May 2013 —

Dearest Leah,

Congratulations to our Master of Food Science and Nutrition! You are going to — we are so *proud* of you.

Love always,
Mom and Dad

Chemistry of Food Additives and Preservatives

Chemistry of Food Additives and Preservatives

Titus A. M. Msagati, B.Sc. (Hons), MSc, Ph.D., CChem, MRSC

Department of Applied Chemistry
University of Johannesburg
Republic of South Africa

WILEY-BLACKWELL

A John Wiley & Sons, Ltd., Publication

Registered office: John Wiley & Sons, Ltd, The Atrium, Southern Gate, Chichester, West Sussex, PO19 8SQ, UK

Editorial offices: 9600 Garsington Road, Oxford, OX4 2DQ, UK
                       The Atrium, Southern Gate, Chichester, West Sussex, PO19 8SQ, UK
                       2121 State Avenue, Ames, Iowa 50014-8300, USA

For details of our global editorial offices, for customer services and for information about how to apply for permission to reuse the copyright material in this book please see our website at www.wiley.com/wiley-blackwell.

Library of Congress Cataloging-in-Publication Data

Chemistry of food additives and preservatives / Titus A. M. Msagati.
        p. cm.
    Includes bibliographical references and index.
    ISBN 978-1-118-27414-9 (hardcover : alk. paper)   1. Food additives.   2. Food preservatives.
3. Food–Analysis.   4. Food–Composition.   I. Msagati, Titus A. M.
    TX553.A3C455 2012
    641.3'08–dc23

                                                                          2012009754

A catalogue record for this book is available from the British Library.

Cover image credit – Top: © iStockphoto.com/Pgiam; Bottom: © iStockphoto.com/mattjeacock
Cover design by Meaden Creative

Set in 10/12 pt Times by Aptara® Inc., New Delhi, India
Printed and bound in Malaysia by Vivar Printing Sdn Bhd

1   2013

Contents

Preface

The incorporation of additives in food preparations has been in practice since time immemorial. Additives are used to perform various functions, for example, to impart or enhance flavour (taste) where it is not sharp enough to meet consumer's demand, to give foodstuffs a desired colour (look/appearance) or to increase the shelf life of the food (preservative role). Some additives perform as essential elements or nutritious supplements to cater for the diet deficiencies of specific groups of people; without such additives these individuals would suffer from some specific nutrient deficiency syndrome or malnutrition.

The tendency to incorporate additives in food products has increased lately, with the advent of many new types of additives on the market. Knowledge regarding food additives, how they are prepared, their compositions and how they work has become very important to those in the food industry and research and academic institutions. This book is therefore intended to address all these aspects of food additives, and is expected to be of interest to all stakeholders in academia and research.

The book covers the chemistry of selected food additives such as their chemical nature, the way in which they are incorporated in foods and the technology involved in their preparations and processing steps. The book also covers the mechanisms or modes of action for the active ingredients in each type and class of food additive and preservative; their physico-chemical characteristics which give them special qualities to be used in food processing; parameters used as indicators for the quality assurance of the products; structure-activity relationships; and their safety to consumers.

There has recently been concern about the possible toxicity of some food additives and food processes. This has led to either a total ban of some additives or maximum limits have been set and strict rules have been enforced to safeguard the health of consumer. This aspect has also been dealt with in this book, and the reported toxic additives are discussed as well as the analytical methods to determine the safety of various food additives. Standard methods for control, monitoring and quality assurance certification for food additives have been set in place by various regulatory bodies such as the European Union (EU) and the American Food and Drug Administration (FDA) to control the legality of use for all the additives. Methods for the monitoring of additives and their metabolites are also discussed.

The classes of food additives that are discussed in this book include: antioxidants and radical scavengers; emulsifiers; stabilisers, thickeners and vegetable gums; sweeteners; fragrances, flavourings and flavour enhancers; food acids and acidity regulators; colourings and colour retention agents; flour treatment/improving agents; anticaking agents; humectants; antifoaming agents; minerals and mineral salts; glazers; preservatives; nutraceuticals, nutrigenomics and nutrigenetics; probiotics; prebiotics; synbiotics and micro (bio) capsules.

This book is expected to be a valuable asset to scholars, especially those enrolled in postgraduate courses and research programs in the areas of food chemistry, food processing and food technology, and also to industrialists and researchers in related areas.

Introduction

Food is one of the main basic human requirements of life and is sourced mainly from plants or animals (and other minor sources such as fungi e.g. mushroom and algae e.g. Spirulina). Generally, human foods are never consumed raw; rather, they undergo special processing treatments with or without heat to make them more palatable. The steps involved in the food processing treatments vary depending on the type of food being prepared. Where necessary, some nutritive additives essential for health are added. The process of adding additives in foods involves mixing together various ingredients before or during a heat-treatment step to give the food the intended flavour, taste, texture or appearance. To attain a balanced diet, it has been necessary to add to certain foodstuffs some ingredients missing in that particular diet such as salt, amino acids and vitamins. In cases where food is processed for future use or where there is a necessity to avoid spoilage by the action of microbes, special treatments such as smoking or salting are used to keep the food safe for long periods of time. The tendency to make foodstuffs more appealing and palatable has paved way for the incorporation of a variety of ingredients or some special treatments to impart a desired quality to foodstuffs. This tendency echoes the saying: 'people first eat with their nose, then with their eyes and finally with their mouths'. Aroma, flavour, taste and appearance are all equally important in the appeal of foods.

Food additives are substances incorporated in edible products in order to perform specific roles and functions, such as preservation of foodstuffs by either increasing shelf life or inhibiting the growth of harmful microbes. Other roles include imparting desired colour, odour or a specific flavour to food. Food additives may have a natural origin in the sense that they may be found existing naturally forming part of the indigenous components of the food, or they may be synthetic but replicas of substances found naturally in foodstuffs. They may also be entirely artificial, which implies that they are synthetically produced and are not copies of any compounds found in nature.

There are a number of additives and preservatives commonly used in foods including antioxidants, acids, acid regulators and salts, emulsifiers, colouring agents, minerals and vitamins, stabilisers, thickeners, gelling agents, sweeteners and preservatives. These different food additives have different roles to play in foods depending on their intended purpose. For instance, emulsifiers tend to give food a good texture as well as good homogeneity such that they make it possible for immiscible items such as water and oils to mix well without any separation, as is the case in ice-creams or mayonnaise (Suman *et al.* 2009). Stabilisers, thickeners and gelling agents provide strong texture and smoothness as well as an increase in viscosity (Quemener *et al.* 2000).

Sweeteners are important as flavours, although there are other types of sweetener flavours which perform an important function in the diets of consumers with health problems such as diabetes (Hutteau *et al.* 1998).

Nutritive additives such as minerals, vitamins, essential amino acids, etc., are added to particular food products where they are missing (Nayak and Nair 2003) or in foodstuffs specifically intended for people with deficiency of such additives, for example milk for babies (Ikem *et al.* 2002). Other additives such as antioxidants are needed for the prevention of fat and oil rancidity in baked foods by inhibiting the effects of oxygen on foods and also preventing the loss of flavour, thereby maintaining food palatability and wholesomeness.

Acids and acidic regulators such as citric acid, vinegar and lactic acid are food additives to control food pH (levels of acidity or alkalinity) and they play an important role in the sharpening of flavours (Populin *et al.* 2007), as preservative (Brul and Coote 1999) and as antioxidants. Some acids and acid regulators tend to release acids only when they are subjected to a heat treatment such as with some bakery products (e.g. acids produced by the leavening agents react with baking soda to make the bakery products rise during the baking process).

Colouring and colour retention agents are added to foods to appease the eye of the consumer or beholder; they are also intended to maintain the colour of food in cases where it may fade (MacDougall 1999).

Generally speaking, the desire for a particular quality of food has resulted in the introduction of numerous additives with wide applications in different cultures and civilisations. Currently, many different types of food additives have been commercialised and are finding their way onto the markets worldwide (Baker 2010). This trend in business has contributed to the speedy growth in food processing and other related industries, where food additives are used *en masse*. The economic success of food additives has further encouraged the advent of new technologies in the processing of foods.

However, these new technologies and additives have brought other unwanted outcomes and are an issue of concern. Despite all the benefits and advantages of food additives and preservatives, there is still a potential danger of chemical adulteration of foods. Additives or preservatives in foods may themselves trigger other hormonal or chemical processes in the body that can generate negative physiological responses. The metabolites produced by additives may also cause side effects, because not all food additives enter the markets after being thoroughly studied to prove their safety (Skovgaard 2004). Although most food additives are considered safe, some are known to be carcinogenic or toxic. For these reasons, many food additives and preservatives are controlled and regulated by national and international health authorities. All food manufacturers must comply with the standards set by the relevant authorities without violating the maximum thresholds stated to ensure the safety of the final product to the consumers. In most cases, food processing industries must seek standard certification before using any new additive or preservative or before using any originally certified additive or preservative in a different way (Pinho et al. 2004; Skovgaard 2004).

REFERENCES

Baker, S. R. (2010) *Maximizing the use of food emulsifiers*. MSc thesis, Kansas State University, Manhattan, Kansas, USA.

Brul, S. & Coote, P. (1999) Preservative agents in foods: Mode of action and microbial resistance mechanisms. *International Journal of Food Microbiology* 50, 1–17.

Hutteau, F., Mathlouthi, M., Portmad, M. O. & Kilcast, D. (1998) Physicochemical and psychophysical characteristics of binary mixtures of bulk and intense sweeteners. *Food Chemistry* 63 (1), 9–16.

Ikem, A. Nwankwoala, A., Odueyungbo, S., Nyavor, K. & Egiebor, N. (2002) Levels of 26 elements in infant formula from USA, UK, and Nigeria by microwave digestion and ICP–OES. *Food Chemistry* 77, 439–447.

MacDougall, D. B. (1999) *Coloring of Food, Drugs, and Cosmetics*. Marcel Dekker, Inc., New York, Basel, USA.

Nayak, B., & Nair, K. M. (2003) In vitro bioavailability of iron from wheat flour fortified with ascorbic acid, EDTA and sodium hexametaphosphate, with or without iron. *Food Chemistry* 80, 545–550.

Pinho, O., Ferreira, I. M. P. L. V. O., Oliveira, M. B. P. P. & Ferreira, M. A. (2000) Quantitation of synthetic phenolic antioxidants in liver pates. *Food Chemistry* 68, 353–357.

Populin, T., Moret, S., Truant, S. & Conte, L. S. (2007) A survey on the presence of free glutamic acid in foodstuffs, with and without added monosodium glutamate. *Food Chemistry* 104, 1712–1717.

Quemener, B., Marot, C., Mouillet, L., Da Riz, V. & Diris, J. (2000) Quantitative analysis of hydrocolloids in food systems by methanolysis coupled to reverse HPLC. Part 1. Gelling carrageenans. *Food Hydrocolloids* 14, 9–17.

Skovgaard, N. (2004) Safety evaluation of certain food additives and contaminants. *International Journal of Food Microbiology* 90, 115–118.

Suman, M., Silva, G., Catellani, D., Bersellini, U., Caffarra, V. & Careri, M. (2009) Determination of food emulsifiers in commercial additives and food products by liquid chromatography/atmospheric-pressure chemical ionization mass spectrometry. *Journal of Chromatography A*, 1216, 3758–3766.

List of Abbreviations

AAPH	2, 2′-azobis (2-amidino-propane) dihydrochloride
ABTS	2, 2′-azino-bis (3-ethylbenzthiazoline-6-sulphonic acid)
ACI	amylose complexing index
AEDA	aroma extract dilution analysis
AMG	acetylated monoglyceride
APCI-MS	atmospheric pressure chemical ionisation mass spectrometry
AV	acid value
BDMS	butyldimethylsilyl
BHA	butylated hydroxyanisole
BMI	body mass index
BR	Brigg–Rauscher
CDG	calcium diglutamate
CE	capillary electrophoresis
CMG	citrate monoglycerides
CSL	calcium stearoyl 2 lactate
CTAB	cetyltrimethylammonium bromide
CTAC	cetyltrimethylammonium chloride
CZE	capillary zone electrophoresis
DAD	diode array detector
DHC	dihydrochalcone
DMPD	N,N-dimethyl-p-phenylenediamine
DPPH	1, 1-Diphenyl-2-picrylhydrazyl
EDTA	ethylenediaminetetraacetic acid
ELISA	enzyme-linked immunosorbent assay
EU	European Union
FACE	fluorophore-assisted carbohydrate electrophoresis
FAO	Food and Agriculture Organization
FDA	Food and Drug Administration
FRAP	ferric-reducing ability of plasma
FT-IR	Fourier transform infrared spectrometry
GC	gas chromatography
GDL	glucano-delta-lactone
GI	glycemic index
GL	glycemic load
GLC	gas liquid chromatography
GPC	gel permeation chromatography
GRAS	generally recognised as safe
HDB	hexadimetrine bromide

HFCS	high-fructose cone syrup
HHP	high hydrostatic pressure
HIL	high-intensity laser
HLB	hydrophilic–lipophilic balance
HORAC	hydroxyl radical antioxidant capacity
HPAEC	high-performance anion exchange chromatography
HPH	high-pressure homogenisation
HPLC	high-performance liquid chromatography
HPU	high-power ultrasound
HVAD	high-voltage arc discharge
LDL	low-density lipoprotein
LOD	limit of detection
LOQ	limit of quantification
MALDI-MS	matrix-assisted laser desorption-ionisation mass spectrometry
MAP	modified atmosphere packaging
MEKC	micellar electrokinetic chromatography
MSG	monosodium glutamate
NNS	non-nutritive sweeteners
OAV	odour activity values
OMF	oscillating magnetic fields
ORAC	oxygen radical absorbance capacity
PCL	photochemiluminescence
PEF	pulsed electrical fields
PG	propylene glycol
PGA	propylene glycol alginate
PGPR	polyglycerol polyricinoleate
PHIL	pulsed high-intensity light
PKU	phenylketonuria
PPO	polyphenol oxidase
PWL	pulsed white light
RMCD	random methylated b-cyclodextrin
RNS	reactive nitrogen species
ROS	reactive oxygen species
-SH	sulphhydryl
SMG	succinylated monoglyceride
SOD	superoxide dismutase
SP	streamer plasma
SWV	square-wave voltammetry
TBARS	thiobarbituric acid reactive substances
TEAC	trolox equivalent antioxidant equivalent
TMS	trimethylsilyl
TRAP	total radical trapping antioxidant parameter
TSS	total soluble solids
UV-Vis	ultraviolet-visible
WHO	World Health Organisation

1 Antioxidants and Radical Scavengers

Abstract: Food antioxidants play an important role in the food industry due to their ability to neutralise free radicals that might be generated in the body. They do that by donating their own electrons to free radicals without becoming free radicals in the process themselves, hence terminating the radical chain reaction. The converted free radical products will then be eliminated from the body before causing any harm; in this regard, antioxidants play the role of scavengers protecting body cells and tissues. In this chapter, the processes which lead to the formation of these reactive species (free radicals) and the different additives used as antioxidants or radical scavengers to counter the effects of free radicals will be discussed. Sources of different types of antioxidants, the various mechanisms by which they work and analytical methods for determination and quality control are also examined.

Keywords: antioxidants; free radical species; ORAC assay; HORAC assay; DPPH assay; FRAP assay; Trolox; TEAC assay; ABTS assay; PCL assay; DMPD assay; DL assay; TBARS assay; Brigg-Rauscher assay

1.1 CHEMISTRY OF FREE RADICALS AND ANTIOXIDANTS

1.1.1 Introduction

From the viewpoint of chemistry, free radicals refer to any molecule with an odd unpaired electron in its outer electronic shell, a configuration responsible for the highly reactive nature of such species. The presence of such highly reactive free radicals in biological systems is directly linked to the oxidative damage that results in severe physiological problems. The free radical species that are of concern in living systems include the reactive oxygen species (ROS), superoxide radicals (SOR), hydroxyl radicals and the reactive nitrogen species (RNS). The oxygen-containing reactive species are the most commonly occurring free radicals in living medium and are therefore of greatest concern. The oxidative damage caused by these free radicals can be prevented by using antioxidants which include enzymatic antioxidant systems such as catalase, glutathione peroxidase and superoxide dismutase (SOD) as well as non-enzymatic antioxidants (Figure 1.1). It should be noted that, in nature, the generation of free radicals which cause oxidative stress and that of antioxidants or radical scavengers is carefully controlled such that there is always a balance between the two (Vouldoukis *et al.* 2004). Examples of non-enzymatic antioxidants include vitamin C (ascorbic acid) which is a sugar acid, vitamin E (α-tocopherol) and β-carotene, bilirubin, propyl gallate (PG, a

Chemistry of Food Additives and Preservatives, First Edition. Titus A. M. Msagati.
© 2013 John Wiley & Sons, Ltd. Published 2013 by John Wiley & Sons, Ltd.

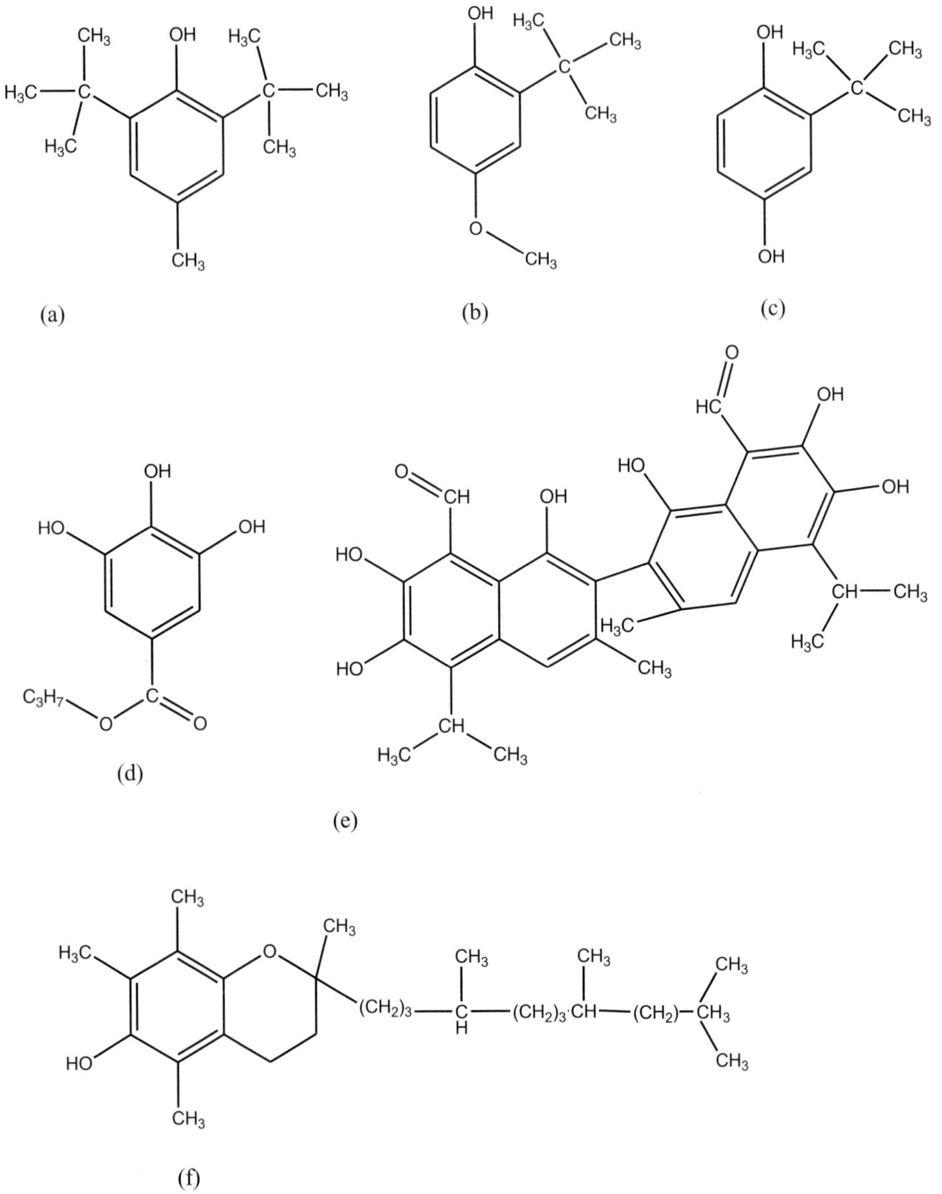

Fig. 1.1   Examples of synthetic antioxidants used in food industries: (a) BHT; (b) BHA; (c) t-BHQ; (d) PG; (e) gossypol; and (f) tocopherol.

condensation ester product of gallic acid and propanol), uric acid, tertiary butylhydroquinone (t-BHQ), butylated hydroxyanisole (BHA), ubiquinone and macromolecules which include ceruloplasmin, albumin and ferritin. Generally, mixtures of different antioxidants provide better protection against attack by free radicals rather than individual antioxidants.

Due to the importance of antioxidant systems, there are a number of quality assessment criteria for the antioxidant performance of these systems. Various assays have been developed to assess the antioxidant capacities, including the oxygen radical absorbance capacity (ORAC) assay, ferric reducing ability of plasma (FRAP), Trolox equivalent antioxidant capacity (TEAC) assay, etc. Antioxidant foods which are dietary nutrients containing antioxidant compounds and non-nutrient antioxidants which are normally added to foods to play the role of antioxidants will be discussed simultaneously in this chapter, unless indicated otherwise.

Further Thinking

Free radicals are undesirable due to their instability caused by the electron deficiencies in their structures. They have a high electronic affinity which makes them attack any molecule in their vicinity, generating a chain of reactions which are detrimental to the body and which instigate disorders, diseases, aging and even death.

1.1.2 The formation of ROS in living systems

Under normal conditions, oxygen is vital in metabolic reactions which are necessary for life. Due to its high reactive nature however, oxygen also causes severe damage to living systems due to the generation of reactive oxygen species (ROS; Davies 1995).

The reactive free radicals are generated as part of the energy generation metabolic processes (Raha and Robinson 2000), and are released as a result of a number of reaction procedures in the electron transport chain as well as in the form of intermediate reduction products (Lenaz 2001). Due to the highly reactive nature of free radicals that are formed as intermediates, they prompt electrons to proceed in a concerted fashion to molecular oxygen and thus generate superoxide anion (Finkel and Holbrook 2000). A similar scenario occurs in plants for example, whereby reactive oxygen species are produced during the process of photosynthesis (Krieger-Liszkay 2005).

Examples of reactive species produced as a result of these metabolic reactions include: superoxide anion (O_2^-), hydrogen peroxide (H_2O_2), hypochlorous acid and hydroxyl radical ($\cdot OH$) (Valko *et al.* 2007). The hydroxyl radicals are known to be unstable; they react spontaneously with other biological molecules in a living medium, causing destructive reactions in foodstuffs and serious physiological damage to consumers (Stohs and Bagchi 1995).

1.1.3 Negative effects of oxidants in food processes and to food consumers

The oxidation process brings about destructive reactions in food items that lead to off-flavour and loss of colour and texture due to the degradation of carbohydrate, protein, vitamins, sterols and lipid peroxidation (Hwang 1991; Pinho *et al.* 2000; Kranl 2004). The consequences to consumers include damage to nucleic acids, cellular membrane lipids and other cellular organelles, carcinogenesis, mental illnesses and disorders, lung diseases, diabetes, atherosclerosis, autoimmune diseases, aging and heart diseases (Finkel and Holbrook 2000; Lachance *et al.* 2001; Ou *et al.* 2002; Yu *et al.* 2005; Nakabeppu et al. 2006).

1.1.4 Reactive oxygen/nitrogen species and aging

There is strong scientific evidence which relates the reactive oxygen/nitrogen species (ROS/RNS) to aging and pathogenesis (Lachance *et al.* 2001; Yu *et al.* 2005; Nakabeppu *et al.* 2006). In addition, facts have also been presented in many scientific reports that ROS such as peroxyl radicals (ROO·), superoxide ion ($O_2·^+$), hydroxyl radicals (HO), etc. play an active role in promoting or inducing numerous diseases such as different types of cancers (Finkel and Holbrook 2000; Ou *et al.* 2002). Unless these adverse reactions are retarded or prohibited, they will result in food deterioration and health problems to consumers. To counter such harmful effects, antioxidants have been incorporated in many foodstuffs to minimise or solve the problem altogether.

Further Thinking

The incorporation of antioxidants in foodstuffs serves a number of purposes, including the prevention of rancidity phenomena as a result of oxidation (which results in bad odour and off-flavour) of food items containing fats and oils. Antioxidants are also essential in the retention of the integrity of food items (mainly fruits, fruit juices and vegetables) because of their particular properties in preventing browning reactions, extending the shelf life of these food items.

1.2 TYPES OF ANTIOXIDANTS

Antioxidants as food additives are used to delay the onset of or slow the pace at which lipid oxidation reactions in food processing proceed. Most of the synthetic antioxidants contain a phenolic functionality with various ring substitutions (monohydroxy or polyhydroxy phenolic compounds) such as butylated hydroxytoluene (BHT), BHA, t-BHQ, PG, gossypol and tocopherol (Figure 1.1). These compounds make powerful antioxidants to protect foodstuffs against oxidative deterioration of the food ingredients. The main chemical attribute that makes them suitable as antioxidants is their low activation energy property, which enables them to donate hydrogen easily and thus put on hold or lower the kinetics of lipid oxidation mechanisms in food systems. The delay to the onset or slowing of the kinetics of lipid oxidation is possible due to the ability of these compounds to either block the generation of free alkyl radicals in the initiation step or temper the propagation of the free radical chain. Due to their positive effects in food processes antioxidants are also known as potential therapeutic agents, thus playing a medicinal role as well. For safety purposes and adherence to quality control standards, the use of any synthetic antioxidant preparation in food processes is expected to meet the following criteria: effective at low concentrations; without any unpleasant odour, flavour or colour; heat stable; non-volatile; and must have excellent carry-through characteristics (Shahidi and Ho 2007).

1.2.1 Natural antioxidants of plant origin

In addition to chemical or synthetic antioxidants, there are also a number of antioxidants that exist naturally in plants and many other herbal materials (Shahidi and Naczk 1995).

Plants that contain natural antioxidants include: carrots, which contain β-carotene and xanthophyll (Chu *et al.* 2002); ginger roots (Halvorsen *et al.* 2002); and citrus fruits with their abundance of flavonoid compounds and ascorbic acid (vitamin C) (King and Cousins 2006). Tomatoes and pink grapefruit contain ascorbic acid and other carotenoid compounds known as lycopenes which are antioxidants (King and Cousins 2006). Grape seeds well as their skin extracts also contain a number of antioxidant substances, mainly proanthocyanidin bioflavonoids and tannins (DerMarderosian 2001). *Saccharomyces cerevisiae*, which is also known as nutritional yeast, has antioxidants superoxide dismutase (SOD) and glutathione (King and Cousins 2006). Green tea is also known to be rich in catechins and other polyphenol antioxidants (Cai *et al.* 2002; Thielecke and Boschmann 2009); vegetable oils such as soybean oil contains radical scavengers such as vitamin E (tocopherols and tocotrienols) (Nesaretnam *et al.* 1992; Beltrán *et al.* 2010); legumes such as soybean are known to be rich in isoflavones (Luthria *et al.* 2007); oil seeds such as canola and mustard contain phenolic acids and phenylpropanoid antioxidants (Shahidi and Wanasundara 1995); and cereals such as wheat contains phenolic and other flavonoid radical scavengers (Shen *et al.* 2009).

Further Thinking

In nature there are many different types of foodstuffs which are known to be rich in antioxidants. Examples include fruits (grape, orange, pineapple, kiwi fruit, grapefruit, etc.), vegetables (cabbage, spinach, etc.), cereals (barley, millet, oats, corn, etc.), legumes (beans, soybeans, etc.) and nuts (groundnuts, peanuts, etc.). Daily intake of a variety of these antioxidant foods may bring significant health benefits to consumers.

1.2.2 Phenolic non-flavonoid antioxidant compounds from natural sources

Polyphenolic non-flavonoid antioxidant compounds include resveratrol and gallic acid which are abundant in plants such as tea, grapes (red wine) and a variety of other fruits (Amakura *et al.* 2000; Rechner *et al.* 2001). Resveratrol, a phenolic non-flavonoid compound extract from wine, has been reported to inhibit low-density lipoprotein oxidation and reduce platelet aggregation, hence playing a direct role in combating atherothrombogenesis (Frankel *et al.* 1995; Pace-Asciak *et al.* 1995; Belguendouz *et al.* 1997). Resveratrol is considered an important agent for the cardio-protective action of wine and also plays an important role in reducing hepatic synthesis of cholesterol and triglyceride, as observed in experiments performed in rats (Arichi *et al.* 1982; Hung *et al.* 2000). It also inhibit the synthesis of eicosanoids and rat leukocytes, interfering arachidonate metabolism (Kimura *et al.* 1985a, b), and inhibits the activity of some protein kinases (Jayatilake *et al.* 1993). All these biological and pharmacological activities of resveratrol are due to its antioxidant property (Rimando *et al.* 2002). The polyphenolic compound gallic acid (3,4,5-trihydroxybenzoic acid) (Figure 1.2), obtained naturally as a product of either alkaline or acid hydrolysis of tannins, and its derivatives is also found abundantly in wine (Aruoma *et al.* 1993).

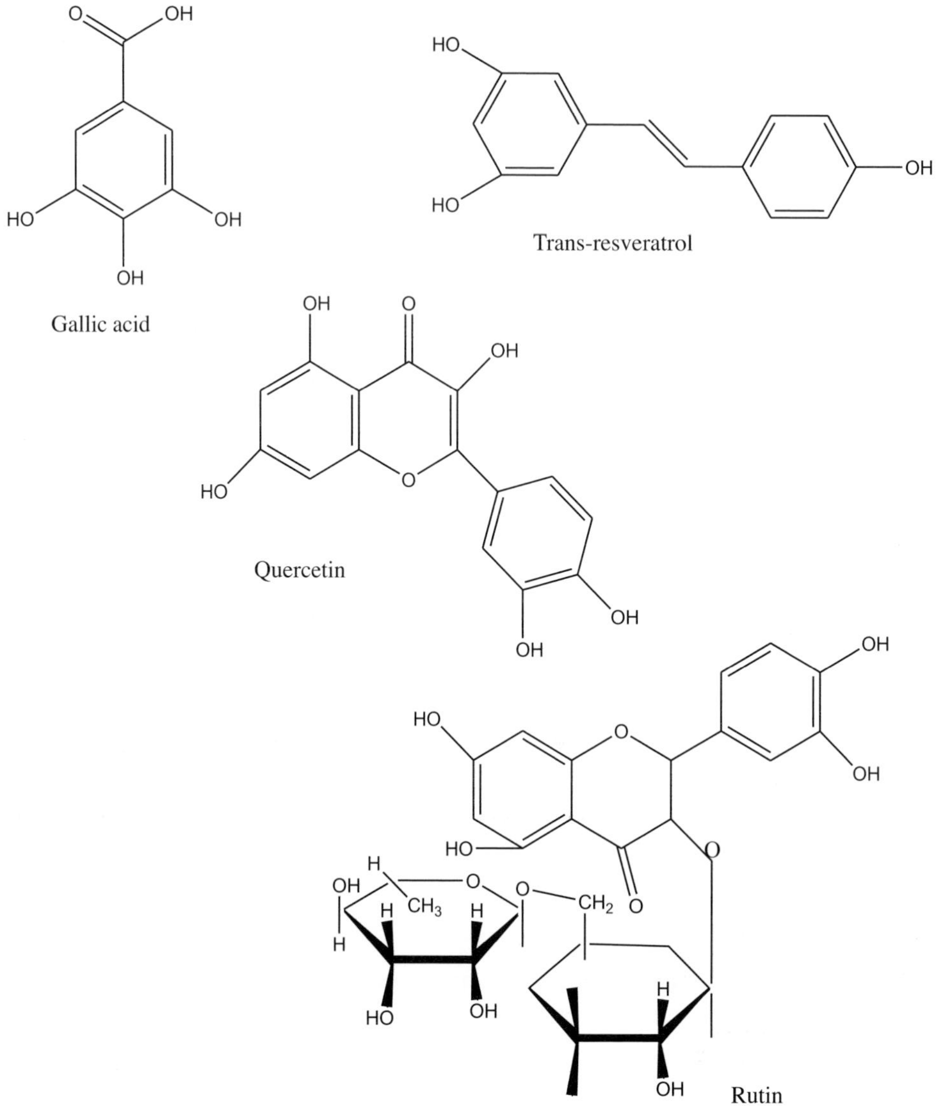

Fig. 1.2   Chemical structures of phenolic non-flavonoid antioxidants.

1.2.3   Phenolic flavonoid antioxidant compounds from natural sources

Antioxidants with flavonoid functionality are low-molecular weight polyphenolics which occur in a variety of vegetables and fruits (Hertog *et al.* 1992). An example of these flavonoid polyphenolic compounds is quercetin, which forms the main aglycone found in many foods (Robards *et al.* 1999). Apart from functioning as antioxidants, various flavonoids also have anti-inflammatory, anti-allergic, anticancer and anti-hemorrhagic properties (Das 1994). The antioxidant properties of flavonoids are responsible for the protective effect of wine and vegetable-rich diets against coronary heart disease (Pearson *et al.* 2001). The majority of phenolic flavonoids extracted from natural sources (for example, gallic acid, trans-resveratrol, quercetin and rutin; Figure 1.2) have demonstrated potential beneficial effects on human health in many ways.

1.2.4 Acidic functional groups responsible for antioxidant activity

The antioxidant activity of certain food plants are due to various functional groups associated with some organic acids such as vanillic, ferulic and p-coumaric acids, found mainly in whole grains. Other acids found in barley grains such as salicylic, p-hydroxybenzoic, protocatechuic, syringic and sinapic acids have functional groups that confer antioxidant activity (Shahidi and Naczk 1995). Generally, corn wheat and barley contain syringic acid, sinapic acid, protocatechuic acid, p-hydroxybenzoic acid, vanillic acid, ferulic acid, salicylic acid and p-coumaric acid as molecules containing antioxidant functional groups (Figure 1.3; Hernández-Borges *et al.* 2005).

Further Thinking

Who needs antioxidants and why?
- *Children need lots of antioxidants (β-carotene, flavonoids, vitamins C and E) as damage caused by free radicals has a much greater effect on their young and tender bodies than compared to adults. Some antioxidants are added to infant formulas (e.g. ascorbyl palmitate, tocopherols and lecithin).*
- *The elderly need antioxidants since the oxidative damage due to free radicals affects the performance of muscles to a greater degree with age, affecting the physical performance and reducing fitness in many areas.*
- *Active sportsmen and those who take part in strenuous exercise or heavy work involving massive physical muscle energy need more antioxidants to protect against the by-products of exercise. This group need extra fatty esters and antioxidants from diets including spices such as from plants of Curcuma longa L. and Zingiberaceae, or collastin supplements which contain natural cyclooxygenase-2 inhibitors that are capable of protecting against cell damage as well as inflammation. Diets with these ingredients as well as some specific antioxidants are essential in maintaining body joints, thus keeping sportsmen fit.*
- *Healthy people need antioxidants as protection from various diseases, illnesses and sicknesses such as cancer, diabetes, etc.*

1.3 EFFICACY OF DIFFERENT ANTIOXIDANTS

The compositions, structural features and chemical structures of antioxidants are important parameters that control their efficacy and also the antioxidant activity (Bors *et al.* 1990a, b). For example, the presence of ortho-dihydroxy functionality in the catechol structure of flavonoid antioxidants has been associated with the increased stability of radicals generated due to the possible formation of hydrogen bonding or the delocalisation of electrons around

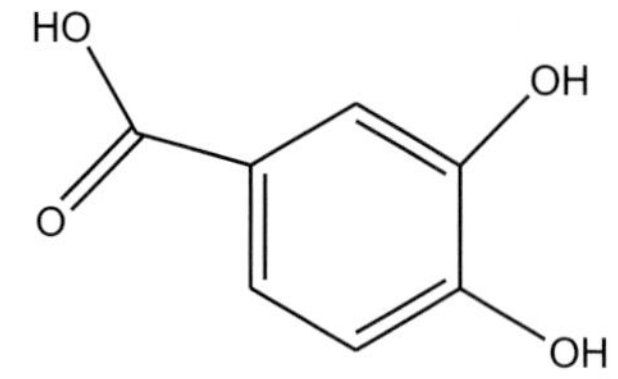

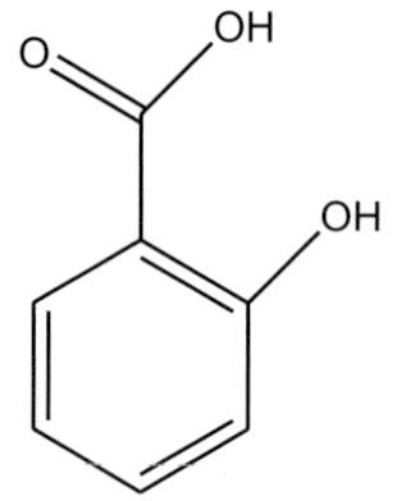

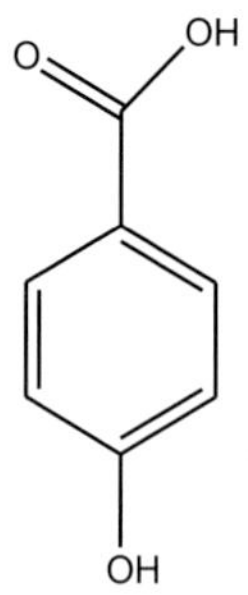

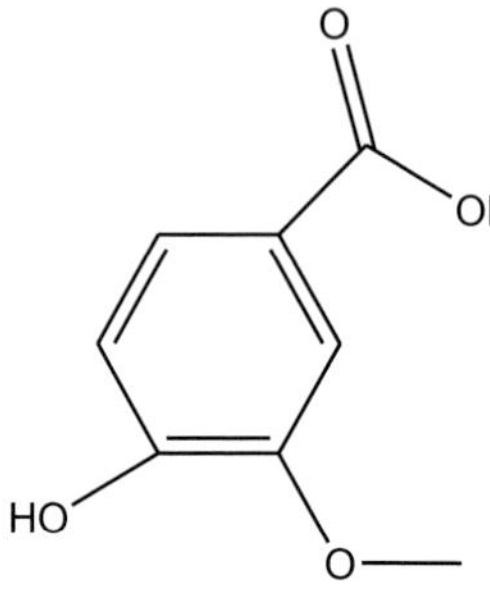

Vanillic acid 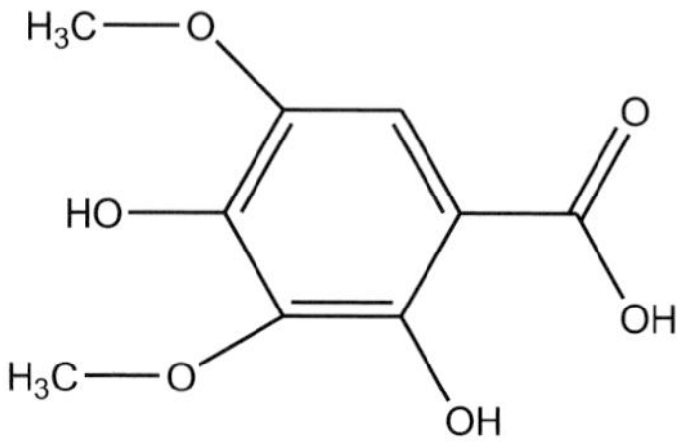   syringic acid

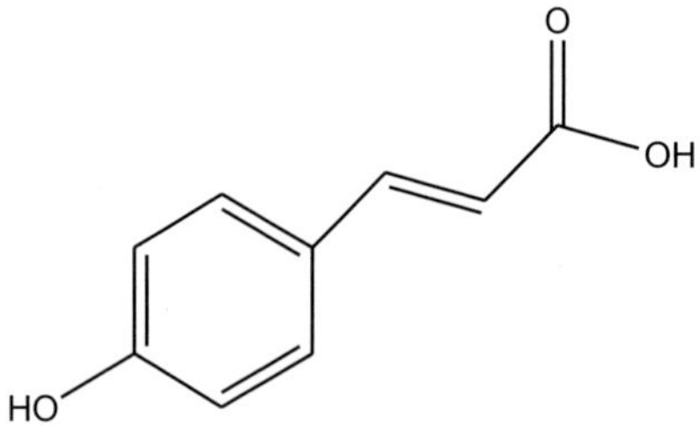

p-coumaric acid 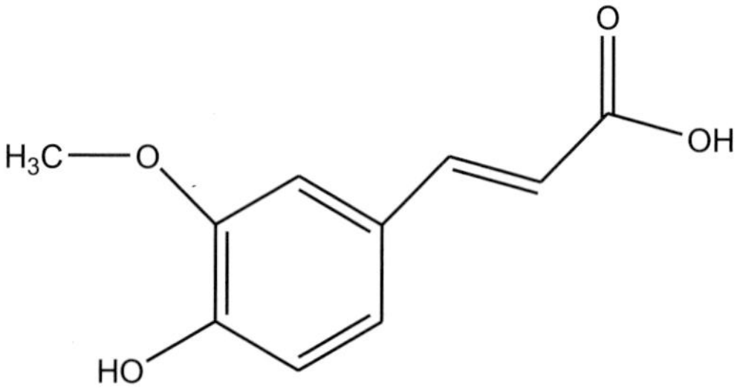   Ferulic acid

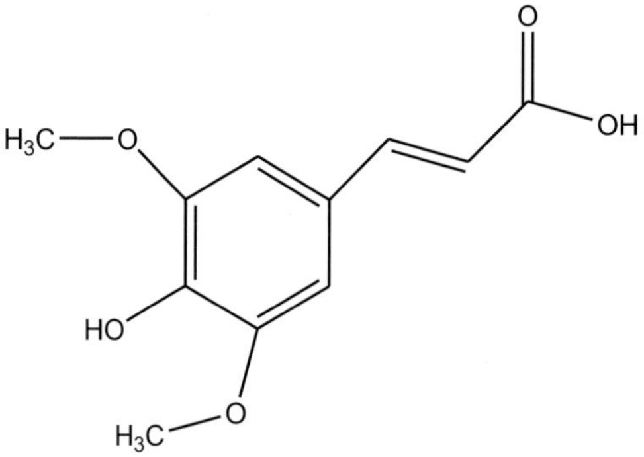

Sinapinic acid

Fig. 1.3   Chemical structures of some antioxidants with acidic functional groups.

the aromatic ring (Apak *et al.* 2007). The presence of hydroxyl groups at positions 3 and 5 of phenolic antioxidants is said to contribute to the stability of antioxidants (Firuzi *et al.* 2005). Phenolic compounds which are dihydroxylated or hydroxylated at position 2 or 4 (ortho or para) or contain a methoxy group are generally more effective than simple phenolics (Van Acker *et al.* 1996; Apak *et al.* 2007; Bracegirdle and Anderson 2010). This is due to the presence of methoxy groups in ortho and para positions of the ring serving as electron-donating groups, thus adding to stability and hence promoting the antioxidant activity (Firuzi *et al.* 2005).

Moreover, phenylpropanoid antioxidants with extended conjugation are known to have enhanced antioxidant activity compared to benzoic acid derivatives because of the resonance stabilisation. The hydrophilicity as well as lipophilicity of the antioxidants is dependent on the correct matching in terms of application of antioxidants; more hydrophilic antioxidants matches is best for use in stabilising bulk oil systems as opposed to oil-in-water emulsions, while the converse is true for the activity of lipophilic antioxidants (Shahidi and Ho 2000).

Further Thinking

Unsaturated and polyunsaturated fats may be preferred over saturated animal fats by many. However, polyunsaturated and saturated fats undergo oxidation easily, hence the problem of rancidity due to the decomposition of fat when they react with oxygen. Peroxides are produced, which result in a bad smell, off-flavour (rancidity) and the soapy texture of food. If oxidation reactions occur in the body system they cause fat deposits to be built up, which may block blood vessels. This necessitates the incorporation of antioxidants in foods which may react with oxygen, hence preventing the formation of peroxides as well as heart problems, cancer diseases, arthritis, tumours etc. Antioxidants also help to preserve the integrity of food items so that they remain fit for human consumption for a long time.

1.4 ACTION MECHANISMS OF ANTIOXIDANTS

From the definition of an antioxidant compound – which refers to a chemicals species capable of suppressing the harmful effects of reactive radicals present in biological systems at low concentration (Gutteridge 1994) – it follows that the mechanisms should involve the protonation by the donor species to the reactive radicals. There are a number of possible mechanisms for antioxidant action and these include: (1) quenching mechanism, which occurs when the radical is in an excited triplet state which makes the antioxidant behave as a quenching agent (Tournaire *et al.* 1993; Anbazhagan *et al.* 2008; Ji and Shen 2008); (2) direct hydrogen transfer mechanism which takes place if the radical is in a doublet state, enabling the direct transfer of the hydrogen atom to the radical (Priyadarsini *et al.* 2003; Luzhkov 2005); (3) charge transfer for doublet radical which yields a closed-shell anion and a radical antioxidant cation (Kovacic and Somanathan 2008; Oschman 2009); and (4) bond-breaking mechanisms, as in the case for vitamin E (Graham *et al.* 1983; Roginsky and Lissi 2005).

1.4.1 Quenching

In this mechanism, which is also known as singlet oxygen scavenging, antioxidants reacts with singlet oxygen ($^1O_2$) to form intermediate compounds such as endoperoxides and final products which are mainly hydroperoxydienones. The final products are responsible for quenching, that is, termination of the propagation process that generates free radicals. Examples of antioxidants which exhibit this phenomenon include vitamin E and carotene.

1.4.2 Hydrogen transfer

A complex is formed between a lipid radical and the antioxidant radical which, in this case, is the free radical acceptor. The processes involve several reactions as depicted in Figure 1.4.

1.4.3 Charge transfer

There are two ways in which the charge transfer antioxidation mechanism takes place, both involving the formation of stable radicals which stops the propagation of reactive species in the biological systems. Firstly, the antioxidation mechanism may occur through

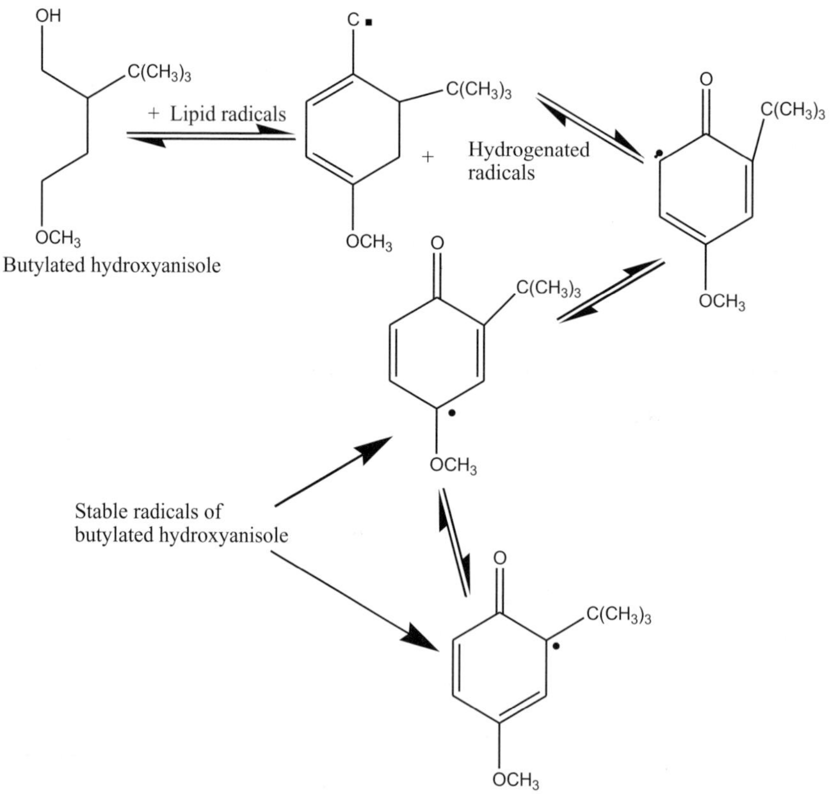

Fig. 1.4   Possible mechanism of butylated hydroxyanisole antioxidants (Lambert *et al.* 1996; Goodman *et al.* 1990)

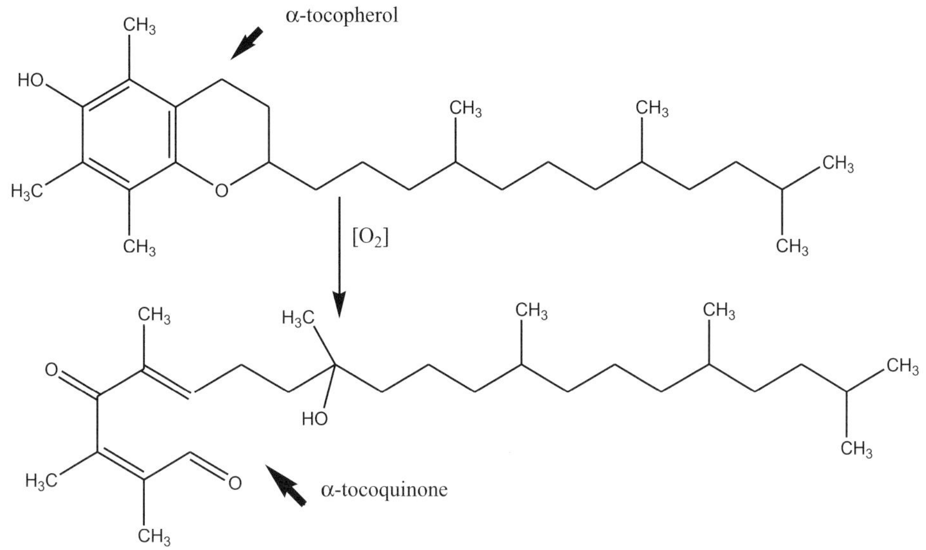

Fig. 1.5 Possible mechanistic reaction of α-tocopherol antioxidant (Herrera and Barbas 2001).

hydrogen transfer processes in which the reactive species themselves abstract a proton from the antioxidant, such that the antioxidant will become a highly stable radical which cannot react with any substrate. The stability of this stable radical is enhanced by resonance effects and hydrogen bonding. The second mechanism is by a one electron transfer process where the antioxidant can donate an electron to the reactive species, making itself a highly stable positively charged radical which cannot undergo any reaction with substrates. Examples of antioxidants which undergo charge transfer mechanisms include flavonoids and other phenolic antioxidants.

1.4.4 Bond-breaking

The α-tocopherol (Figure 1.5) is a hydrophobic antioxidant which plays an important role in protecting the cytoplasmic membranes against oxidation reactions caused by lipid radicals. It protects cell membranes by reacting with the lipid radicals, thus terminating the chain propagation reactions due to the reactive species that would otherwise have continued oxidation reactions with the cell membrane (Herrera and Barbas 2001).

1.5 STRUCTURE–ACTIVITY RELATIONSHIP OF ANTIOXIDANTS

1.5.1 Polyphenol antioxidants

With the phenolic antioxidants it has been established that the presence of o-dihydroxy structure in the B ring (Figure 1.6) contributes significantly to the higher stability of the radical; it also plays a significant role in electron delocalisation, necessary for the antioxidant activity. Moreover, the 3- and 5-OH groups with 4-oxo function in the A and C rings have

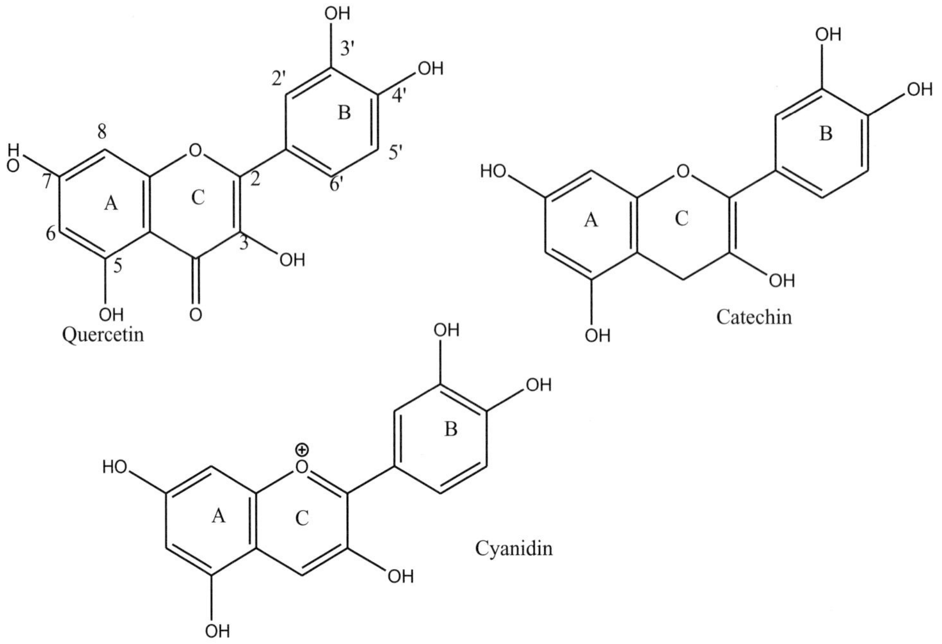

Fig. 1.6   Structure of polyphenol antioxidants.

been reported as necessary for efficient antioxidant activity (Rice-Evans *et al.* 1996). The position and degree of hydroxylation is another aspect that has been reported as essential for the antioxidant activity of phenols and particularly the o-dihydroxylation of the B ring, the carbonyl at position 4, and a free hydroxyl group at positions 3 and/or 5 in the C and A rings, respectively.

1.5.2   Flavonoid antioxidants

The activity of flavonoid antioxidants (for example flavones, isoflavones and flavanones) against peroxyl and hydroxyl radicals (pro-oxidants) was studied by Cao *et al.* (1997). They found that the pro-oxidant activities of these flavonoid antioxidants were strongly influenced by the number of hydroxyl substitutions in their backbone structure, which lacked both the antioxidant as well as the pro-oxidant property. It was evident that the greater the number of hydroxyl substitutions, the stronger the antioxidant and pro-oxidant activities. It was also concluded that those flavonoids with multiple hydroxyl substitutions had higher antiperoxyl radical activities compared to others such as α-tocopherol. Another important observation was that the presence of a single hydroxyl substitution at position 5 as well as the conjugation between rings A and B (Figures 1.7a–c) provided no activity at all, but the di-OH substitution at 3′ and 4′ (Figure 1.7b) proved to be essential for the peroxyl radical absorbing activity of a flavonoid. Cao *et al.* also studied the effect of O-methylation of the hydroxyl substitutions and found that it resulted in the inactivation of both the antioxidant and the pro-oxidant activities of the flavonoids (Cao *et al.* 1997).

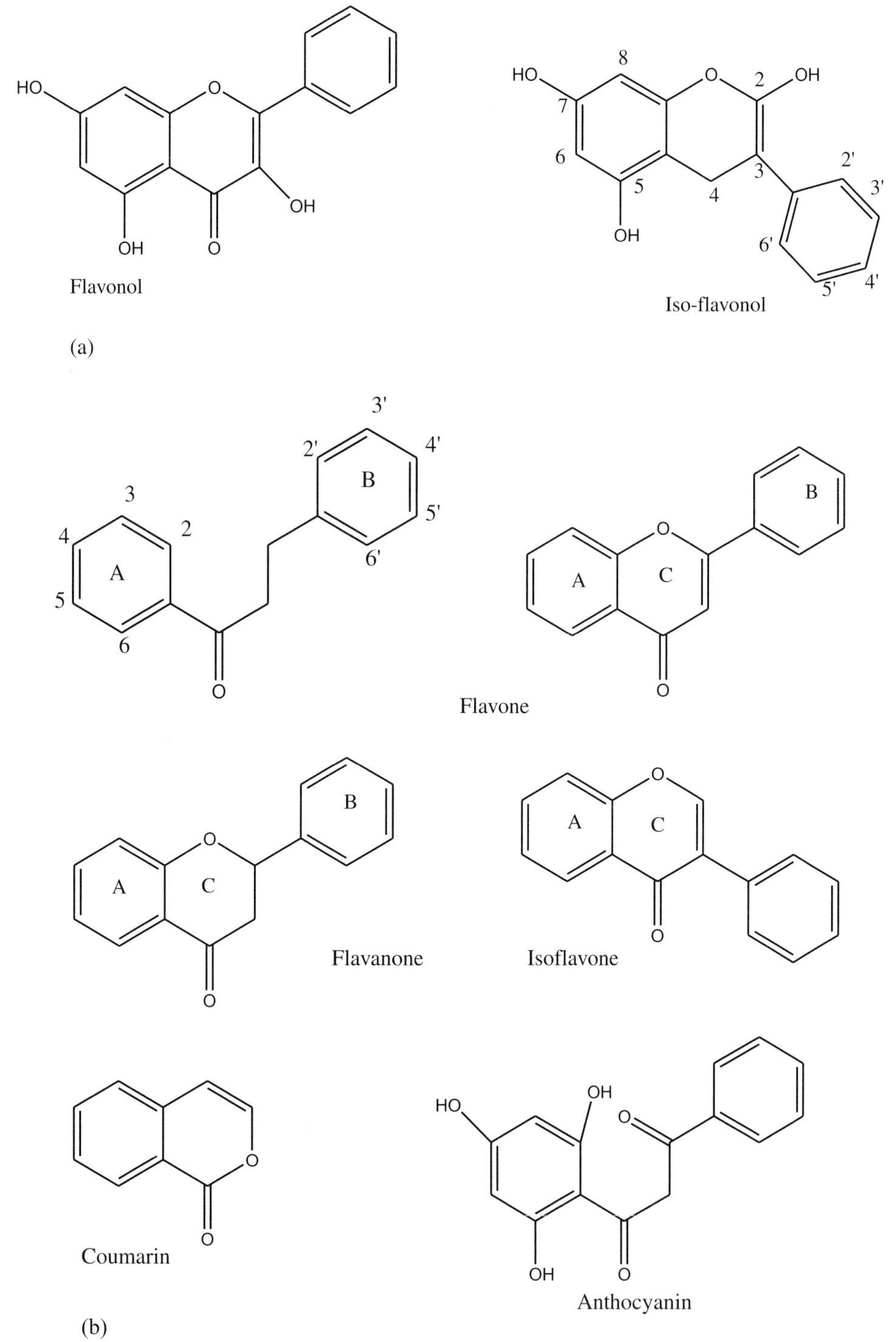

Flavonol

Iso-flavonol

(a)

Flavone

Flavanone

Isoflavone

Coumarin

Anthocyanin

(b)

Fig. 1.7 (a) General structures of the main classes of flavonoid antioxidants and flavonoid-related compounds; (b) basic structure of flavonoids; and (c) possible mechanism of flavonoid antioxidants with radical scavengers (R·) (Pereira and Das 1990).

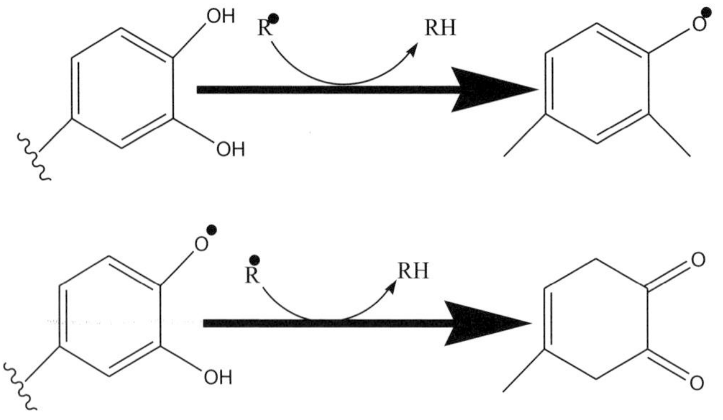

(c)

Fig. 1.7 (Continued)

1.5.3 Mechanism of reactions of flavonoid antioxidants with radical scavengers

Pereira and Das (1990) have reported that the presence of carbonyl group at C-4 and a double bond between C-2 and C-3 are important features for high antioxidant activity in flavonoids (see the basic structure of flavonoids, Figures 1.7b and c).

1.6 FACTORS AFFECTING ANTIOXIDANT ACTIVITY

There are a number of physical factors that influence the activity of the antioxidant, discussed in the following sections.

1.6.1 Temperature

Temperature catalyses the acceleration of the initiation reactions, which results in a decrease in the activity of the already-available or introduced antioxidants (Pokorny 1986). Because of this, the variations in the temperature normally influence the manner in which some oxidants work; note that these variations are not the same for all antioxidants (Yanishlieva 2001). For instance, the effect of temperature variations on the activity of different antioxidants in fats and oils over a large temperature range was that the α-tocopherol activity increased as the working temperature increased throughout the whole temperature range (20–100°C) (Marinova and Yanishlieva 1992, 1998; Yanishlieva and Marinova 1996a, b). Another observation on the effect of temperature variation on the antioxidant activity was that some of the tested antioxidants were found to be sensitive to either concentration or the stabilised substrate (Marinova and Yanishlieva 1992, 1998).

1.6.2 Activation energy and redox potential

Different antioxidants will have different activation energies as well as oxidation-reduction potentials. These properties mean that antioxidants have a varying ability to donate an electron easily.

1.6.3 Stability

Antioxidants have a varying degree of optimal performance with respect to pH. When the antioxidant is in a high-pH medium, it will undergo deprotonation. Its radical scavenging capacity will be enhanced since it will have the ability to donate an electron much easier (Lemaska *et al.* 2001).

Further Thinking

Note that in this chapter antioxidant foods in the sense of (1) foodstuffs containing antioxidant compounds as well as (2) non-nutrition antioxidant compounds which can be added to foods to play the role of radical scavenging have been discussed simultaneously. In the following section, the term antioxidant will however be restricted to the non-nutrient antioxidants (e.g. polyphenols, catechins, etc.) which show antioxidant activity in vitro and allow the artificial index of antioxidant strength to be determined.

1.7 QUALITY ASSESSMENT OF DIETARY ANTIOXIDANTS

Because of the importance of the role played by antioxidants, it is imperative to assess and evaluate their antioxidant capacity or activity. There is generally a variety of chemistries within the antioxidant classes; some are hydrophilic while others are lipid-soluble molecules, implying that they are hydrophobic. All these different functionalities of antioxidants display a multiplicity of antioxidant pathways; there therefore is a need to quantitatively measure the total antioxidant capacity or antioxidant power in food products.

A number of methods and techniques (referred to as assays) have been established for the measurement of total antioxidant capacity in food products, and are discussed in the following sections.

Further Thinking

There are special qualities that antioxidants must possess to be suitable for human consumption. These attributes include solubility in fats and oils and they should maintain the integrity of foods in the sense that they should not in any case impart any unnatural colour, odour or flavour in the foods, even after prolonged periods of storage. Their stability and usability must prove to be effective for at least a year at room temperature. During food processing, they must prove to be stable to the processing heat without affecting the integrity of the final product in any way. Moreover, they must be easy to incorporate in foods and effective especially at low concentrations.

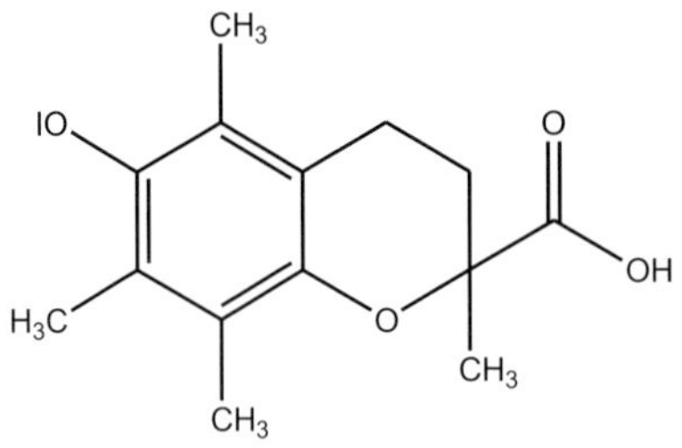

Fig. 1.8   Chemical structure of Trolox.

1.7.1   Total radical trapping antioxidant parameter/oxygen radical absorbing capacity

The oxygen radical absorbing capacity (ORAC) assay measures the extent of oxidative degradation of either β-phycoerythrin or fluorescein following the reaction with azo-initiator compounds, the source of the free peroxy-radicals (Cao *et al.* 1993). In some cases however, the AAPH (2, 2′-azobis (2-amidino-propane) dihydrochloride) has been used as the sole free-radical generator. The reaction is monitored by measuring the rate of the degeneration (or decomposition) of fluorescein as the presence of the antioxidant slows the fluorescence disappearance (decay) with time (Cao *et al.* 1993; Ou *et al.* 2001). The decay curves of fluorescence intensity against time are plotted, and the area under the curve calculated. The extent of the antioxidant-mediated protection is quantified against a standard antioxidant known as Trolox, which actually is a variant of tocopherols (vitamin E) (Huang *et al.* 2005).

The total radical trapping antioxidant parameter (TRAP) which refers to the moles of peroxyl radical trapped by a litre of fluid is calculated using 6-hydroxy-2,5,7,8-tetramethylchroman-2-carboxylic acid (Trolox) as a standard (Figure 1.8). The stoichiometric factor between the peroxyl radical per Trolox molecule is 2.

The ORAC assay is the assay mostly used for the determination of antioxidant activities, and has therefore been reported for many applications such as the determination of antioxidants in fruits and fruit juices (Wang *et al.* 1996); in fruits and vegetables (Wang *et al.* 1997); in tea extracts (Cao *et al.* 1996); in green and black tea (Serafijni 1996); and in a variety of herbs (Zheng and Wang 2001), and in the investigation of the influence of beer on the antioxidant activity (Ghiselli *et al.* 2000).

The wide application of the ORAC assay is due to its advantages, which include the fact that it can work effectively for samples with either slow- or fast-acting antioxidants or for mixed phases (Cao *et al.* 1993). However, ORAC assays are known to only work against peroxyl radicals, and there is no evidence that these radicals do form or even that the radicals are involved in the reactions as the damaging reactions cannot be characterised by ORAC. Due to these limitations of ORAC, a number of other ORAC-modified methods have been proposed and reported with the majority utilising the same principle (i.e. measurement of 2, 2′-azobis (2-amidino-propane) dihydrochloride (AAPH)-radical mediated damage of fluorescein). One of these ORAC-modified method is the ORAC-electron paramagnetic resonance (EPR), which actually gives a direct measurements of the decrease of AAPH-radical level by the scavenging action of the antioxidant substance (Kohri *et al.* 2009).

The higher ORAC magnitude of a certain food, typically given as ORAC units, the higher the level of antioxidants is in that particular food (Ou *et al.* 2001; Huang *et al.* 2002, 2005; Yu *et al.* 2005).

The ORAC assay is mostly suitable for hydrophilic and lipophilic antioxidants. Other methods such as the randomly methylated β-cyclodextrin (RMCD) have been developed, and are used as a molecular species to enhance the solubility of hydrophobic antioxidants (Huang *et al.* 2002). RMCD has been reported to be efficient at solubilising vitamin E compounds (among other hydrophilic antioxidants), though it cannot be applied to others such as carotenoids (Huang *et al.* 2002).

1.7.2 Hydroxyl radical antioxidant capacity (HORAC)

A hydroxyl radical antioxidant capacity (HORAC) assay is a complement to the ORAC assay and utilises the oxidation reaction of fluorescein by hydroxyl radicals via a classic hydrogen atom transfer (HAT) mechanism to generate free hydroxyl radicals by hydrogen peroxide (H_2O_2) (Luo *et al.* 2009). These free radicals will then be used to suppress the fluorescence of fluorescein over time. In the presence of antioxidants, a blockage of the hydroxyl radicals formed will initiate and proceed until all of the antioxidant activity in the sample is completely exhausted, leaving the H_2O_2 radicals to react with the fluorescence of fluorescein. The area under the fluorescence diminishing plot allows the total hydroxyl radical antioxidant activity in a sample to be calculated and compared to a standard curve (normally that of polyphenolic compounds such as gallic acid).

The advantage of this assay is that it gives a more direct measurement of antioxidant capacity for hydroxyl radicals. Unlike the ORAC which is validated for the determination of peroxyl radical absorbance capacity, the HORAC analyses the hydroxyl radical prevention capacity.

1.7.3 DPPH

This assay is based on the scavenging of DPPH (1,1-diphenyl-2-picrylhydrazyl) free radical (Om and Bhat 2009). The DPPH is a stable free radical of red colour and has an absorbance band at 515 nm. If free radicals have been scavenged by an antiradical compound DPPH will change colour to yellow, which also causes its absorption to disappear. The DPPH has a lone electron which causes a strong absorption maximum at 515 nm; when this lone electron is paired with another electron from an antioxidant, the absorption strength decreases causing a change of colour from red to yellow (Figure 1.9). The colour change is known to be stoichiometric relative to the number of electrons captured.

The decrease in absorbance is normally monitored at a wavelength band of 515 nm before the commencement of the reaction (time $= 0$ minutes), then at constant time intervals until the reaction plateaus. Antioxidant activity is then calculated as the amount of oxidant required to decrease the initial amount of DPPH by half (50%). The efficiency concentration is referred as EC_{50} (mol/L of AO divided by mol/L of DPPH). The antiradical power (ARP) is defined as the reciprocal of EC_{50}, i.e. $1/EC_{50}$. From these mathematical relationships, it follows that the larger the ARP value the more efficient the antioxidant (Brand-Williams *et al.* 1995).

1.7.4 Ferric reducing antioxidant power

The ferric reducing antioxidant power (FRAP) assay measures the reducing ability of antioxidants and, unlike many other assays, it does not make use of any radical; it only measures the reducing ability, and not even the radical quenching capacity (Benzie and Strain 1999).

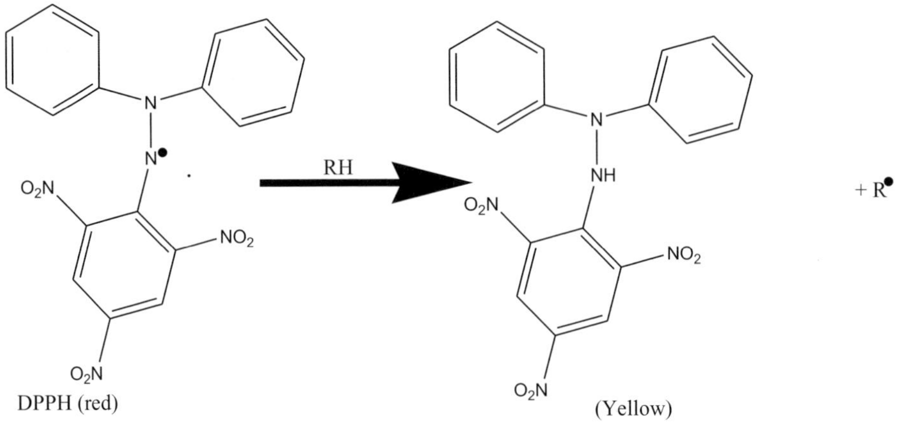

Fig. 1.9   Proposed reactions in the DPPH assay for antioxidant quality assurance (Om and Bhat 2009).

This test system uses antioxidants as reductants in a redox-linked colourimetric method, applying easily reduced oxidant species. At acidic pH, reduction of ferric tripyridyl triazine (Fe III TPTZ) complex to blue ferrous species can be monitored by measuring the change in absorption at 593 nm. The change in absorbance is directly proportional to the combined or total reducing power of the electron-donating antioxidants present in the reaction mixture.

1.7.5   Trolox equivalent antioxidant capacity (TEAC)

The chemical/scientific name for Trolox is 6-hydroxy-2, 5, 7, 8-tetramethylchroman-2-carboxylic acid, a hydrophilic compound which is a derivative of tocopherol and is widely used in biological and biochemical research to slow the oxidative stress and oxidative damage caused by the free radicals (Re *et al.* 1999). Trolox is the standard upon which the measurement of the Trolox equivalent antioxidant activity (TEAC) strength is based. The units for TEAC assays are in Trolox Equivalents (TE) and it is most often measured using ABTS (2, 2′-azino-bis (3-ethylbenzthiazoline-6-sulphonic acid), a chemical compound (Figure 1.10) used to monitor the decolourisation progress (Re *et al.* 1999).

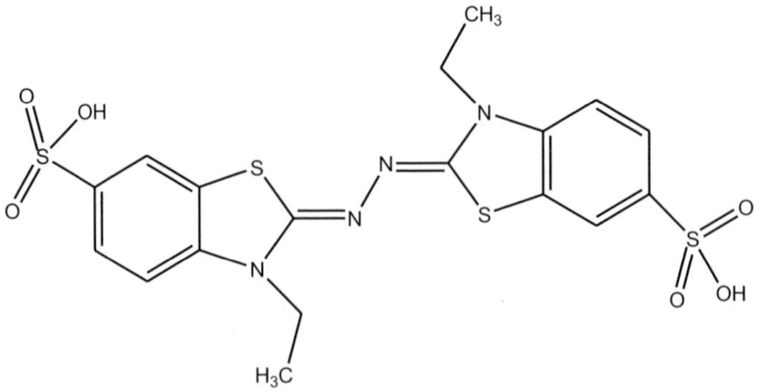

Fig. 1.10   The chemical structure of ABTS (2, 2-azino-bis (3-ethylbenzthiazoline 6-sulphonic acid)).

The TEAC assay has been used as an *in vitro* assay to ascertain the antioxidant capacity of foods and beverages (Huang *et al.* 2005). There are other antioxidant capacity assays that employ Trolox as a standard, including the ORAC, DPPH and FRAP assays.

1.7.6 ABTS

The ABTS assay is among the most widely employed technique for the measurement of antioxidant activity, and is based on the decolourisation of ABTS radical cation (2, 2-azino-bis (3-ethylbenzthiazoline 6-sulphonic acid)) (Rice-Evans and Miller 1994; Ivekovic *et al.* 2005).

The ABTS assay gives a measure of the overall antioxidant capacity within a given sample. The assay is based on the ability of antioxidants to inhibit the oxidation reactions of ABTS (Re *et al.* 1999). The assay itself involves the oxidation of ABTS to a product with an intensely coloured nitrogen-centred radical cation, ABTS·+ (Figure 1.11), which has an absorption maxima at 734 nm. Since most food extracts are also highly coloured but do not absorb light at 734 nm, this assay is a very useful tool for testing such foods.

The advantage of the ABTS system is that it yields the cumulative effect of all antioxidants present in the sample; more meaningful information can therefore be deduced, compared to the measurement of individual antioxidants (Re *et al.* 1999). It is also viable for both aqueous and lipophilic types of systems.

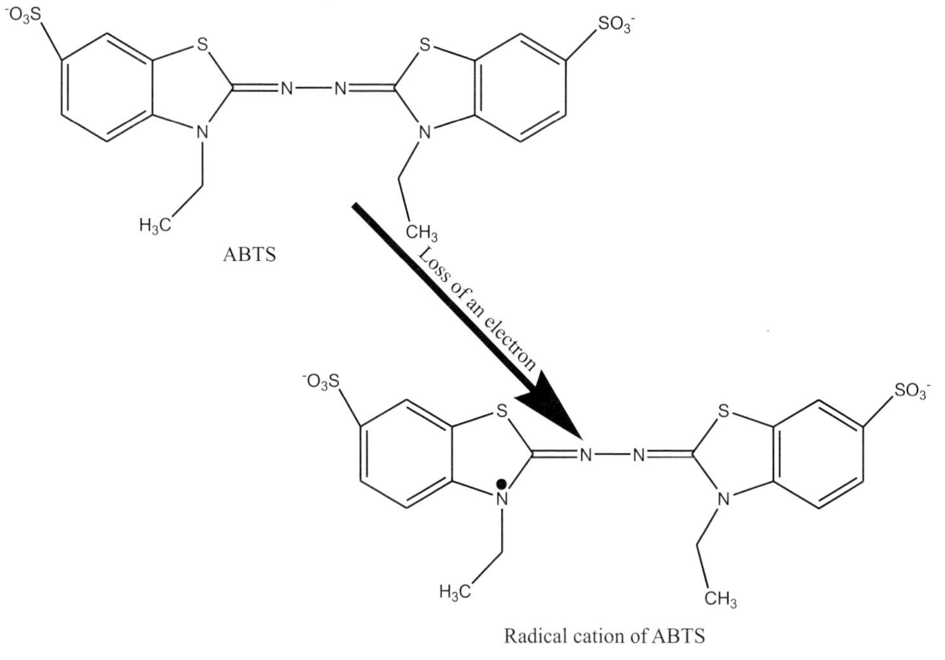

Fig. 1.11 Oxidation of ABTS to ABTS·+ radical (Re *et al.* 1999)

1.7.7 Copper (Cu^{2+}) reduction

The Cu^{2+} reduction assay measures antioxidant capacity by the simple principle of the reduction of the cupric ion to cuprous ions (i.e. Cu^{2+} to Cu^{+}) (Campos *et al.* 2009). Matrices containing antioxidants are mixed with Cu^{2+} solution and the Cu^{2+} ions will be reduced by antioxidants in the matrices to Cu^{+}, which will then react with chromatic solution (bathocuproine). The reaction with chromatic solution can be monitored by measuring absorbance at a range of wavelengths from 480 to 490 nm, and the antioxidant capacity can be easily calculated. The advantage of this assay is that it can be used to measure the antioxidant capacity of both hydrophilic antioxidants such as vitamin C and glutathione and hydrophobic antioxidants such as vitamin E (Proudfoot *et al.* 1997).

1.7.8 Photochemiluminescence (PCL)

In the photoluminescence (PCL) assay the process of photochemical generation of free radicals is coupled to the detection step, which is by means of chemiluminescence. The mechanism of this process is based on the photo-induced antioxidation inhibition of luminol (which works as a photosensitiser as well as the O$_2$ radical determination reagent) by antioxidants, mediated from the radical anion superoxide O$_2\cdot^{-}$ (Besco *et al.* 2007). The process is described by the equation (Popov and Lewin 1999):

$$\text{Luminol} + \text{light} + O_2 \rightarrow (\text{Luminol}^*O_2)\backslash \rightarrow \text{Luminol} \cdot^{+} + O_2\cdot^{-} \qquad (1)$$

The photochemical generation reaction is initiated by the optical excitation of a photosensitiser such as luminal, which then generates superoxide radical O$_2\cdot^{-}$ (Popov and Lewin 1999). The assay is mostly suitable for the measurements of radical antioxidation properties of a single antioxidant and also for more complex systems at very low concentrations. The antioxidant potential capacity is obtained by plotting the lag phase at various ranges of concentrations using a Trolox calibration curve, reported as mmol equivalent in antioxidant activity of Trolox.

1.7.9 Chemiluminescence

The antioxidant capacity can also be ascertained by monitoring the ability of antioxidants to quench chemiluminescence (Frei *et al.* 1988). In this assay, lipid hydroperoxide and isoluminol/microperoxidase reagent are used as the source to generate chemiluminescence. During the generation of chemiluminescence, lipid hydroperoxide reacts with microperoxidase to form an oxyradical (LO·) which then reacts with isoluminol to form a semiquinone radical, which will oxidise oxygen to O$_2\cdot^{-}$. The chemiluminescence is derived from isoluminol endoperoxide. Using a constant amount of lipid hydroperoxide (oxyradical donor: cumene hydroperoxide) the ability of antioxidants can be estimated as the decrease of chemiluminescence.

1.7.10 Fluorometric

This assay is normally used to measure the antioxidant power of the aqueous as well as the lipid antioxidants (Rimet *et al.* 1987; Brenan and Parish 1988). A lipid soluble radical initiator such as MeO-AMVN (2,2′-azobis(4-methoxy-2,4-dimethyl-valeronitrile)) is used together with a lipophilic fluorescence probe CII-BODIPY 581/591(4,4'-difluoro-5-(4-phenyl-1,3-butadienyl)-4-bora-3a,4a-diaza-5-indacene-3-undecanoic acid) to monitor the lipid compartment plasma oxidation. The red fluorescence decrease (excitation wavelength $\lambda_{ex} = 580$ nm, emission wavelength $\lambda_{em} = 600$ nm) of BODIPY and the green fluorescence increase ($\lambda_{ex} = 500$ nm, $\lambda_{em} = 520$ nm) of the oxidation are measured.

1.7.11 N, N-dimethyl-p-phenylenediamine

N,N-dimethyl-p-phenylenediamine (DMPD) is used to generate a radical cation which results in a coloured solution. When reacted with an antioxidant, the colour formation is inhibited (Locatelli *et al.* 2009). The coloured radical is formed by adding ferric chloride to the DMPD solution (Fe^{3+}:DPMD ratio 1:10) and the absorbance of this solution is measured at 505 nm. It may be stable (constant) up to 12 hours at room temperature. 50 µL of the antioxidant solution is added to 1 mL DMPD·$^+$ solution. The absorbance at 505 nm is measured after 10 min at 25°C under continuous stirring. Antioxidant activity is calculated as the percentage of the uninhibited radical solution according to the equation:

$$\text{Antioxidant activity } (\%) = [1 - (E_{505}\text{sample}/E_{505}\text{DMPD·}^+)] \times 100 \qquad (2)$$

This antioxidant activity can be expressed in terms of Trolox (Fogli *et al.* 1999).

1.7.12 Low-density lipoprotein (LDL)

This assay monitors the kinetics of the oxidation of low-density lipoprotein (LDL), in which polyunsaturated fatty acids of low-density lipoproteins are oxidised to form various products (Sakaue *et al.* 2000). The kinetic process is monitored continuously by observing changes of 234 nm diene absorbance which develops in LDL as the oxidation proceeds, resulting in the generation of conjugated fatty acid hydroperoxides. The formation of the dienes is directly proportional to the generation of lipid hydroperoxide.

The LDL assay has some major disadvantages, however. The ultracentrifugation, which is the most widely used procedure for LDL isolation, is a time-consuming step and preservatives such as ethylenediaminetetraacetic acid (EDTA) are regularly included in the high-salt solutions to limit oxidation. Isolated LDLs are often extensively dialysed in order to remove these compounds prior to the oxidation assay. Findings by Scheek *et al.* (1995) indicated that 56–65% in the concentrations of β-carotene, lycopene and α-tocopherol were due to dialysis. Due to this, Puhl *et al.* (1994) proposed the option of using gel filtration as a reliable alternative to dialysis.

1.7.13 Thiobarbituric acid reactive substances (TBARS)

This assay is used for the detection of lipid peroxidation where malondialdehyde is formed as a result of lipid peroxidation. The malondialdehyde reacts with barbituric acid to generate a pink pigment that has an absorbance maximum at 532 nm (Dawn-Linsley *et al.* 2005).

However, a shortcoming of the TBARS assay is that the reaction is not specific; many other substances including alkanals, proteins, sucrose and urea may react with thiobarbituric acid to form coloured species that can interfere with the assay. To counter this shortcoming and enhance the specificity of the assay, the use of high-performance liquid chromatography (HPLC) for the separation of the complex formed prior to measurements has been proposed and used successfully. Other approaches, including the extraction of malondialdehyde prior to the formation of chromogen and/or derivative spectrophotometry, have also been used with success.

1.7.14 Brigg-Rauscher

The Brigg-Rauscher (BR) assay is a procedure to monitor the relative activity of antioxidants according to the inhibitory effects exerted by each of the free radical scavengers, as measured by the oscillations of the BR mechanistic processes (Cervellati *et al.* 2001). The BR oscillating system is generated through the iodination and oxidation of malonic acid and related substrates using acidic iodate, with hydrogen peroxide and manganese ions (Mn^{2+}) serving as catalysts. The antioxidant leads to the immediate cessation of the oscillation, but after the so-called inhibition time the oscillation behaviour is regenerated. The BR reaction shows good amplitude, frequency and duration of oscillation at pH c. 2.

1.7.15 Electrochemical

The electrochemical assay is normally performed in a flow injection analysis fashion using an electrochemical detector and a glassy carbon electrode running amperometrically at a constant potential (normally at +0.5V; Buratti *et al.* 2001). Flow injection experiments are performed amperometrically under the principles of either oxidation or reduction of an electro-active compound at the glassy carbon working electrode at a constant applied potential. The measured current is a direct measurement of the electrochemical reaction rate.

1.7.16 β-carotene bleaching

This assay follows the principles of the coupled oxidation of β-carotene and linoleic acid. The bleaching of β-carotene resulting from oxidation by degeneration products of linoleic acid is measured.

1.7.17 Comparison of different assays for dietary total antioxidant capacity

Due to the large number of various analytical test assay systems, an evaluation of their performance for the purposes of comparison would be useful. Prior and Cao (1999) made a comparative study of FRAP, TEAC, TRAP and β-carotene bleaching assays, in terms of their reaction principles; however, no conclusive details resulted from their study. Other researchers (Wiseman *et al.* 1997; Cao and Prior 1998; Rice-Evans 2000; Protoggente *et al.* 2002) have compared the above-mentioned assay systems and shown that their performance was similar. Vinson *et al.* 2001 included LDL in the comparison, and reported similar results. Ou *et al.* 2002 however demonstrated different antioxidant activity trends for 927 freeze-dried fruits, although they reported an irregular relationship between ORAC and FRAP values. Schlesier *et al.* 2002 compared TEAC, TRAP, DPPH, DMPD, PCL and FRAP assays for gallic, uric acid, ascorbic acid and Trolox; results showed that TEAC indicated gallic as the strongest antioxidant while DMPD indicated ascorbic acid.

1.8 HOW SAFE ARE FOOD ANTIOXIDANTS?

The overwhelming application of antioxidants in foods (especially processed) in this era is certainly alarming, and is an issue of concern to health practitioners due to the possible health risks associated with the many antioxidants used. Antioxidants of chemical or synthetic origin generate the most concern. Some of these synthetic radical scavengers, for example monomeric antioxidants, have been associated with a number of pathological effects. They are potential carcinogens and may interact negatively with enzymes to have undesirable effects on health and reproduction (Gower 1988; Sun 1990).

Due to the low concentration levels at which they are used, the majority of antioxidants are however expected to be non-toxic in food production practices (Daniel 1986). Excessive application of antioxidants to food has the potential to promote lipid peroxidation in cooking-ware made of copper and iron, however (Reddy and Lokesh 1992). Phenolic antioxidants such as BHA and BHT have been associated with the worsening of diseases such as urticaria (Goodman *et al.* 1990). Generally, an overdose of BHT is very harmful to human beings (Shlian and Goldstone 1986). Propyl gallate is another phenolic flavonoid which has been listed as a human carcinogen (van der Heijden *et al.* 1986).

Further Thinking

Despite the fact that antioxidants in foods provide health benefits, their safety always needs to be established and verified scientifically. Some conditions need to be fulfilled for an antioxidant to be certified as fit for human consumption, such as their LD50 values (the lethal dose at which 50% of test species die) not exceeding 1000 mg/kg body weight. Their toxicities should be proved not to cause any significant physiological effects to experimental organisms (e.g. rats) when tested over a long period at 100 times the concentration levels expected to be used in foodstuffs for human consumption. Moreover, an antioxidant must demonstrate that it is not toxic (not mutagenic, teratogen or carcinogenic).

Due to the possible health hazards of some of the residues of antioxidants used in foods, there are a number of guidelines that have been set by international authorities such as the European Union with regard to the use of food supplements (Directive 2002/46/EC). It has been legislated that the total concentration of permitted antioxidants incorporated singly or in a mixture should be below 200 parts per million by weight when measured in fats (Directive 2002/46/EC).

Adherence to the legislation is monitored and a number of methods including: electrochemical detection (Brainina *et al.* 2007; Milardovic *et al.* 2007; Kamel *et al.* 2008; Ragubeer *et al.* 2010); spectrophotometric (Szydłowska-Czerniaka *et al.* 1994); fluorometric (López *et al.* 2003; Ribeiro *et al.* 2010); capillary electrophoresis (Herrero-Martinez *et al.* 2004; Hernández-Borges *et al.* 2005); liquid chromatography (Zhang *et al.* 2005; Celik *et al.* 2010); gas chromatography (Caceres *et al.* 1963); and chromatographic methods hyphenated to mass spectrometric detection (Bravo *et al.* 2007) have been reported. See the following sections for descriptions of these methods.

1.8.1 Electrochemical

One of the mechanisms of action for antioxidants involves the donation of electrons; this allows the possibility of electrochemical methods to be applied in the determination of such molecules (Chevion *et al.* 1997).

An example of an electrochemical method is cyclic voltammetry (CV), which has been reported in the determination of antioxidant capacity of various food products (Chevion *et al.* 2000). Cyclic voltammetry is also useful in the measurement of the ability of a number of other molecules with regard to their ability to donate electrons (Huang *et al.* 2004). It has been reported that most of the low molecular weight antioxidants are excellent reducing agents, related to their high capabilities in terms of donating electrons (i.e. strong electro-active species). The magnitude of half-wave potential ($E_{1/2}$ value), defined as the potential at half the height of the peak of the anodic current wave, is used as an indicator of the reducing power of antioxidants. The square wave voltammetry (SWV) is also used for the determination of antioxidants.

Electro-analytical methods for the determination of antioxidants are generally attractive because of the fact that they are: easy to control; not affected by turbid solutions of analytes; and can be used to analyse radical species in organic or aqueous solvents (Buratti *et al.* 2001; de Abreu *et al.* 2002).

1.8.2 High-performance liquid chromatography (HPLC)

Methods involving the use of HPLC in the determination of antioxidants in foods are attractive because the technique itself is known for its versatility, precision and relatively low cost (Escarpa and Gonzalez 2000, 2001; Tsao and Yang 2003). In most cases, liquid chromatography for analysis of antioxidants is performed either under reversed-phase or ion-exclusion conditions using a variety of packing stationary phases such as C18 columns, mobile phases consisting of acidified water and polar organic solvents (e.g. acetonitrile or methanol) and diode array detection (DAD) (Merken and Beecher 2000; Robards 2003).

1.8.3 Capillary electrophoresis

The two most widely used modes of capillary electrophoresis (CE) for the determination of antioxidants are: (1) capillary zone electrophoresis (CZE) and (2) micellar electrokinetic chromatography (MEKC) (Pietta *et al.* 1998; Pomponio *et al.* 1998; Sheu *et al.* 2001; Chen *et al.* 2001; Pomponio *et al.* 2002). For quick electrophoretic separations of anionic antioxidant species, the electro-osmotic flow (EOF) is normally reversed and the cationic surfactants included in the buffering electrolyte medium (Masselter and Zemann 1995; Volgger *et al.* 1997).

1.8.4 Mass spectrometry

Most of the methods described here, and especially those involving mass spectrometry, are sensitive enough to monitor these compounds to very low detection limits and are suitable for both routine analyses as well as for confirmation. Spectrometric methods for the determination of antioxidants are mainly hyphenated to chromatographic methods, with either a liquid chromatograph or a gas chromatograph being coupled to a mass spectrometer (Choy *et al.* 1963).

1.8.5   Spectroscopy

Methods developed for measuring antioxidant capacity are based on either inhibition or non-inhibition principles (Ronald and Guohua 1999), but all lie within the major class of spectrophotometry.

1.9   SUMMARY

The main function performed by food antioxidants is to either control or slow down the auto-oxidation processes that are always undesirable in foods; they are responsible for rancidity phenomena, spoilage and off-flavours. There are many processes such as photo-oxidation, oxidation triggered by enzymes such as lipo-oxygenase and thermal-induced oxidation which all result in food quality deterioration. There will always be a need to control and retard such processes to ensure food quality; the presence of antioxidants is therefore of huge importance in foods.

REFERENCES

Amakura, Y., Okada, M., Tsuji, S. & Tonogai, Y. (2000) Determination of phenolic acids in fruit juices by isocratic column liquid chromatography. *Journal of Chromatography A*, 891 (1), 183–188.

Anbazhagan, V., Kalaiselvan, A., Jaccob, M., Venuvanalingam, P. & Renganathan, R. (2008) Investigations on the fluorescence quenching of 2, 3-diazabicyclo (2.2.2) oct-2-ene by certain flavonoids. *Journal of Photochemistry and Photobiology B* 91 (2–3), 143–150.

Apak, R., Güçlü, K., Demirata, B., Özyürek, M., Çelik, S. E., Bektaşoğlu, B., Berker, K. I. & Özyurt, D. (2007) Comparative evaluation of various total antioxidant capacity assays applied to phenolic compounds with the CUPRAC Assay. *Molecules*, 12, 1496–1547.

Arichi, H., Kimura, Y., Okuda, H., Baba, K., Kozawa, M. & Arichi, S. (1982) Effects of stilbene components of the roots of polygonum-cuspidatum on lipid-metabolism. *Chemical & Pharmaceutical Bulletin* 30 (5), 1766–1770.

Aruoma, O. I., Murcia, A., Butler, J. & Halliwell, B. (1993) Evaluation of the antioxidant and prooxidantactions of gallic acid and its derivatives. *Journal of Agriculture & Food Chemistry* 41 (11), 1880–1885.

Belguendouz, L., Fremont, L. & Linard, A. (1997) Resveratrolinhibits metal ion-dependent and independent peroxidation of porcine low-density. *Biochemical Pharmacology* 53 (9) 1347–1355.

Beltrán, G., Jiménez, A., del Rio, C., Sánchez, S., Martínez, L., Uceda, M. & Aguilera, M. P. (2010) Variability of vitam in E in virgin olive oil by agronomical and genetic factors. *Journal of Food Composition and Analysis* 23 (6), 633–639.

Benzie, F. F. & Strain, J. J. (1999) Ferric Reducing/Antioxidant Power assay: Direct measure of total antioxidant activity of biological fluids and modified version for simultaneous measurement of total antioxidant power and ascorbic acid concentration. *Methods in Enzymology* 299, 15–23.

Besco, E., Braccioli, E., Vertuani, S., Ziosi, P., Brazzo, F., Bruni, R., Sacchetti, G. & Manfredini, S. (2007). The use of photochemiluminescence for the measurement of the integral antioxidant capacity of baobab products. *Food Chemistry* 102 (4), 1352–1356.

Bors, W., Hellers, W., Michel, C., & Saran, M. (1990a) Radical chemistry of flavonoid antioxidants. In: *Antioxidants in Therapy and Preventive Medicine*, Emerit, I. (ed.), Plenum Pub Corp., 1, 165–170.

Bors, W., Werner, H., Michel, C. & Saran, M. (1990b) Flavonoids as antioxidants – determination of radical-scavenging efficiencies. *Methods in Enzymology* 186, 343–355.

Bracegirdle, S. & Anderson, E. A. (2010) Arylsilane oxidation—new routes to hydroxylated aromatics. *Chemical Communications* 46, 3454–3456.

Brainina, Kh. Z., Ivanova, A. V., Sharafutdinova, E. N., Lozovskaya, E. L. & Shkarina, E. I. (2007) Potentiometry as a method of antioxidant activity investigation. *Talanta* 71, 13–18.

Brand-Williams, W., Cuvelier, M. E. & Berset, C. (1995) Use of a free radical method to evaluate antioxidant activity. *Lebensmittel-Wissschaft und Technologie* 28, 25–30.

Bravo, L., Goya, L. & Lecumberri, E. (2007) LC/MS characterization of phenolic constituents of mate (*Ilex paraguariensis*, St. Hil.) and its antioxidant activity compared to commonly consumed beverages. *Food Research International* 40, 393–405.

Brenan, M. & Parish, C. R. (1988) Automated fluorometric assay for T cell cytotoxicity. *Journal of Immunology Methods*, 112 (1), 121–131.

Buratti, S., Pellegrini, N., Brenna, O. V. & Mannino, S. (2001) Rapid electrochemical method for the evaluation of the antioxidant power of some lipophilic food extracts. *Journal of Agriculture & Food Chemistry* 49 (11), 5136–5141.

Caceres, C. A., Calatayud, J. B., Orvis, H. H., Fawal, I. A., Thomas, R., Kelser, G. A. Jr, Abraham, S. & Anderson, A. (1963) An evaluation of clinical and laboratory findings in male subjects on long-term, low-fat, low-protein diets. *The New England Journal of Medicine* 269, 550–555.

Cai, Y-J., Ma, L-P., Hou, L-F., Zhou, B., Yang, L. & Liu, Z.-L. (2002) Antioxidant effects of green tea polyphenols on free radical initiated peroxidation of rat liver microsomes. *Chemistry & Physics of Lipids*, 120 (1–2), 109–117.

Campos, C., Guzmán, R., López-Fernández, E. & Casado, Á. (2009) Evaluation of the copper (II) reduction assay using bathocuproinedisulphonic acid disodium salt for the total antioxidant capacity assessment: The CUPRAC–BCS assay. *Analytical Biochemistry* 392 (1), 37–44.

Cao, G. H. & Prior, R. L. (1998) Comparison of different analytical methods for assessing total antioxidant capacity of human serum. *Clinical Chemistry* 44 (6), 1309–1315.

Cao, G., Alessio, H. & Cutler, R. (1993) Oxygen-radical absorbance capacity assay for antioxidants. *Free Radical Biology & Medicine* 14 (3), 303–11.

Cao, G. H., Sofic, E. & Prior, R. L. (1996) Antioxidant capacity of tea and common vegetables. *Journal of Agricultural and Food Chemistry* 44 (11), 3426–3431.

Cao, G., Sofic, E. & Prior, R. L. (1997) Antioxidant and prooxidant behavior of flavonoids, Structure-activity relationships. *Free Radical Biology & Medicine* 22 (5), 749–760.

Celik, S. E., Ozyurek, M., Guclu, K. & Apak, R (2010) Determination of antioxidants by a novel on-line HPLC-cupric reducing antioxidant capacity (CUPRAC) assay with post-column detection. *Analytica Chimica Acta* 674 (1), 79–88.

Cervellati, R., Honer, K., Furrow, S. D., Neddens, C. & Costa, S. (2001) The Briggs-Rauscher reaction as a test to measure the activity of antioxidants. *Helvetica Chimica Acta* 84 (12), 3533–3547.

Chen, Z. L., Krishnamurti, G. S. & Naidu, R. (2001) Separation of phenolic acids in soil and plant tissue extracts by co-electroosmotic capillary electrophoresis with direct UV detection. *Chromatographia* 53, 179–184.

Chevion, S., Berry, E. M., Kitrossky, N. & Kohen, R. (1997) Evaluation of plasma low molecular weight antioxidant capacity by cyclic voltammetry. *Free Radical Biology & Medicine* 22 (3), 411–421.

Chevion, S., Roberts, M. A. & Chevion, M. (2000) The use of cyclic voltammetry for the evaluation of antioxidant capacity. *Free Radical Biology & Medicine* 28 (6), 860–870.

Choy, T. K., Quattrone Jr. J. J. & Alicino, N. J. (1963) A gas chromatographic method for the determination of the antioxidants BHA, BHT and ethoxyquin in aqueous and in hydrocarbon soluble samples. *Journal of Chromatography A*, 12, 171–177.

Chu, Y. F., Sun, J., Wu, X. & Liu, R. H. (2002) Antioxidant and antiproliferative activities of common vegetables. *Journal of Agriculture & Food Chemistry* 50 (23), 6910–6916.

Daniel, J. W. (1986) Metabolic aspects of antioxidants and preservatives. *Xenobiotica*, 16, 1073–1078.

Das, D. K. (1994) Naturally-occurring flavonoids:structure, chemistry, and high-performance liquid-chromatography methods for separation and characterization. *Methods in Enzymology* 234, 410–420.

Davies, K. (1995) Oxidative stress: the paradox of aerobic life. *Biochemical Society Symposia* 61, 1–31.

Dawn-Linsley, M., Ekinci, F. J., Ortiz, D., Rogers, E. & Shea, T. B. (2005) Monitoring thiobarbituric acid-reactive substances (TBARs) as an assay for oxidative damage in neuronal cultures and central nervous system. *Journal of Neuroscience Methods*, 141 (2), 219–222.

de Abreu, F. C., de L. Ferraz, P. A. & Goulart, M. O. F. (2002) Some applications of electrochemistry in biomedical chemistry. Emphasis on the correlation of electrochemical and bioactive properties. *Journal of Brazilian Chemical Society* 13 (1), 19–35.

DerMarderosian, A. (2001) *The Review of Natural Products. Facts and Comparisons*. Lippincott, Williams and Wilkins, St Louis, USA.

Directive 2002/46/EC of the European Parliament and of the Council of 10 June 2002 on the approximation of the laws of the Member States relating to food supplements. Commission of the European Parliament and the Council of the European Union, Brussels, Belgium.

Escarpa, A. & Gonzalez, M. C. (2000) Optimization strategy and validation of one chromatographic method as approach to determine the phenolic compounds from different sources. *Journal of Chromatography A*, 897, 161–170.

Escarpa, A. & Gonzalez, M. C. (2001) Approach to the content of total extractable phenolic compounds from different food samples by comparison of chromatographic and spectrophotometric methods. *Analytica Chimica Acta*, 427, 119–127.

Finkel, T. & Holbrook, N. J. (2000) Oxidants, oxidative stress and the biology of ageing. *Nature* 408 (6809), 240–247.

Firuzi, O., Lacanna, A., Petrucci, R., Marrosu, G. & Saso, L. (2005) Evaluation of the antioxidant activity of flavonoids by 'Ferric Reducing Antioxidant Power' assay and cyclic voltammetry. *Biochimica et Biophysica Acta*, 1721, 174–184.

Fogli, S., Danesi, R., Innocenti, F., Di Paolo, A., Bocci, G., Barbara, C. & Del Tacca, M. (1999) An improved HPLC method for therapeutic drug monitoring of daunorubicin, idarubicin, doxorubicin, epirubicin, and their 13-dihydro metabolites in human plasma. *Therapeutic Drug Monitoring* 21, 367–375.

Frankel, E. N., Waterhouse, A. L. & Teissedre, P. L. (1995) Principal phenolic phytochemicals in selected california wines and their antioxidant activity in inhibiting oxidation of human low-density lipoproteins. *Journal of Agriculture & Food Chemistry* 43 (4), 890–894.

Frei, B., Yamamoto, Y., Niclas, D. & Ames, B. N. (1988) Evaluation of an isoluminol chemiluminescence assay for the detection of hydroperoxides in human blood plasma. *Analytical Biochemistry* 175 (1), 120–130.

Ghiselli, A., Natella, F., Guidi, A., Montanari, L., Fantozzi, P. & Scaccini, C. (2000) Beer increases plasma antioxidant capacity in humans. *Journal of Nutritional Biochemistry* 11 (2), 76–80.

Goodman, D. L., McDonnell, J. T., & Nelson, H. S. (1990) Chronic urticaria exacerbated by the antioxidant food preservatives, butylated hydroxyanisole (BHA) and butylated hydroxytoluene (BHT). *Journal of Allergy & Clinical Immunology* 86, 570–575.

Gower, J. D. (1988) A role for dietary lipids and antioxidants in the activation of carcinogens. *Free Radical Biology & Medicine* 5 (2), 95–111.

Graham, S., Haughey, B., Marshall, J., Priore, R., Byers, T., Rzepka, T., Mettlin, C. & Pontes, J. E. (1983) Diet in the epidemiology of carcinoma of the prostate-gland. *Journal of the National Cancer Institute* 70 (4), 687–692.

Gutteridge, J. M. C. (1994) Free radicals and aging. *Reviews in Clinical Gerontology* 4, 279–288.

Halvorsen, B. L., Holte, K., Myhrstad, M. C., Barikmo, I., Hvattum, E., Remberg, S. F., Wold, A. B., Haffner, K., Baugerod, H., Andersen, L. F., Moskaug, O., Jacobs, D. R. Jr. & Blomhoff, R. A. (2002) Systematic screening of total antioxidants in dietary plants. *Journal of Nutrition* 132 (3), 461–471.

Hernández-Borges, J., González-Hernández, G., Borges-Miquel, T. & Rodríguez-Delgado, M. A. (2005) Determination of antioxidants in edible grain derivatives from the Canary Islands by capillary electrophoresis. *Food Chemistry* 91, 105–111.

Herrera, E. & Barbas, C. (2001) Vitamin E: action, metabolism and perspectives. *Journal of Physiology & Biochemistry* 57 (2), 43–56.

Herrero-Martinez, J. M., Schoenmakers, P. J. & Kok, W. T. (2004) Determination of the amylose-amylopectin ratio of starches by iodine-affinity capillary electrophoresis. *Journal of Chromatography A*, 1053 (1–2), 227–234.

Hertog, M. G. L., Hollman, P. C. H. & Katan, M. B. (1992) Content of potentially anticarcinogenic flavonoids of 28 vegetables and 9 fruits commonly consumed in the Netherlands. *Journal of Agriculture & Food Chemistry* 40 (12), 2379–2383.

Huang, D., Ou, B., Hampsch-Woodill, M., Flanagan, J. A. & Deemer, E. K. (2002) Development and validation of oxygen radical absorbance capacity assay for lipophilic antioxidants using randomly methylated â-cyclodextrin as the solubility enhancer. *Journal of Agriculture & Food Chemistry* 50, 1815–1821.

Huang, D., Ou, B. & Prior, R. (2005) The chemistry behind antioxidant capacity assays. *Journal of Agriculture & Food Chemistry* 53 (6), 1841–1856.

Huang, T., Gao, P. & Hageman, M. J. (2004) Rapid screening of antioxidants in pharmaceutical formulation development using cyclic voltammetry: potential and limitations. *Current Drug Discovery Technologies* 1 (2), 173–179.

Hung, L. M., Chen, J. K., Huang, S. S., Lee, R. S. & Su, M. J. (2000) Cardioprotective effect of resveratrol, a natural antioxidant derived from grapes. *Cardiovascular Research* 47 (3), 549–555.

Hwang, P. L. (1991) Biological-activities of oxygenated sterols – physiological and pathological implications. *BioEssays* 13 (11), 583–589.

Ivekovic, D., Milardovic, S., Roboz, M. & Grabaric, B. S. (2005) Evaluation of the antioxidant activity by flow injection analysis method with electrochemically generated ABTS radical cation. *Analyst* 130 (5), 708–714.

Jayatilake, G. S., Jayasuriya, H., Lee, S. S., Koonchanok, N. M., Geahlen, R. L., Ashendel, C. L., McLaughlin, J. L. & Chang, C. J. (1993) Kinase inhibitors from polygonum-cuspidatum. *Journal of Natural Products* 56 (10), 1805–1810.

Ji, H.-F. & Shen, L. (2008) Mechanisms of singlet and triplet state riboflavin quenching by 3-hydroxypyridine and 4-hydroxypyridine. A quantum chemical study. *Journal of Molecular Structure: Theochem* 865 (1–3), 25–27.

Kamel, A. H., Moreira, F. T. C., Delerue-Matos, C., Goreti M. & Sales, F. (2008) Electrochemical determination of antioxidant capacities in flavored waters by guanine and adenine biosensors. *Biosensors and Bioelectronics* 24, 591–599.

Kimura, Y., Okuda, H. & Arichi, S. (1985a) Effects of stilbenes on arachidonate metabolism in leukocytes. *Biochimica et Biophysica Acta* 834 (2), 275–278.

Kimura, Y., Okuda, H. & Arichi, S. (1985b) Effects of stilbene derivatives on arachidonate metabolism in leukocytes. *Biochimica et Biophysica Acta* 837 (2), 209–212.

King, J. C. & Cousins, R. J. (2006) *Zinc in Modern Nutrition in Health and Disease*, 10th edition. Lipponcott Williams and Wilkins, Philadelphia, 271–285.

Kohri, S., Fujii, H., Oowada, S., Endoh, N., Sueishi, Y., Kusakabe, M., Shimmei, M. & Kotake, Y. (2009) An oxygen radical absorbance capacity-like assay that directly quantifies the antioxidant's scavenging capacity against AAPH-derived free radicals. *Analytical Biochemistry* 386 (2), 167–171.

Kovacic, P. & Somanathan, R. (2008) Ototoxicity and noise trauma: Electron transfer, reactive oxygen species, cell signaling, electrical effects, and protection by antioxidants: Practical medical aspects. *Medical Hypotheses* 70 (5), 914–923.

Kranl, K. (2004) Comparing antioxidative food additives and secondary plant products. *Food Chemistry* 93, 171–175.

Krieger-Liszkay, A. (2005) Singlet oxygen production in photosynthesis. *Journal of Experimental Botany* 56 (411), 337–346.

Lachance, P. A., Nakat, Z. & Jeong, W.-S. (2001) Antioxidants: an integrative approach. *Nutrition* 17, 835–838.

Lambert, C. R., Black, H. S. & Truscott, T. G. (1996) Reactivity of butylated hydroxytoluene. *Free Radical Biology and Medicine* 21 (3), 395–400.

Lemaska, K., Szymusiak, H., Tyrakowska, B., Zieliski, R., Soffers, E. M. F. & Rietjens, I. M. C. M. (2001) The influence of pH on antioxidant properties and the mechanism of antioxidant action of hydroxyflavones. *Free Radical Biology & Medicine* 31 (7), 869–881.

Lenaz, G. (2001) The mitochondrial production of reactive oxygen species: mechanisms and implications in human pathology. *IUBMB Life* 52 (3–5), 159–164.

Locatelli, M., Gindro, R., Travaglia, F., Coïsson, J.-D., Rinaldi, M. & Arlorio, M. (2009) Study of the DPPH·-scavenging activity: Development of a free software for the correct interpretation of data. *Chemistry* 114 (3), 889–897.

Lopez, M., Martinez, F., Del Valle, C., Ferrit, M. & Luque, R. (2003) Study of phenolic compounds as natural antioxidants by a fluorescence method. *Talanta* 60 (2–3), 609–616.

Luo, Y., Xao-rong, W., Liang-liang, J. & Yan, S. (2009) EPR detection of hydroxyl radical generation and its interaction with antioxidant system in *Carassius auratus* exposed to pentachlorophenol. *Journal of Hazardous Materials* 171 (1–3), 1096–1102.

Luthria, D. L., Biswas, R. & Natarajan, S. (2007) Comparison of extraction solvents and techniques used for the assay of isoflavones from soybean. *Food Chemistry* 105 (1), 325–333.

Luzhkov, V. B. (2005) Mechanism of antioxidant activity: The DFT study of hydrogen abstraction from phenol and toluene by the hydroperoxyl radical. *Chemical Physics* 314 (1–3), 211–217.

Marinova, E. M. & Yanishlieva, N. V. (1992) Effect of temperature on the antioxidative action of inhibitors in lipid autoxidation. *Journal of the Science of Food and Agriculture* 60, 313–318.

Marinova, E. M. & Yanishlieva, N. V. (1998) Antioxidative action of quercetin and morin in triacylglycerols of sunflower oil at ambient and high temperatures. *Seifen Öle Fette Wachse* 124, 10–16.

Masselter, S. M. & Zemann, A. J. (1995) Influence of organic solvents in co-electroosmotic capillary electrophoresis of phenols. *Analytical Chemistry* 67, 1047–1053.

Merken, H. M. & Beecher, G. R. (2000) Measurement of food flavonoids by high-performance liquid chromatography: A review. *Journal of Agriculture & Food Chemistry* 48, 577–599.

Milardovic, S., Kerekovic, I. & Rumenjak, V. (2007) Analytical, nutritional and clinical methods a flow injection biamperometric method for determination of total antioxidant capacity of alcoholic beverages using bienzymatically produced ABTS. *Food Chemistry* 105, 1688–1694.

Nakabeppu, Y., Sakumi, K., Sakamoto, K., Tsuchimoto, D., Tsuzuki, T. & Nakatsu, Y. (2006) Mutagenesis and carcinogenesis caused by the oxidation of nucleic acids. *Biology & Chemistry* 387 (4), 373–379.

Nesaretnam, K., Khor, H. T., Ganeson, J., Chong, Y. H., Sundram, K. & Gapor, A. (1992) The effect of vitamin E tocotrienols from palm oil on chemically induced mammary carcinogenesis in female rats. *Nutrition Research* 12 (7), 879–892.

Om, P. & Bhat, T. K. (2009) DPPH antioxidant assay revisited. *Food Chemistry* 113 (4), 1202–1205.

Oschman, J. L. (2009) Charge transfer in the living matrix. *Journal of Bodywork and Movement Therapy* 13 (3), 215–228.

Ou, B., Hampsch-Woodill, M. & Prior, L. (2001) Development and validation of an improved oxygen radical absorbance capacity assay using fluorescein as the fluorescent probe. *Journal of Agriculture & Food Chemistry* 49 (10), 4619–4626.

Ou, B., Hampsch-Woodill, M., Flanagan, J., Deemer, E. K., Prior, R. L. & Huang, D. (2002) Novel fluorometric assay for hydroxyl radical prevention capacity using fluorescein as the probe. *Journal of Agriculture & Food Chemistry* 50, 2772–2777.

Pace-Asciak, C. R., Hahn, S. E., Diamandis, E. P., Soleas, G. & Goldberg, D. M. (1995) The red wine phenolics trans-resveratrol and quercetin block human platelet-aggregation and eicosanoid synthesis: implications for protection against coronary heart-disease. *Clinica Chimica Acta* 235 (2), 207–219.

Pearson, D. A., Schmitz, H. H., Lazarus, S. A. & Keen, C. L. (2001) Inhibition of in vitro low-density lipoprotein oxidation by oligomeric procyanidins present in chocolate and cocoas: Flavonoids and other polyphenols. *Methods in Enzymology* (Book Series) 335, 350–360.

Pereira, T. A. & Das, N. P. (1990) The effects of flavonoids on the thermal autoxidation of palm oil and other vegetable oils determined by differential scanning calorimetry. *Thermochimica Acta*, 165 (1), 129–137.

Pietta, P., Mauri, P. & Bauer, R. (1998) MEKC analysis of different Echinacea species. *Planta Medica* 64 (7), 649–652.

Pinho, O., Ferreira, I. M. P. L. V. O., Oliveira, M. B. P. P. & Ferreira, M. A. (2000) Quantification of synthetic phenolic antioxidants in liver pates. *Food Chemistry* 68 (3), 353–357.

Pokorny, J. (1986) Addition of antioxidants for food stabilization to control oxidative rancidity. *Czech Journal of Food Science* 4, 299–307.

Pomponio, R., Gotti, R., Hudaib, M. & Cavrini, V. (2002) Analysis of phenolic acids by micellar electrokinetic chromatography: application to Echinacea purpurea plant extracts. *Journal of Chromatography A*, 945, 239–247.

Popov, I. & Lewin, G. (1999) Photochemiluminescent detection of antiradical activity. VI. Antioxidant characteristics of human blood plasma, low density lipoprotein, serum albumin and amino acids during in vitro oxidation. *Luminescence* 14 (3), 169–174.

Prior, R. L. & Cao, G. H. (1999) In vivo total antioxidant capacity: Comparison of different analytical methods. *Free Radical Biology and Medicine* 27 (11–12), 1173–1181.

Priyadarsini, K. I., Maity, D. K., Naik, G. H., Kumar, M. S., Unnikrishnan, M. K., Satav, J. G. & Mohan, H. (2003) Role of phenolic O-H and methylene hydrogen on the free radical reactions and antioxidant activity of curcumin. *Free Radical Biology & Medicine* 35 (5), 475–484.

Protoggente, A. R., Pannala, A. S., Paganga, G., Van Buren, L., Wagner, E., Wiseman, S., Van De Put, F., Dacombe, C. & Rice-Evans, C. A. (2002) The antioxidant activity of regularly consumed fruit and vegetables reflects their phenolic and vitamin C composition. *Free Radical Research* 36, 217–233.

Proudfoot, J. M., Croft, K. D., Puddey, I. B. & Beilin, L. J. (1997) The role of copper reduction by -tocopherol in low-density lipoprotein oxidation. *Free Radical Biology & Medicine* 23 (5), 720–728.

Puhl, H., Waeg, G. & Esterbauer, H. (1994) Methods to determine oxidation of low-density lipoproteins. *Methods in Enzymology* 233, 425–441.

Ragubeer, N., Beukes, D. R. & Limson, J. L. (2010) Critical assessment of voltammetry for rapid screening of antioxidants in marine algae. *Food Chemistry* 121, 227–232.

Raha, S. & Robinson, B. (2000) Mitochondria, oxygen free radicals, disease and ageing. *Trends in Biochemical Science* 25 (10), 502–508.

Re, R., Pellegrini, N., Pannala, A., Yang, M. & Rice-Evans, C. (1999) Antioxidant activity applying an improved ABTS radical cation decolorization assay. *Free Radical Biology & Medicine* 26, 1231–1237.

Rechner, A. R., Spencer, J. P. E., Kuhnle, G., Hahn, U. & Rice-Evans, C. A. (2001). Novel biomarkers of the metabolism of caffeic acid derivatives in vivo. *Free Radical Biology & Medicine* 30 (11), 1213–1222.

Reddy, A. C. P. & Lokesh, B. R. (1992) Studies on spice principles as antioxidants in the inhibition of lipid-peroxidation of rat-liver microsomes. *Molecular & Cellular Biochemistry* 111 (1–2), 117–124.

Ribeiro, J. P. N., Magalhaes, L. M., Segundo, M. A., Reis, S. & Lima, J. L. F. C. (2010) Fully automatic flow method for the determination of scavenging capacity against nitric oxide radicals. *Analytical and Bioanalytical Chemistry* 397 (7), 3005–3014.

Rice-Evans, C. A. (2000) Measurement of total antioxidant activity as a marker of antioxidant status in vivo: Procedures and limitations. *Free Radical Research* 33 (Supplement: S), S59–S66.

Rice-Evans, C. & Miller, N. J. (1994) Total antioxidant status in plasma and body fluids. *Methods in Enzymology* 234, 279–293.

Rice-Evans, C. A., Miller, N. J. & Paganga, G. (1996) Structure-antioxidant activity relationships of flavonoids and phenolic acids. *Free Radical Biology & Medicine* 20 (7), 933–956.

Rimando, A. M., Cuendet, M., Desmarchelier, C., Mehta, R. G., Pezzuto, J. M. & Duke, S. O. (2002) Cancer chemopreventive and antioxidant activities of pterostilbene, a naturally occurring analogue of resveratrol. *Journal of Agriculture & Food Chemistry* 50 (12), 3453–3457.

Rimet, O., Chauvet, M., Bourdeaux, M. & Briand, C. (1987) A novel fluorometric assay for quantitative analysis of dihydrofolate reductance activity in biological samples. *Journal of Biochemical & Biophysical Methods* 14 (6), 335–342.

Robards, K. (2003) Strategies for the determination of bioactive phenols in plants, fruit and vegetables. *Journal of Chromatography A* 1000, 657–691.

Robards, K., Prenzler, P. D., Tucker, G., Swatsitang, P. & Glover, W. (1999) Phenolic compounds and their role in oxidative processes in fruits. *Food Chemistry* 66 (4), 401–436.

Roginsky, V. & Lissi, E. A. (2005) Review of methods to determine chain-breaking antioxidant activity in food. *Food Chemistry* 92 (2), 235–254.

Ronald, L. P. & Guohua, C. (1999) In vivo total antioxidant capacity: comparison of different analytical methods. *Free Radical Biology & Medicine* 27, 1173–1181.

Sakaue, T., Hirano, T., Yoshino, G., Sakai, K., Takeuchi, H. & Adachi, M. (2000) Reactions of direct LDL-cholesterol assays with pure LDL fraction and IDL: comparison of three homogeneous methods. *Clinica Chimica Acta* 295 (1–2), 97–106.

Scheek, L. M., Wiseman, S. A., Tijburg, L. B. M. & Vantol, A. (1995) Dialysis of isolated low-density-lipoprotein induces a loss of lipophilic antioxidants and increases the susceptibility to oxidation in-vitro. *Atherosclerosis* 117 (1), 139–144.

Schlesier, K., Harwat, M., Bohm, V. & Bitsch, R. (2002) Assessment of antioxidant activity by using different in vitro methods. *Free Radical Research* 36 (2), 177–187.

Serafijni, M., Ghiselli, A. & FerroLuzzi, A. (1996) In vivo antioxidant effect of green and black tea in man. *European Journal of Clinical Nutrition* 50 (1), 28–32.

Shahidi, F. & Naczk, M. (1995) *Food Phenolics: Sources, Chemistry, Effects, Applications.* Technomic Publishing Co., Lancaster, Basel.

Shahidi, F. & Wanasundara, U. (1995) Effect of natural antioxidants on the stability of canola oil. *Developments in Food Science* 37 (1), 469–479.

Shahidi, F. & Ho, C.-T. (eds) (2000) *Phytochemicals and Phytopharmaceuticals.* AOCS Press, Champaign, IL, USA.

Shahidi, F. & Ho, C.-T. (Eds) (2007) *Antioxidant Measurement and Applications*, ACS Symposium Series 956, American Chemical Society, Washington, DC, USA.

Shen, Y., Jin, L., Xiao, P., Lu, Y. & Bao, J. (2009) Total phenolics, flavonoids, antioxidant capacity in rice grain and their relations to grain color, size and weight. *Journal of Cereal Science* 49 (1), 106–111.

Sheu, S. J., Chieh, C. L. & Weng, W. C. (2001) Capillary electrophoretic determination of the constituents of Artemisiae Capillaris Herba. *Journal of Chromatography A*, 911, 285–293.

Shlian, D. M. & Goldstone, J. (1986) Toxicity of butylated hydroxytoluene. *New England Journal of Medicine* 314, 648–649.

Stohs, S. & Bagchi, D. (1995) Oxidative mechanisms in the toxicity of metal ions. *Free Radical Biology & Medicine* 18 (2), 321–336.

Sun, Y. (1990) Free radicals, antioxidant enzymes, and carcinogenesis. *Free Radical Biology & Medicine* 8 (6), 583–599.

Szydłowska-Czerniaka, A., Dianoczkib, C., Recseg, K., Karlovits, G. & Szłyk, E. (2008) Determination of antioxidant capacities of vegetable oils by ferric-ion spectrophotometric methods. *Talanta* 76, 899–905.

Thielecke, F. & Boschmann, M. (2009) The potential role of green tea catechins in the prevention of the metabolic syndrome. *Phytochemistry* 70 (1), 11–24.

Tournaire, C., Croux, S., Maurette, M.-T., Beck, I., Hocquaux, M. l., Braun, A. M. & Oliveros, E. (1993) Antioxidant activity of flavonoids: Efficiency of singlet oxygen ($^1\Delta_g$) quenching. *Journal of Photochemistry & Photobiology B* 19 (3), 205–215.

Tsao, R. & Yang, R. (2003) Optimization of a new mobile phase to know the complex and real polyphenolic composition: towards a total phenolic index using high-performance liquid chromatography. *Journal of Chromatography A*, 1018, 29–40.

Valko, M., Leibfritz, D., Moncol, J., Cronin, M., Mazur, M. & Telser, J. (2007) Free radicals and antioxidants in normal physiological functions and human disease. *International Journal of Biochemistry & Cell Biology* 39 (1), 44–84.

Van Acker, S. A. B. E., van den Berg, D.-J., Tromp, M. N. J. L., Griffioen, D. H., van Bennekom, W. P., van der Vijgh, I. J. G. H. & Bast, W. J. F. (1996) Structural aspects of antioxidant activity of flavonoids. *Free Radical Biology & Medicine* 20, 331–342.

van der Heijden, C. A., Janssen, P. J. & Strik, J. J. (1986) Toxicology of gallates: a review and evaluation. *Food and Chemical Toxicology* 24, 1067–1070.

Vinson, J. A., Jang, J. H., Yang, J. H., Dabbagh, Y., Liang, X. Q., Serry, M., Proch, J. & Cai, S. H. (2001) Vitamins and especially flavonoids in common beverages are powerful in vitro antioxidants which enrich lower density lipoproteins and increase their oxidative resistance after ex vivo spiking in human plasma. *Journal of Agricultural and Food Chemistry* 49 (9), 4520–4520.

Volgger, D., Zemann, A. J., Bonn, G. K. & Antal, M. J. (1997) High speed separation of carboxylic acids by co-electroosmotic capillary electrophoresis with direct and indirect UV detection. *Journal of Chromatography A* 758, 263–276.

Vouldoukis, I., Conti, M., Krauss, P., Kamate, C., Blazquez, S., Tefit, M., Mazier, D., Calenda, A. & Dugas, B. (2004) Supplementation with gliadin-combined plant superoxide dismutase extract promotes antioxidant defenses and protects against oxidative stress. *Phytotherapy Research* 18 (12), 957–962.

Wang, H., Cao, G. H. & Prior, R. L. (1996) Total antioxidant capacity of fruits. *Journal of Agricultural and Food Chemistry* 44 (3), 701–705.

Wang, H., Cao, G. H. & Prior, R. L. (1997) Oxygen radical absorbing capacity of anthocyanins. *Journal of Agricultural and Food Chemistry* 45 (2), 304–309.

Wiseman, S. A., Balentine, D. A. & Frei, B. (1997) Antioxidants in tea. *Critical Reviews in Food Science and Nutrition* 37 (8), 705–718.

Yanishlieva, N. V. (2001) Inhibiting oxidation. In: *Antioxidants in Food – Practical Applications*, Pokorny, J., Yanishlieva, N. V. & Gordon, M. H. (eds), Woodhead Publishing, Cambridge (UK), pp. 22–70.

Yanishlieva, N. V. & Marinova, E. M. (1996a) Antioxidative action of some flavonoids at ambient and high temperatures. *Rivista Italiana delle Sostanze Grasse* 73, 445–449.

Yanishlieva, N. V. & Marinova, E. M. (1996b) Antioxidative effectiveness of some natural antioxidants in sunflower oil. *Zeitschrift Lebensmittel Untersuchung und Forschung* 203, 220–223.

Yu, L. L., Zhou, K. K. & Parry, J. (2005) Antioxidant properties of cold-pressed black caraway, carrot, cranberry, and hemp seed oils. *Food Chemistry* 91 (4), 723–729.

Zhang, J., Cui, M., He, Y., Yu, H. & Guo, D. (2005) Chemical fingerprint and metabolic fingerprint analysis of Danshen injection by HPLC-UV and HPLC-MS methods. *Journal of Pharmaceutical and Biomedical Analysis* 36 (5), 1029–1035.

Zheng, W. & Wang, S. Y. (2001) Antioxidant activity and phenolic compounds in selected herbs. *Journal of Agriculture and Food Chemistry* 49 (11), 5165–5170.

FURTHER READING

Burton, G. W., Joyce, A. & Ingold, K. U. (1983) Is vitamin E the only lipid-soluble, chain-breaking antioxidant in human blood plasma and erythrocyte membranes? *Archives of Biochemistry & Biophysics* 221 (1), 281–290.

Duarte, A. (1995) *Health Alternatives*. Megasystems, Morton Grove (IL).

Janeiro, P. & Brett, A. M. O. (2004) Catechin electrochemical oxidation mechanisms. *Analytica Chimica Acta*, 518 (2), 109–115.

Kubow, S. (1992) Routes of formation and toxic consequences of lipid oxidation-products in foods. *Free Radical Biology & Medicine* 12 (1), 63–81.

McCann, D., Barrett, A., Cooper, A., Crumpler, D., Dalen, L., Grimshaw, K., Kitchin, E., Lok, K., Porteous, L., Prince, E., Sonuga-Barke, E., Warner, J. O. & Stevenson, J. (2007) Food additives and hyperactive behaviour in 3-year-old and 8-to-9-year-old children in the community: a randomized, double-blinded, placebo-controlled trial. *The Lancet*, Sept 2007.

Milardovic, S., Ivekovic, D. & Grabaric, B. S. (2005) A novel amperometric method for antioxidant activity determination using DPPH free radical. *Bioelectrochemistry* 68, 180–185.

Samotyja, U. & Malecka, M. (2007) Effects of blackcurrant seeds and rosemary extracts on oxidative stability of bulk and emulsified lipid substrates. *Food Chemistry* 104 (1), 317–323.

Santos, S. A. O., Pinto, P. C. R. O., Silvestre, A. J. D. & Neto, C. P. (2010) Chemical composition and antioxidant activity of phenolic extracts of cork from *Quercus suber* L. *Industrial Crops and Products*, 31 (3), 521–526.

Schmitt-Koplin, P., Garrison, A. W., Perdue, E. M., Freitag, D. & Kettrup, A. (1998) Capillary electrophoresis in the analysis of humic substances: Facts and artifacts. *Journal of Chromatography A*, 807, 101–109.

Sharma, O. P., Bhat, B. (2009) DPPH antioxidant assay revisited. *Food Chemistry* 113 (4), 1202–1205.

Whitehead, T. P., Robinson, D., Allaway, S., Syms, J. & Hale, A. (1995) Effect of red wine ingestion on the antioxidant capacity of serum. *Clinical Chemistry* 41, 32–35.

Yang, B., Kotani, A., Arai, K. & Kusu, F. (2001) Estimation of antioxidant activities of flavonoids from their oxidation potentials. *Analytical Science* 17 (5), 599–604.

Zhao, Y. & Lunte, L. E. (1999) pH-mediated field amplification on column preconcentration of anions in physiological samples for capillary electrophoresis. *Analytical Chemistry* 71, 3985–3991.

2 Emulsifiers

Abstract: In food industries, forming a homogeneous mixture of food components which are totally immiscible (such as oil and water) is a challenge. Food emulsifiers are chemical molecules characterised by the presence of a hydrophilic and a hydrophobic part. The hydrophobic component is made up of fatty acid, while the hydrophilic portion consists of either glycerol or one of its ester derivatives generated from the reaction with organic acids such as lactic, citric, acetic or tartaric acid. As food additives, emulsifiers play a very important role in enabling hydrophilic components (e.g. water) and hydrophobic substances such as oils to mix together to form a stable continuous homogeneous product, or an emulsion. Examples of emulsion food products include mayonnaise, ice-cream and homogenised milk, which are composed of hydrophilic and hydrophobic substances.

Keywords: acacia gums; acetylated monoglycerides; cellulose alkyl esters; emulsifiers; glycol alginates; HLB; lactylated derivatives; sorbitan monoesters; stearoyl lactylates; succinylated derivatives

2.1 MECHANISMS OF FOOD EMULSIFIERS

By definition, emulsions are heterogeneous colloid mixtures of small molecule droplets of one component suspended in another component immiscible to it (Stampfli and Nerden 1995). This is possible because the surface tension that exists between the immiscible components is reduced by the action of emulsifiers which enable the immiscible phases to form one stable homogeneous phase known as an emulsion (Dziezak 1988). In the case of liquids, the term emulsion refers to the dispersion of two liquids that, under normal circumstances, are not miscible (e.g. oil and water). When mixed, an emulsion will form as tiny droplets of one phase dispersed into a continuous phase of another (Krog 1990; Stampfli and Nerden 1995). The availability of both the hydrophobic and hydrophilic affinities within one and the same material encourages the emulsifier molecules to position themselves accordingly at the junction between one phase and the continuous phase; the hydrophobic part of the molecule will be directed towards the oil phase while the water-loving hydrophilic part will orient itself towards the aqueous phase (Krog 1990; Stampfli and Nerden 1995). This mechanism prevents the droplets from lumping with other droplets which could result in the collapse of the emulsion (Figure 2.1a, b). This means that, in the real sense, emulsifiers do not create emulsions; instead, mechanical energy creates the emulsion and emulsifiers simply lock the emulsion that has been created.

Chemistry of Food Additives and Preservatives, First Edition. Titus A. M. Msagati.
© 2013 John Wiley & Sons, Ltd. Published 2013 by John Wiley & Sons, Ltd.

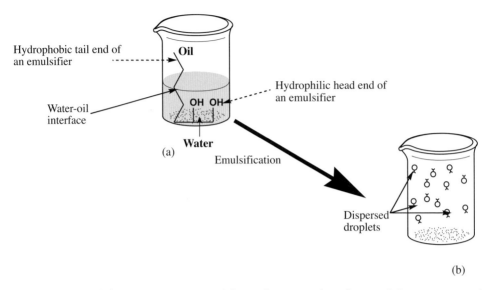

Fig. 2.1   The emulsification process: (a) emulsifier at the water–oil interface; and (b) suspension and dispersion of droplets to the other phase (water or oil).

Further Thinking

Emulsifiers in foodstuffs are very important as they help to mix together food components that are immiscible and that would otherwise collapse to form distinct layers or separate into their respective phases.

Among those emulsifiers used by the food industry are the monoglycerides, which are prepared using fats and oils as raw materials (Flack 1987). Fats are made up of a hydrophilic backbone of triglyceride with three fatty acid molecules attached to it, making the whole molecule lipophilic in nature (Lal *et al.* 2006). When the triglyceride molecule is cleaved, the lipophilic nature of the tail of the fatty acids balances the hydrophilic properties of the glyceride head to stabilise the emulsion (Lal *et al.* 2006).

Figure 2.1a shows how an emulsifier forms a film when it is adsorbed at the interface between the oil and water phases. The hydrophilic head end of the fatty acids in the emulsifiers points towards the aqueous phase while the lipophilic end points towards the oil phase in an air–water or oil–water interface, in order to lower the surface tension or interfacial tension (Jaynes 1985; Krog *et al.* 1985; Friberg and Larsson 1990). By forming a film at the interface, an emulsifier converts one of the two immiscible phases (water and oil in this case) into drops, which become suspended and dispersed over the other phase (Figure 2.1b). The emulsifier stabilises the droplets by encircling them to completely isolate them from each other; the immiscible phases will therefore not separate.

Emulsifiers are therefore liaisons between the two immiscible components and serve to stabilise such mixtures. In terms of their physical appearances and texture, emulsions are thick and normally find application in foods as well as in pharmacy, where they stabilise pharmaceutical products and formulation (Lal *et al.* 2006).

2.2 THE ROLE OF EMULSIFIERS IN FOODS

The primary role of emulsifiers as food additives is to improve food palatability, maximise the volume and aeration of food items, reduce the stickiness, enhance food flavour, improve the textural properties of foods and impart foam stability (Dieffenbacher and Martin 1987; Suman *et al.* 2009).

Further Thinking

Emulsifiers enhance the appeal of foods; they maintain its quality and freshness and prevent the growth of microbes such as fungi (which cannot grow in an environment with oils or fats).

Apart from the role of assisting oil and water to remain in stable emulsions, a property useful in the preparation of many food products such as salad dressings, emulsifiers have many other roles that they play in the food industry. Their unique molecular structures enable them to perform a variety of other roles in the improvement of the quality of a wide variety of food products (Baker 2010). This may explain why many processed foods consist to some extent of emulsions and, in some cases, the whole food may be an emulsion (or the food may have been in an emulsified state at some stage of the processing).

Emulsifiers are also attractive for use as food additives due to their safety record; they can be present in a human body up to 125 mg kg^{-1} body weight without causing any health problems (FAO 1963; FAO/WHO 1963).

2.2.1 Emulsification

Emulsification (the process of maintaining the emulsions) is one of the primary roles of emulsifiers in foods. The choice of the emulsifier is largely dependent upon the type of food material which forms the dispersed phase and the continuous phase. For example, if oil is the continuous phase, the emulsifier must be more lipophilic; if it is an aqueous system such as water then a hydrophilic emulsifier will be the best choice for use.

2.2.2 Starch complexing

Another role of emulsifiers in food is that of starch complexing, which has a wide range of applications. Starch granules comprise a linear polymer polysaccharide of water-insoluble D-glucose sugar units known as amylose (Figure 2.2a) and another highly branched water-soluble glucose polymer molecule known as amylopectin (Figure 2.2b). When starch is dispersed in water and heated the granules tend to absorb water and swell. They become gelatinised in the sense that the starch molecules attain a viscous state, forming a gel structure. When the product is cooled the starch molecules will tend to be closer to one another, squeezing the absorbed water out. This causes the starch to recrystallise in a process known as retrogradation.

In products where retrogradation takes place (e.g. bread), emulsifiers are incorporated to retard this process and maintain the softness of the product. During gelatinisation, the linear

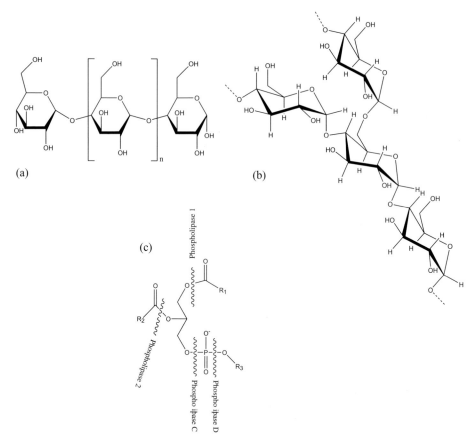

(a)

(b)

(c)

Fig. 2.2   Chemical structure of (a) amylase; (b) amylopectin (Baker 2010); and (c) phospholipase B.

water-insoluble amylose molecule forms a helical structure, the inside of which tends to possess a mild lipophilic property. When the emulsifiers complex with the amylose through the attachment of their hydrophilic tail ends inside the helical structure, it automatically physically inhibits the amylose molecule from retrograding.

Since there is a diverse range of emulsifiers, there are differences in the extent at which different emulsifiers complex to starch molecules and the shape of the molecule has a great effect on whether it fits within the helical structure. For instance, the fatty acid has to be either fully saturated or trans-oleic fatty acid for it to fit inside the helix. If it exists in cis isomeric form or in a polyunsaturated configuration, the structure with an optical bend on the fatty acid will make it impossible to fit into the helix; the molecular structure and shape must be straight. Structures such as lecithin which contain two sets of fatty acids will not be effective starch complexing agents unless they undergo some chemical modifications by cleaving one of the fatty acids using a phospholipase enzyme (Figure 2.2c).

The ability of food emulsifiers to form complexes with starch components, especially amylose, is evaluated using the amylose complexing index (ACI). The ACI takes a value between 0 and 100; 100 represents no iodine affinity (IA), implying that complete amylose-emulsifier complex formation has been attained. Iodine affinity (IA) for iodine complexes is defined as the fraction or ratio of amylase to starch.

2.2.3   Foam stabilisation and aeration

Food emulsifiers also play the important role of foam stabilisation and provide aeration in baked and dairy products by improving the mechanism by which air is incorporated and retained. Air is important in the process of baking to give the required texture; air sacks in the baking batter expand due to the carbon dioxide generated from the leavening of the baked products, resulting in the form and texture of baked products. Without this process, the products tend to harden.

In other products such as ice-cream, emulsifiers stabilise the foam by destabilising the emulsion of the product as well as the indigenous proteins present in ice-creams; the emulsion is then stabilised by employing hydrophobic binding interactions to the triglycerides on the surface of fat molecules. This mechanism prevents fat agglomeration and forms clusters which are important for creating foam; these clusters have the ability to coat the surface of air sacks, thereby stabilising them.

Another role of emulsifiers in food is that of interactions with gluten proteins in baked products. This interaction results in the formation of an elastic network which actually shapes the dough into a desired structure and prevents the possible loss of leavening gases, thus maintaining the volume of the final product. Ionic emulsifiers such as stearoyl lactylates are known to be very effective in the interaction with gluten proteins due to their ability in forming hydrophobic or ionic interactions or in hydrogen bonding.

2.3   CLASSIFICATION OF EMULSIFIERS

Due to a diverse range of chemistries that exist within so many different types of food emulsifiers, it follows that the choice of appropriate emulsifier for a particular food product can be difficult. Emulsifiers can however be classified in many different ways to provide an indicator of performance. These classifications are based on the hydrophilic-lipophilic balance (HLB), the ionic charge and the crystal stability.

2.3.1   Hydrophilic-lipophilic balance

The hydrophilic-lipophilic balance (HLB) measurement is important in determining how effective an emulsifier will be in a particular medium, especially for simple foods (Schmidts *et al.* 2010). The HLB has values ranging from 0 to 20, indicative of the emulsifier's oil or water affinity property. If an emulsifier exhibits a low HLB value, this indicates that it is strongly lipophilic; high HLB implies is it hydrophilic (Griffin 1949, 1954; Davies 1957). In other words, if the continuous phase is formed of a high proportion of oil, the best emulsifier is one with lower value of HLB number (Schmidts *et al.* 2010). The converse is true in cases where water forms the continuous phase.

The HLB values can be used to describe important characteristics of emulsifiers. For instance, an emulsifier is a good antifoaming agent if the HLB value ranges between 0 and 3; it has good water-in-oil emulsifying properties if the HLB values are between 4 to 6; and it is a good wetting agent if HLB values are between 7 and 9. Emulsifiers with HLB value in the range 8–18 are better suited as oil-in-water emulsifiers and also form good hydrolysers.

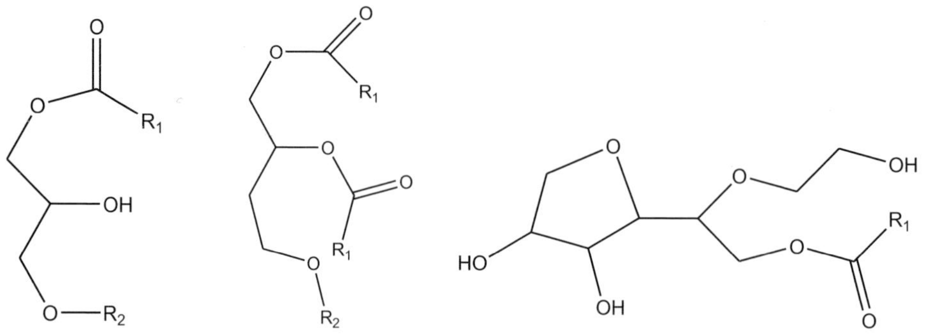

(a)   (b)   (c)

Fig. 2.3   Chemical structure of (a) monoglyceride; (b) diglyceride; and (c) polysorbates (where R_1 may either be stearic or oleic acid) (Suman *et al.* 2009; Baker 2010).

2.3.2   Ionic charge

Another criterion for the choice of an emulsifier for a particular food product is ionic charge. There are certain types of emulsifiers which can form anions when in aqueous media (e.g. stearoyl lactylates) as well as those which possess carboxylic acid functionality (e.g. diacetyl tartaric acid esters of monoglycerides).

2.3.3   Crystal stability

A number of emulsifiers also possess polymorphic properties which give them flexibility in that they can exist in different crystal forms such as alpha, beta, etc. Emulsifiers may crystallise in one form initially and then transform to another. Examples of emulsifiers with this type of transformation tendency include acetic and lactic acid esters, polyglycerol esters, propylene glycol esters and sorbitan esters.

2.4   TYPES OF FOOD EMULSIFIERS

2.4.1   Main classification

Food emulsifiers may be classified as one of two main groups: synthetic and naturally occurring. The synthetic types of emulsifiers used in the food industry include low-molecular-weight (mono- and di-glycerides) and high-molecular-weight (polymeric, such as polysorbates, alginates, carrageenans, gums and gelatins; Figure 2.3a–c) emulsifiers. The low-molecular-weight mono- and di-glycerides are highly lipophilic, with HLB values within the range 1–10. The majority of synthetic emulsifiers are chemical modifications of monoglycerides such as acetylated monoglycerides or organic acid esters of monoglycerides (O'Brien 2004; Baker 2010).

Synthetic food emulsifiers are partial esters of either fatty acids or polyols and water-soluble organic acids, and therefore contain both hydrophilic and hydrophobic functionalities within the same molecule. These molecules have the structures required to maintain the stability of the emulsion, that is, they have both hydrophilic and hydrophobic ends enabling them to stabilise emulsions by increasing their kinetic energy (Figure 2.4a–e).

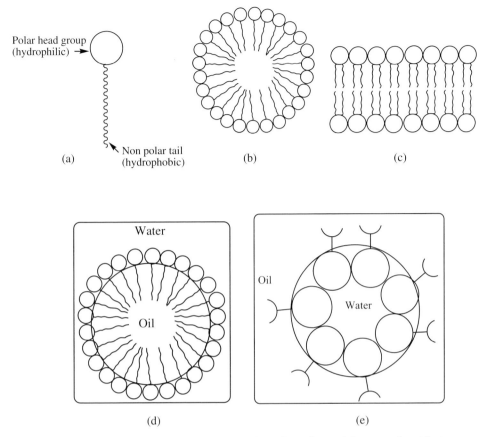

Fig. 2.4 Structures of micelles: (a) micelle; (b) non-spherical micelles; (c) bilayer micelles; (d) oil-in-water emulsion and (e) water-in-oil emulsion (Mariani and Fedeli 1985).

Further Thinking

Food emulsions can either be oil-in-water emulsion, implying that oil droplets are dispersed in the water phase, or water-in-oil emulsion whereby water droplets are dispersed in the oil phase. Since the two media are immiscible, a third phase which is an emulsifier is always needed to facilitate the stable mixing.

Normally the hydrophobic moiety of the emulsifiers is formed by fatty acids (Figure 2.4a, b) while the hydrophilic component of it consists of either propane-1, 2, 3-triol (commonly known as glycerol mono-, di- or tri-acylglycerol; Figure 2.5c–h) or ester derivatives of the same glycerol (Mariani and Fedeli 1985). The structure of stearic fatty acid as shown in Figure 2.5a has 18 carbons. Butyric acid has 4 carbons (Figure 2.5b); the end with two oxygen atoms is polar, but the rest of the molecules are completely non-polar.

Although shorter fatty acids with carbon chains 12 (palmitic acid) or less demonstrate good emulsification properties, they are avoided as they easily undergo hydrolysis to form

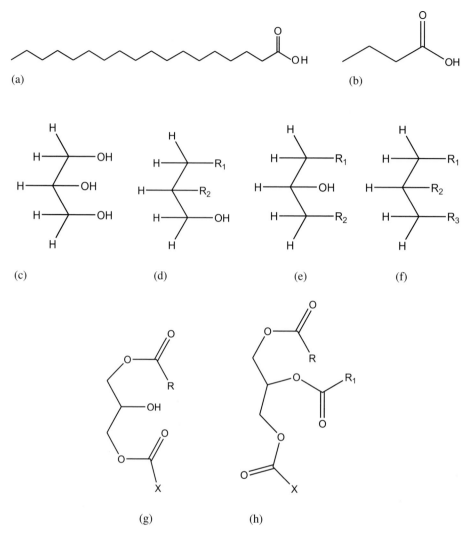

Fig. 2.5   (a) Stearic acid (a long chain fatty acid); (b) butyric acid (a short chain fatty acid); (c) glycerol; (d) 1, 2-diacylglycerol; (e) 1, 3-diacylglycerol; (f) tri-acylgrlycerol; and (g, h) chemical structures of mono- and diacylglycerol esters of fatty acid emulsifiers (Jaynes 1985; Krog *et al.* 1985; Friberg and Larsson 1990; Suman *et al.* 2009; Baker 2010). Typically, R: lauric acid, myristic acid, palmitic acid, oleic acid, and stearic acid and x:diacetyl tartaric acid, acetic acid, lactic acid, and citric acid.

soapy products which cause off-flavours (Jaynes 1985; Krog *et al.* 1985; Friberg and Larsson 1990). In addition, some of the unsaturated fatty acids such as linoleic acid (C18:2) have a tendency to undergo oxidation; this affects the flavour of products and hence linoleic acid is also avoided (Jaynes 1985; Krog *et al.* 1985; Friberg and Larsson 1990). The oxidation of these fatty acids is due to the presence of iron and lipid hydroperoxides. Iron-hydroperoxide interactions cause the decomposition of fatty acids, leading to the development of rancidity.

As well as synthetic types, natural emulsifiers such as lecithin, casein, mustard and cayenne pepper are also used. Lecithin is derived from soybean oil (Pyler 1988; Baker

2010), and its functional component as an emulsifier is a mixture of phospholipids. These have a hydrophilic polar head as well as a lipophilic portion of the molecule, and two fatty acid tails. In the same way as for monoglyceride emulsifiers, lecithin can be modified to many other chemical patterns to perform different functions.

2.4.2 Low-molecular-weight glyceride emulsifiers

Monoglycerides are prepared by the interesterification process, which involve triglycerides and glycerol (Scheme 2.1). Glycerin fatty acid esters are prepared from glycerin and animal and plant oils/fats or their fatty acids. In order to improve the properties, a highly purified and distilled monoglyceride is required. The process of molecular distillation is therefore introduced for the production of this high-quality product. The reaction described by Scheme 2.1 can be represented by the chemical equation depicted in Figure 2.6.

The chemical structure of glycerin (Figure 2.6) shows that it has three hydroxyl groups. If only one of the three OH groups is esterified with a fatty acid, the resulting ester is called monoglyceride. The di-and tri-glycerides have two and three fatty acid groups esterified at hydroxyl group, respectively.

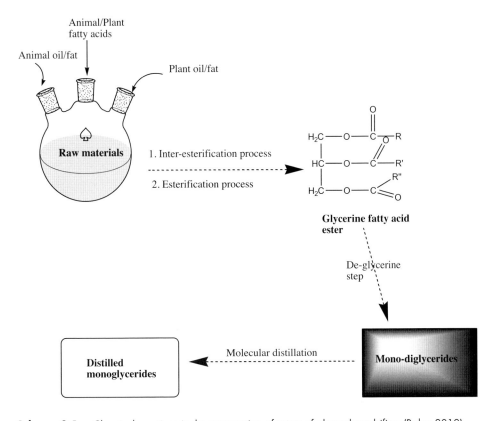

Scheme 2.1 Chemical reactions in the preparation of esters of glycerol emulsifiers (Baker 2010).

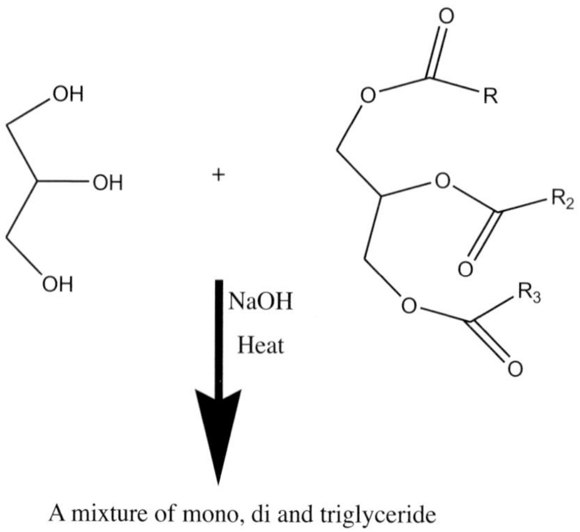

A mixture of mono, di and triglyceride

Fig. 2.6 Esterification of glycerol by fatty acid (Baker 2010)

2.4.2.1 Specific applications

Monoglycerides used as emulsifiers in the food industry have different physico-chemical properties and characteristics according on the type and the composition of fatty acid used as the starting material during its preparation. They are therefore used for a range of different purposes such as emulsifying, foaming and anti-foaming agent, aerators and crystal stabilisers in margarines, whipped toppings and peanut butter. The diacetyl tartaric acid esters of monoglycerides perform the role of film formers in baked products, dairy product analogues and in confectionaries and that of lubricants and releasing agents in dairy products and soft candy etc. The type of monoglyceride selected is therefore dependent upon its intended role.

2.4.2.2 Ester derivatives of acylglycerol

Ester derivatives of acylglycerols are composed of mainly mono- and di–forms. They have strong lipophilicity due to the fact that their hydroxyl micelle head components are relatively small and are non-ionic, a property which tends to enhance emulsification ability (Figure 2.5).

Examples of monoacyl glycerol derivatives include acetate monoacyl glycerols (CH_3COO^-), lactate monoacyl glycerols ($CH_3CH(OH)COO^-$), citrate monoacyl glycerols ($-OOCH_2CC(COO^-)(OH)H_2CCOOH$) and succinate monoacyl glycerols, to mention a few.

2.4.2.3 Acetylated monoglyceride

As their name suggests, acetylated glyceride derivatives contain an acetyl group that is linked to a monoglyceride moiety (Figure 2.7a–c). Despite the fact that they lack appreciable activities as food emulsifiers, some members of this class are soft enough to be used where expansion at conditions of increasing tension is needed. They also have high oil stability, their measure of peroxide value is intact even at higher temperatures (e.g. even at the water

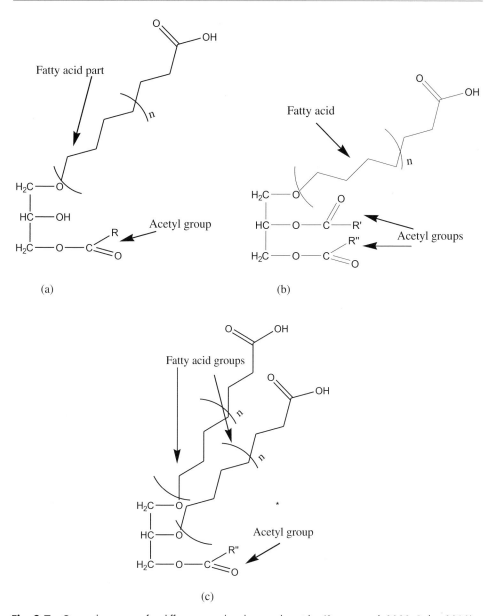

Fig. 2.7 General structures for different acetylated monoglycerides (Suman *et al.* 2009; Baker 2010).

boiling temperatures) and they remain stable even after being exposed to high temperatures for long periods of time.

Acetyl derivatives are normally required where there is a need to reduce hydrophilicity functionalities, and they are easily obtained by performing an acetylation process. This replaces hydroxyl groups with an acetyl group in the monoacetyl glycerol molecules (Jaynes 1985; Krog *et al.* 1985; Friberg and Larsson 1990). The acetylation process may be accomplished by treating monoacetyl glycerols with acetic anhydride in the presence of a

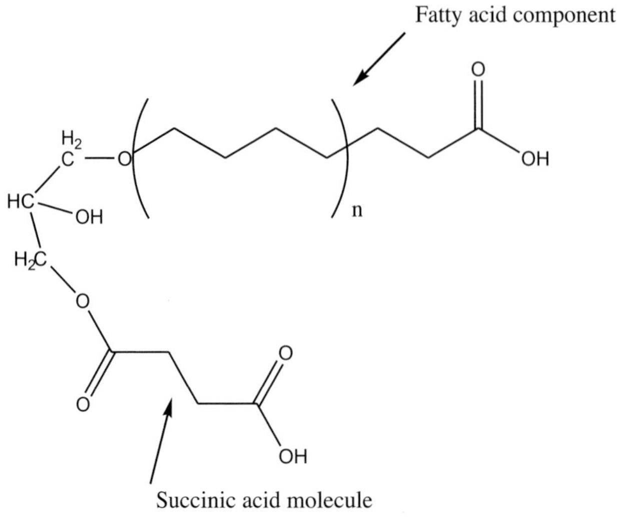

Fig. 2.8 General structure of succinylated glyceride derivative (Hadeball *et al.* 1986; Sax and Lewis 1989).

concentrated mineral acid to produce both an acid and an ester which can easily be separated out by distillation. Alternatively, the acetylation may proceed via the reaction between monoacyl glycerols and glyceryl triacetate in the presence of a basic catalyst to form the di- and tri-acetates of glycerol.

Acetylated monoglycerides are used as film foamers and moisture barriers in fruits, nuts and pizza (Jaynes 1985; Krog *et al.* 1985; Friberg and Larsson 1990).

2.4.2.4 Succinic acid esters of monoglycerides (SMG)

The succinylated monoglycerides are characterised by the presence of the succinic acid molecule attached to the monoglyceride molecule (Figure 2.8). The important properties of succinylated monoglycerides in their role as emulsifiers include the fact that they are not soluble in cold water and they form dispersions in hot water; however, they show appreciable solubility in organic media such as hot alcohol, fats and oils.

The preparation of succinylated monoacyl derivatives involves the use of succinic anhydride (as opposed acetic anhydride used in the preparation of acetylated derivatives of monoacyl glycerols; Hadeball *et al.* 1986; Sax and Lewis 1989) (Figure 2.9). The differences between the succinylated and the acetylated derivatives include the fact that one carboxyl group by-product in the former is not removed as in the latter, and that the hydrophilicity becomes relatively larger with succinylated derivatives with an anionic character stable at an appropriate optimal pH value (Hadeball *et al.* 1986; Sax and Lewis 1989). In comparison, the acetylated monoacyl glycerol derivatives have the advantage of flexibility in terms of the diversity of their alkyl chain, and therefore form better emulsifying agents than the succinylated derivatives (Guillard *et al.* 2004). The succinylated derivatives are mostly used as dough modifying agents and as emulsifiers for shortening, and they also have roles as both softener as well as conditioners in a variety of food products (Jaynes 1985; Krog *et al.* 1985; Friberg and Larsson 1990).

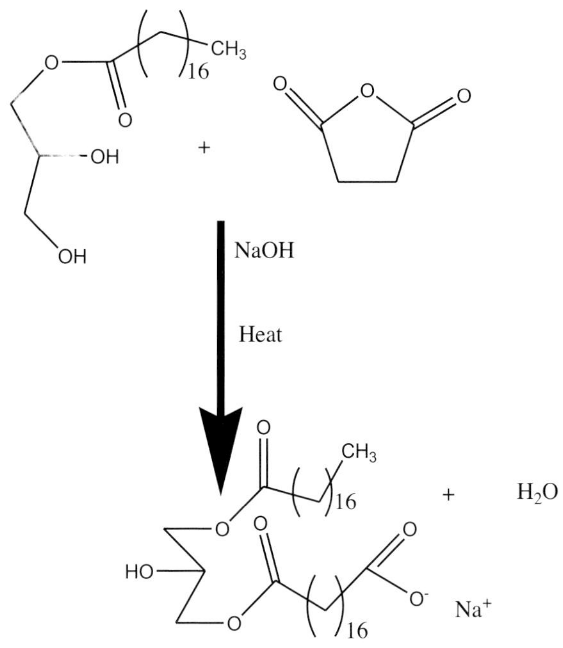

Fig. 2.9   Preparation of succinylated derivatives (Hadeball *et al.* 1986; Sax and Lewis 1989).

2.4.2.5   *Lactic acid esters of monoglycerides (LMG)*

In this class of monoglyceride derivatives a lactic acid molecule is attached to a mono-glyceride group (Figure 2.10a, b). The lactated monoglyceride derivatives are known for their high foaming efficiency which surpasses their emulsifying ability. They are used in shortening for cakes, desserts and foaming for cream.

The lactylated monoacyl derivatives are normally prepared through a condensation re-action that involves the carboxyl and hydroxyl groups of the monoacylglycerol molecules (Figure 2.11; Schmidt *et al.* 1976b). The resulting lactylated monoacylglycerol formed has a positive effect on the size of the hydrophilicity as it tends to enlarge it without affecting the integrity of the non-ionic property of the molecule itself (Schmidt *et al.* 1976b). Lactylated monoacylglycerol derivatives are normally considered to have advantages over acetylated derivatives as the reaction mixture of the former is relatively less corrosive and flammable, hence safer (Hasenhuettl and Hartel 2008). Lactylated monoglycerides are used not only as emulsifiers but also as plasticisers and surface active agents in baked products and whipped toppings (Jaynes 1985; Krog *et al.* 1985; Friberg and Larsson 1990).

2.4.2.6   *Stearoyl lactylates (calcium stearoyl-2-lactate derivatives)*

Calcium stearoyl-2-lactate (CSL) emulsifiers (Figure 2.12), which have ionic and hydrophilic properties, are prepared from a reaction involving a number of lactic acid molecules with stearic acid. The product is partially neutralised with calcium (or sometimes sodium), thus forming an anionic emulsifier with a strong ability to bind proteins (Figures 2.11 and 2.12). This means they are also lactic acid esters of monoglyceride with sodium or calcium and are used as dough modifiers for flour foods such as bread, and the stearoyl lactylates of calcium

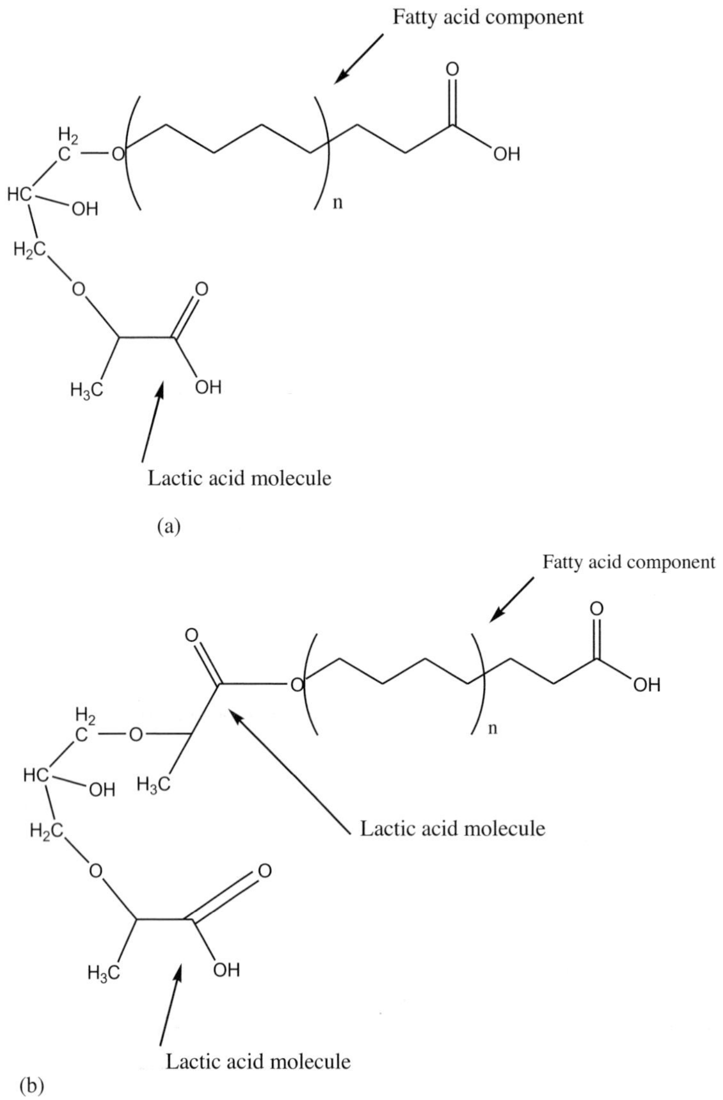

(a)

(b)

Fig. 2.10   General structures of lactylated monoglycerides (Baker 2010).

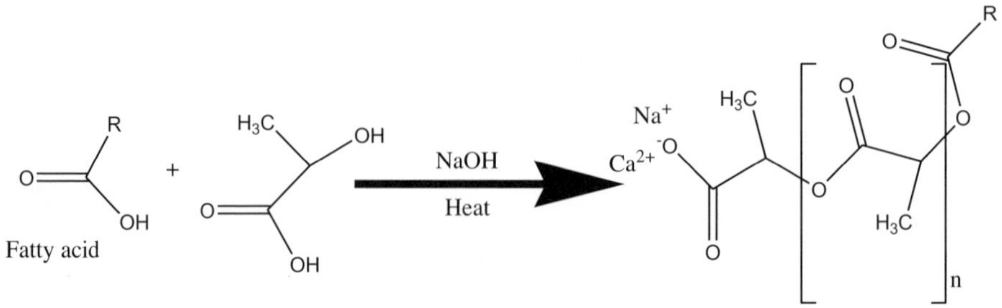

Fig. 2.11   Preparation of stearoyl-2-lactylate (used as softener and conditioner) (Schmidt 1976b).

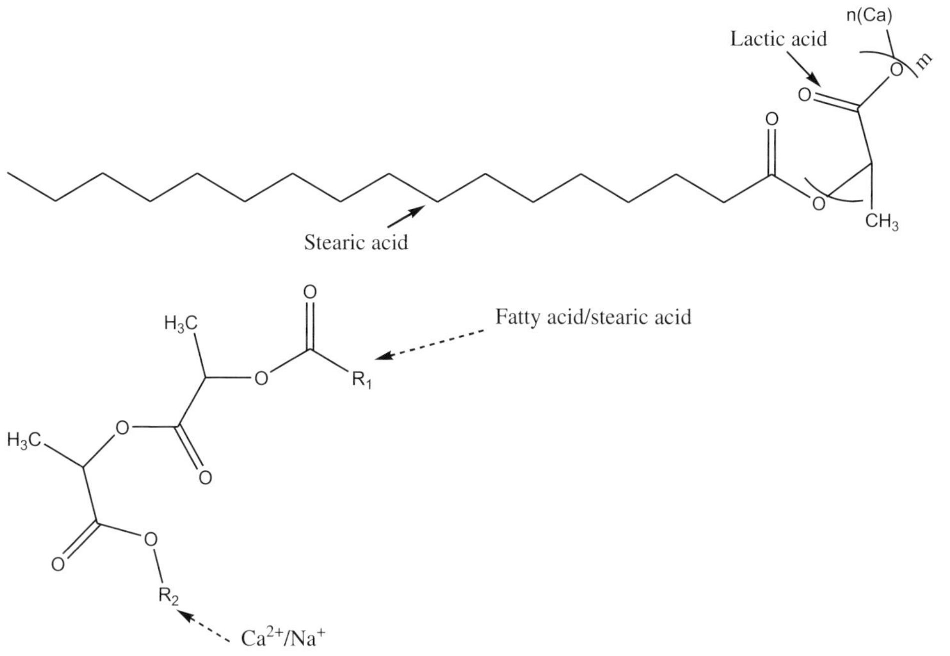

Fig. 2.12 General structures of stearoyl lactylates (Schmidt 1976b).

and sodium are used as conditioners and whipping agents in coffee whiteners, icing and dehydrated potatoes. They are also known to form a strong complex with gluten in starch and are therefore valuable in baked products.

2.4.2.7 Monoglyceride citrate ester derivatives

As the name suggests, citric acid ester derivatives of monoglyceride contain citric acid molecules attached to a monoglyceride part (Figure 2.13) and are known to be highly hydrophilic emulsifiers. Structurally, they are stable α-crystals which have applications in the preparation of margarine, coffee whitener and cream. They can also be applied as emulsion stabilisers for mayonnaise.

Monoacylglycerol citrate ester derivatives, of varying functions, can normally be prepared through the condensation reaction involving the monoacylglycerols and either citric acid or citric acid anhydride (Figure 2.13). They therefore have a hydrophilic head group which is expanded and diversified (Figure 2.13).

2.4.3 High-molecular-weight (polymeric) emulsifiers

2.4.3.1 Propylene glycol (PG) esters of fatty acids

Propylene glycol (PG) ester derivatives of fatty acids are a product of interesterification of propylene glycol and fatty acid (Figure 2.14a–c). Since the interesterification produces a mixture of monoesters and diesters, an isolation step is needed to extract a high-purity monoester product. This product has desirable surface-active effects and can be obtained through

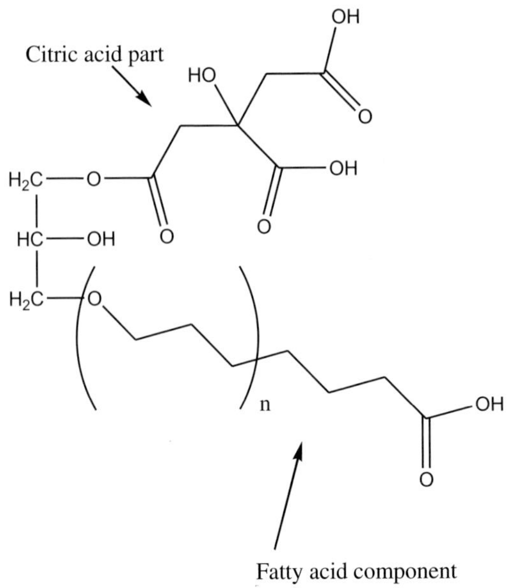

Fatty acid component

Fig. 2.13   General structure of citric acid ester derivative (Baker 2010).

the molecular distillation process involving the monoglyceride. The monoester product generated is known to have little emulsifying action, although it has qualities to maintain the integrity of its α-crystal structure (Hasenhuettl and Hartel 2008). The monoester derivatives are in most cases produced in powdered form, and are used together with monoglyceride as powder-foaming agents in the preparation of cakes and desserts as well as liquid shortening products.

2.4.3.2   *Sucrose ester derivatives of fatty acids*

Sucrose ester derivatives of fatty acids contain sucrose and fatty acid (Figure 2.15) and are known to have a wide range of hydrophilic-lipophilic balance (HLB) values of 1–16. This property makes them suitable for many different roles in the food industry, such as an emulsifying or dispersing agent and as bactericidal agents for dairy products.

2.4.3.3   *Glycol alginate and cellulosic alkyl ester derivatives*

Propylene glycol alginate and cellulose alkyl esters are prepared differently to those discussed above. Propylene glycol alginate is prepared from the esterification reaction of alginic acid and propylene glycol or propylene oxide (Figure 2.16). Alginic acid is composed of hydroxyl groups and is comprised of repeating units of mannuronic acid and guluronic acid unit (Figure 2.16). Sodium alginates and calcium alginates are known gelling agents and are important food additives, playing a role as thickeners (Jaynes 1985; Krog et al. 1985; Friberg and Larsson 1990). However, both sodium and calcium alginate lack an effective surface activity to be used as effective emulsifiers in food processes. An esterification of alginic and propylene glycol is therefore necessary to produce an emulsifier of the desired quality (Figures 2.16 and 2.17).

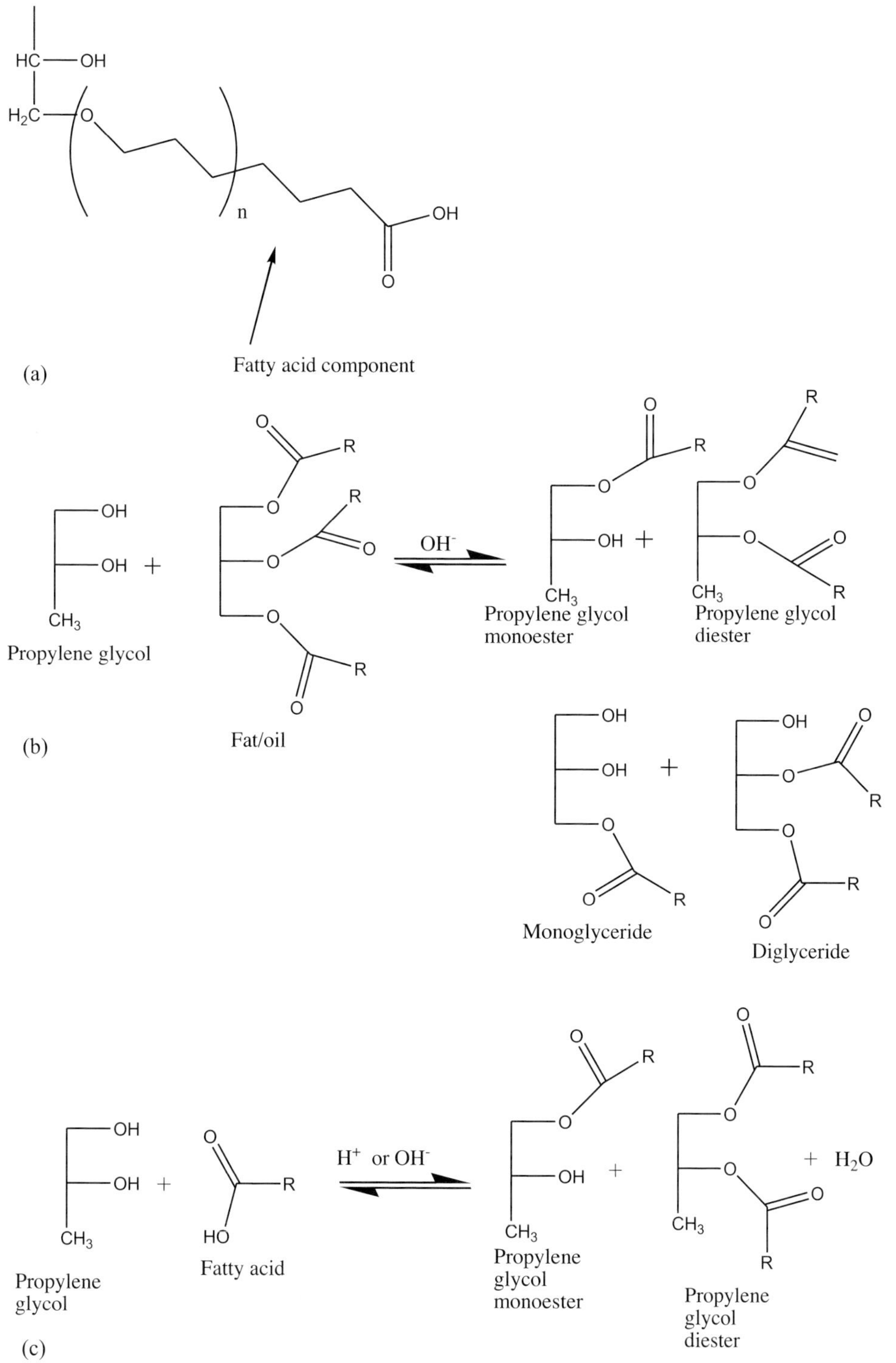

Fig. 2.14 (a) General chemical structure of the propylene glycol ester of fatty acids; (b) preparation of propylene glycol monoesters by interesterification; and (c) preparation of propylene glycol monoesters by direct esterification (Stauffer 1999; Hasenhuettl and Hartel 2008; Baker 2010).

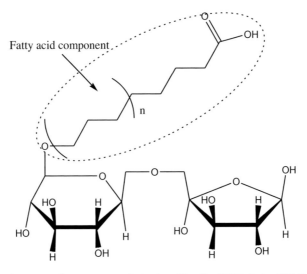

Fig. 2.15   General structure of sucrose ester derivatives (Stauffer 1999; Baker 2010).

Sucrose esters of fatty acids finds their use as emulsifiers, texturisers and film formers in baked food items, fruit coatings and confectionaries, while propylene glycol esters are used as emulsifiers and aerators in cake mixing and whipped toppings (Jaynes 1985; Krog *et al.* 1985; Friberg and Larsson 1990). Polyglycerol esters are also used as emulsifiers, aerators and cloud inhibitors in salad oils, peanut butter and fillings.

The cellulose alkyl ester derivatives are prepared by the reaction of alkyl chlorides and cellulose. This reaction is possible since cellulose contains OH groups which may be activated,

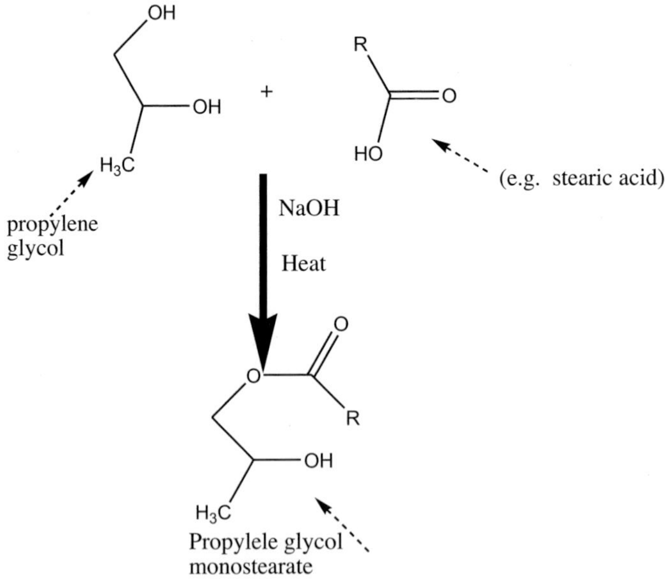

Fig. 2.16   Preparation of propylene glycol monostearate by esterification of propylene glycol with stearic acid (Stauffer 1999; Hasenhuettl and Hartel 2008; Baker 2010).

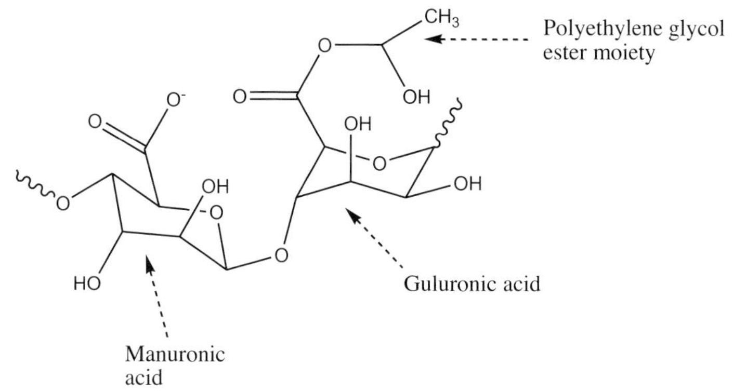

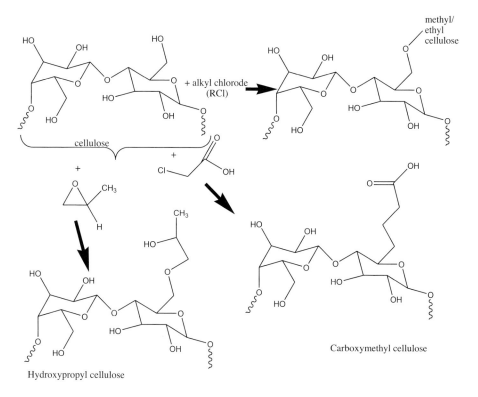

CH₃ ········ Polyethylene glycol
ester moiety

Guluronic acid

Manuronic
acid

Fig. 2.17 General structure of alginic acid, highlighting where esterification reactions take place (Jaynes 1985; Krog *et al.* 1985; Friberg and Larsson 1990).

thereby acquiring a higher ability for water absorption and swelling as well as increased lipophilicity which results in higher surface activity (Belitz *et al.* 2004; Figure 2.18).

Polyglyceryl ester derivatives are a product of esterification of polyglycerine with fatty acids (Figure 2.19) which are soluble in oils, but can only be dispersible in water. The polymer lengths tend to vary for different derivatives, affecting its HLB values which lie

methyl/
ethyl
cellulose

+ alkyl chloride
(RCl)

cellulose

Hydroxypropyl cellulose

Carboxymethyl cellulose

Fig. 2.18 Structures of various cellulose that may react with alkyl chloride to form alkyl ester derivatives (Belitz et al. 2004b).

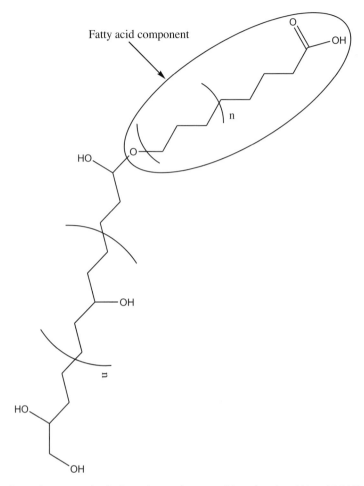

Fatty acid component

Fig. 2.19 General structure of polyglyceryl ester derivative (Hasenhuettl and Hartel 2008).

within the range 3–13. This derivative is used in the food industry as an emulsifier as well as a modifier to regulate fat crystallisation.

2.4.3.4 Polyglycerol polyricinoleate

Polyglycerol polyricinoleate derivative is somewhat similar to polyglyceryl ester derivative; the main difference is that it has two sets of fatty acid groups, one of which has a variable length (Figure 2.20a, b). Polyglycerol polyricinoleate derivatives are known to be strong lipophilic water-in-oil (W/O) emulsifiers and they are highly viscous and insoluble in water and ethanol, but soluble in fats and oils (Hasenhuettl and Hartel 2008). They have applications in the chocolate industry, where they play a role as viscosity-reducing agents.

2.4.3.5 Sorbitan esters of fatty acids

Sorbitan esters are a mixture of sorbitol ester and sorbide ester produced by an esterification reaction involving sorbitol (Figure 2.21a) and fatty acids (Figure 2.21b–d). Different fatty

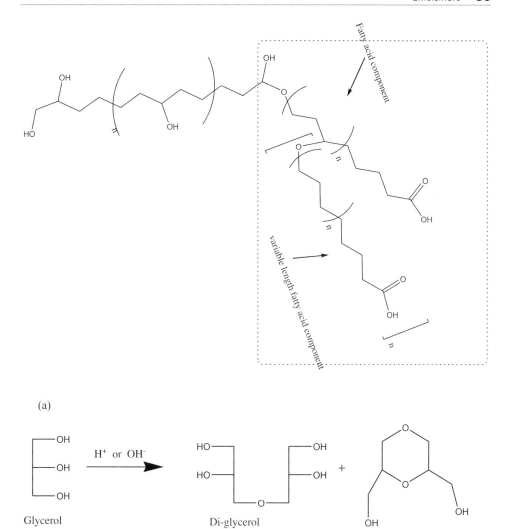

Fig. 2.20 (a) General structure of polyglycerol polyricinoleate derivatives; and (b) polymerisation of glycerol (Hasenhuettl and Hartel 2008).

acids are used for this purpose, hence producing different types of sorbitan ester derivatives with a different extent of esterification.

Sorbitan esters are generally used as an emulsifier for cream. They are also used in other food types (such as ice-creams and chocolate) as major emulsifying agents.

2.4.3.6 *Sorbitan monostearate*

These food emulsifiers have an HLB value of 4.7 and are usually used in conjunction with polysorbates in oil as well as cake mixes.

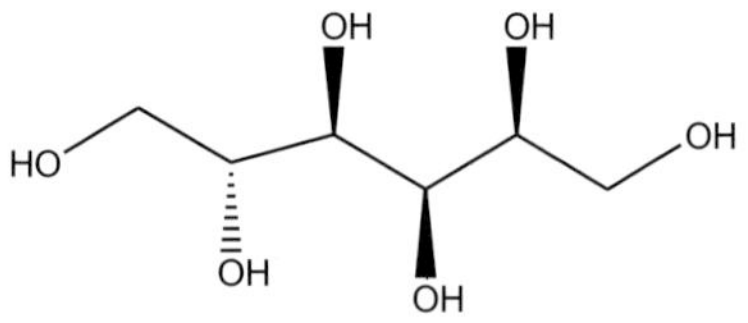

(a)

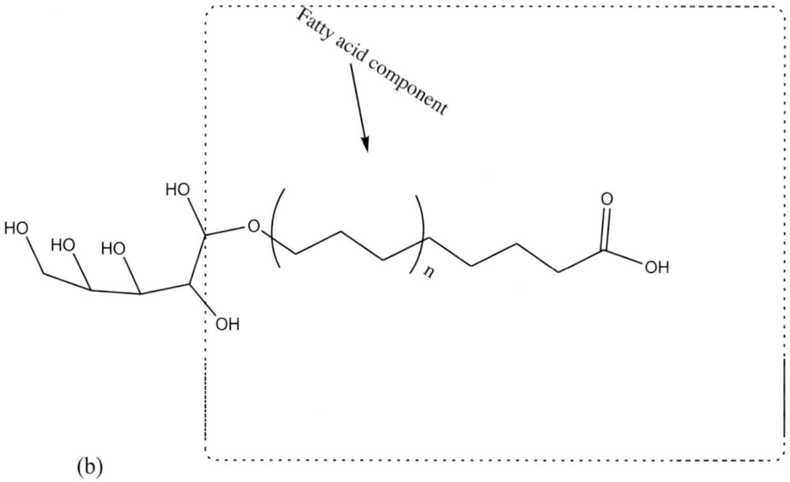

(b)

Fig. 2.21   Chemical structure of (a) sorbitol (Stauffer 1999; Baker 2010); (b) sorbitol monoester (Stauffer 1999); (c) sorbide monoester (Stauffer 1999); and (d) sorbitan monoester (Stauffer 1999; Baker 2010).

2.4.4   Natural emulsifiers

Natural emulsifiers used in food preparations include lecithin, casein, mustard and cayenne pepper. Lecithin is widely used, while casein is mainly a natural emulsifier in dairy products. Lecithin (Figure 2.22a, b) has a chemical structure containing phospholipids, ether of either choline, ethanolamine, inositol or serine and fatty acids. Phospholipids normally form the major component of the molecule. Lecithins are widely found in nature from both plant (from soybeans, corn and rapeseed) and animal sources (yolk lecithin made by excluding the phospholipid). Fractionated lecithin is obtained by isolation from special components of the raw materials (Pyler 1988; Baker 2010).

2.4.4.1   Lecithin

Lecithin may have other derivatives which are obtained by some specialised enzymatic treatments such as enzyme treatment or enzyme digestion processes (Figure 2.23a, b). In this process, lecithin properties which make it useful as emulsifiers are enhanced by improving the hydrophilicity by a reaction with phospholipase enzyme.

   Lecithin is used as emulsifiers in many food preparations including butter and mayonnaise. Chemically, lecithin is composed of a mixture of phospholipids including phosphatidylcholine (actual lecithin), phosphatidylethanolamine (cephalin), phosphatidylinositol

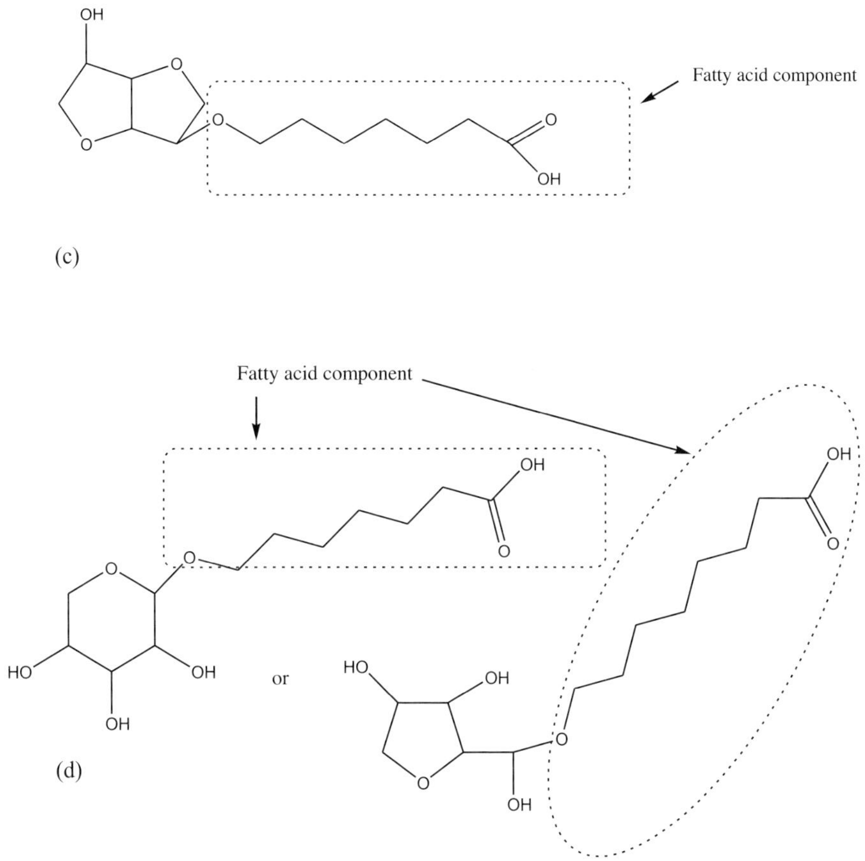

(c)

(d)

Fig. 2.21 (Continued)

and phosphatidylserine. The commercial preparations of lecithin are known to contain trigle-cerides, fatty acids, pigments, carbohydrates and sterol (Figure 2.24).

The structure of lecithin indicates that saturated fatty acids are attached to the α position of phosphatidylcholine and to the β position of phosphatidylethanolamine. Other observations of the structure of lecithin reveal that the various phosphatides contain both lipophilic and hydrophilic functions, a property which demonstrates their ability to play roles as emulsifying agents.

Lecithin is not only used in food products as an emulsifier, but it has other specific roles such as an anti-spattering agent in margarine. Monoglycerides are normally included to play the role of stabilisation. The stabilisation phenomenon is important because, during heating, margarine will tend to melt which results in the spattering of oil; the presence of lecithin will however cause water droplets to coalesce into larger droplets which will eventually evaporate, halting the spattering of the heated oil.

Lecithin is also used as a dispersion agent in cocoa powder which contains a film of cocoa butter on the surface. This mixture cannot disperse in either water or milk, especially at temperatures below the melting point of cocoa butter. When lecithin is used, its lipophilic and hydrophilic properties assist the cocoa powder to come to solution.

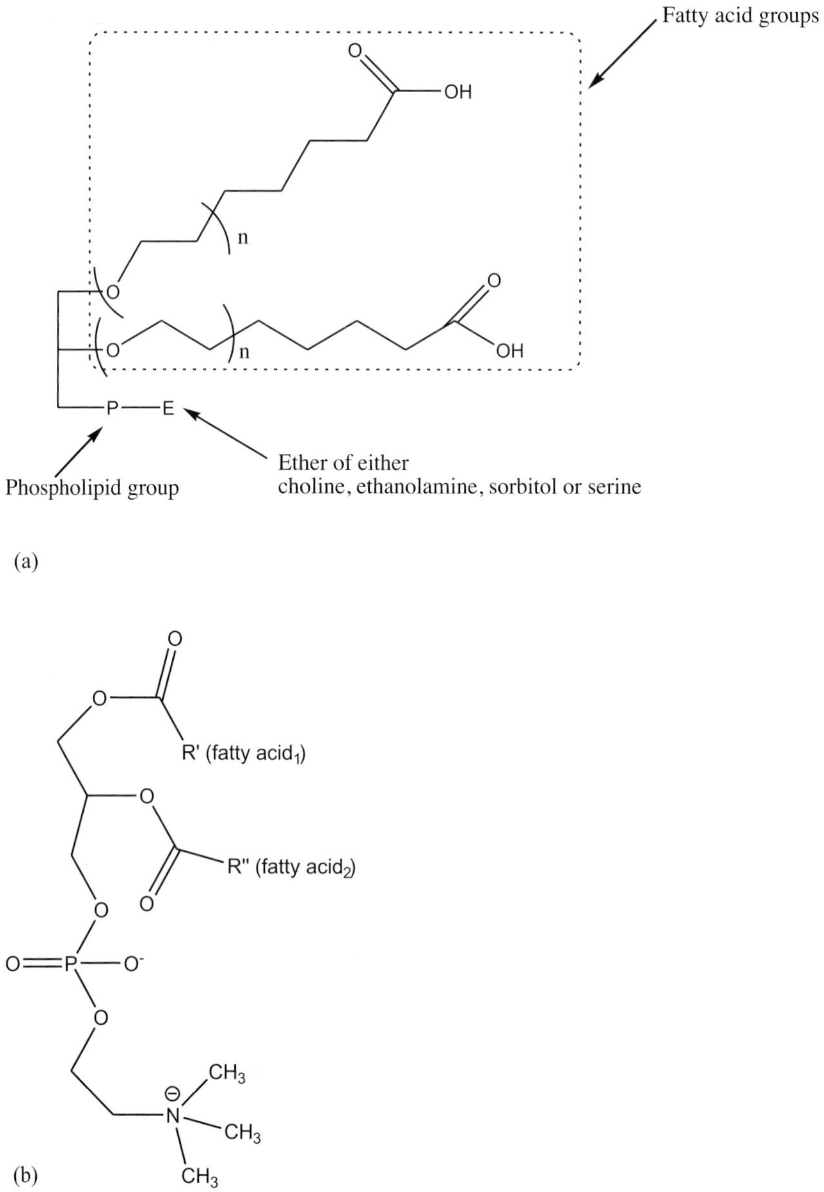
Fatty acid groups
O
OH
n
O
O
n
O
OH
P
E
Phospholipid group
Ether of either
choline, ethanolamine, sorbitol or serine
(a)
O
O
R' (fatty acid1)
O
R'' (fatty acid2)
O
O
O
P
O-
O
CH3
N
CH3
CH3
(b)

Fig. 2.22 General structure of (a) lecithin, showing where various groups are attached; and (b) lecithin (Pyler 1988; Szuhaj 2005; Baker 2010).

2.4.4.2 *Acacia gums*

Acacia gums contain oils, calcium, magnesium and potassium salts of arabic acid whereby the oil exists in the emulsion form (in same way fat exists in milk). In cases where acacia gum is used as an emulsifier, particles are coated with a film of gum which is in molecular association with the oil. This prevents coalescing, and the viscosity produced in the mixture prevents separation.

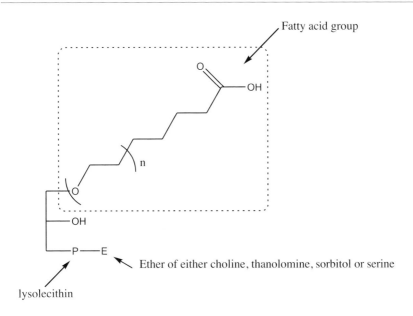

(a)

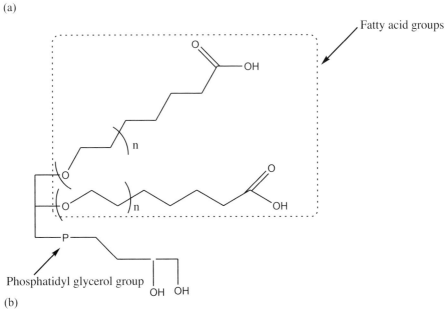

(b)

Fig. 2.23   Chemical structure of (a) enzyme digested lecithin; and (b) enzyme treated lecithin (Pyler 1988; Szuhaj 2005; Baker 2010).

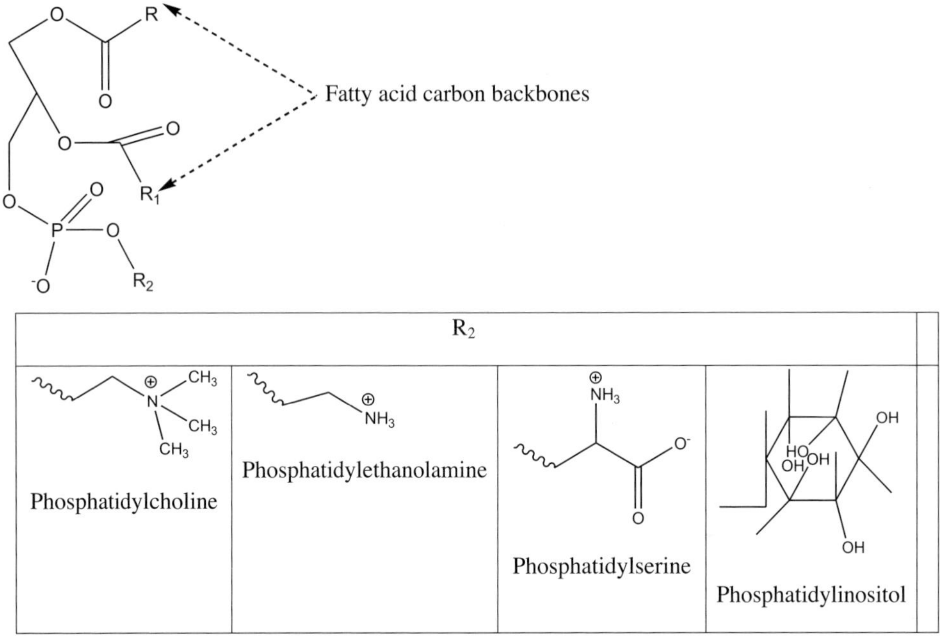

Fatty acid carbon backbones

Fig. 2.24 Chemical structures of the α-forms of commercial lecithin (Pyler 1988; Szuhaj 2005; Baker 2010).

2.4.4.3 Casein

Casein is a protein-like emulsifier prepared from milk. It is used as a stabiliser in dairy product.

2.4.4.4 Mustard seeds

Mustard seeds contain proteins, carbohydrates and oil. When the seeds are crushed to release the contents, the proteins and carbohydrates tend to associate chemically with the oil component. This therefore works as a mechanism to prevent the seed components (proteins, carbohydrates and oils) repelling each other, to stabilise and maintain the emulsion. Mustard is used as an emulsifier in foods such as lemon juice and mayonnaise.

2.5 QUALITY AND ANALYSIS OF FOOD EMULSIFIERS

Emulsifiers have a vast number of applications in the food industry and are therefore subject to certain procedural performance checks (Dieffenbacher and Martin 1987). These restrictions mean that food products containing emulsifiers are monitored to ensure regulations are adhered to. For this reason, a number of analytical methods have been been developed and reported for the determination of emulsifying molecules such as mono-, di- and tri-glycerides.

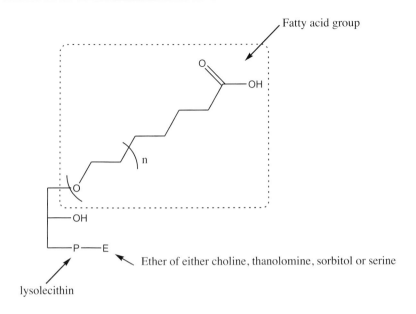

Fatty acid group

Ether of either choline, thanolomine, sorbitol or serine

lysolecithin

(a)

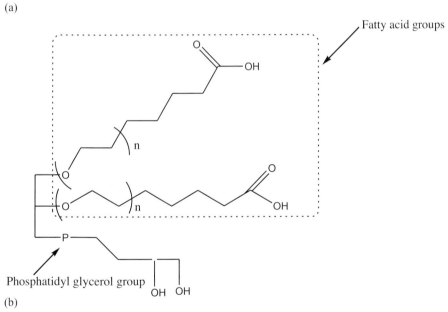

Fatty acid groups

Phosphatidyl glycerol group

(b)

Fig. 2.23 Chemical structure of (a) enzyme digested lecithin; and (b) enzyme treated lecithin (Pyler 1988; Szuhaj 2005; Baker 2010).

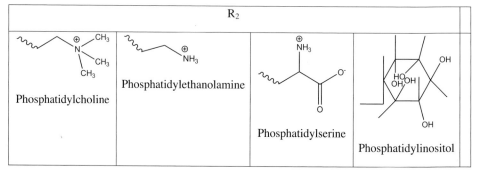

Fig. 2.24   Chemical structures of the α-forms of commercial lecithin (Pyler 1988; Szuhaj 2005; Baker 2010).

2.4.4.3   Casein

Casein is a protein-like emulsifier prepared from milk. It is used as a stabiliser in dairy product.

2.4.4.4   Mustard seeds

Mustard seeds contain proteins, carbohydrates and oil. When the seeds are crushed to release the contents, the proteins and carbohydrates tend to associate chemically with the oil component. This therefore works as a mechanism to prevent the seed components (proteins, carbohydrates and oils) repelling each other, to stabilise and maintain the emulsion. Mustard is used as an emulsifier in foods such as lemon juice and mayonnaise.

2.5   QUALITY AND ANALYSIS OF FOOD EMULSIFIERS

Emulsifiers have a vast number of applications in the food industry and are therefore subject to certain procedural performance checks (Dieffenbacher and Martin 1987). These restrictions mean that food products containing emulsifiers are monitored to ensure regulations are adhered to. For this reason, a number of analytical methods have been been developed and reported for the determination of emulsifying molecules such as mono-, di- and tri-glycerides.

The most common method of determination used for monoglycerides in foods is iodimetric titration, a method based on the oxidation of 1-monoacylglyceride using periodic acid which results in the production of formaldehyde (EU Commission 2003). The main limitations of this method are that it is only useful for the determination of 1-monoacylglycerol with the assumption that the other isomers (the 1-mono- and 2-monoacylglycerol) are in equilibrium. Further, the method is not very economical and the use of huge amounts of (possibly poisonous) solvents is dangerous.

The products formed from monoacylglycerol derivatives are generally characterised according to either their peroxide value; saponification value; hydroxyl value; acid number; iodine value; water-insoluble combined lactic acid (WICLA) value; or by chromatography techniques (Schmidt 1976a), as described in the following sections.

2.5.1 Peroxide value

The peroxide value is used as an indicator of the extent of oxidative rancid reactions that might have taken place in oils/fats, and hence reflects the palatability of these products. Fats and oils tend to undergo auto-oxidative reactions at the points of unsaturation (the double bonds) found in fat acid chains, resulting in the formation of free radicals and hydroperoxides. This leads to the deterioration of fats/oils, exemplified by off-flavour or a bad odour (Gotoh and Wada 2006).

2.5.2 Saponification value

Saponification refers to the reaction which occurs in the carboxylic acid ester group, resulting in the formation of an alcohol and the corresponding salt of the carboxylic acid. The saponification value of surfactants (emulsifiers) gives a measure of the mean molecular weight or chain length of all the fatty acids that are present.

2.5.3 Hydroxyl value

Hydroxyl value gives a measure of unesterified hydroxyl groups left after the esterification processing of polyols to produce surfactant emulsifiers. It is therefore an indicator of the hydrophilic character of the emulsifiers; the higher the hydroxyl value, the higher the HLB value.

2.5.4 Acid number

Acid number (also referred to as acid value or AV or neutralisation number) refers to the number of carboxylic acid groups present in a single compound (or even in a mixture of compounds), and also gives a measure of the freshness of oil. It has been reported that the AV of hydrogenated oil is an indicator of the oxidisation tendency of the oil; a low AV is highly desirable in the food industry (Koseoglu and Lucas 1990; Fu *et al.* 2008). The AV is also used to measure the quality of emulsifiers by comparing with the standard characteristics for fatty acids. For example, for a C12 (lauric) the standard acid number is 2.81, for C16 (palmitic) the standard acid number is 2.19 and for C18 (stearic or oleic) it is 1.99 (Hasenhuettl and

Hartel 2008). The acid number for synthetic or extracted emulsifiers should conform to the standard values.

2.5.5   Iodine value

Also referred to as iodine adsorption value, iodine number or iodine index measurement, the iodine value is a measure of the amount (in grams) of iodine that is used up 100 grams of a chemical species. The iodine value gives a measure of the extent of unsaturation (number of double bonds) contained in fatty acids. The higher the iodine number, the greater the number of unsaturated fatty acid (double bonds) present in a fat (Bosworth and Brown 1933).

2.5.6   Chromatography

Chromatographic techniques are also very useful in the determination of emulsifiers, including techniques such as high-performance liquid chromatography (HPLC; Aitzetmuller 1977; Soe 1983); gas chromatography (GC; Soe 1983); and thin-layer chromatography (Soe 1983). Of the chromatographic techniques, gas chromatography is the most suitable for the determination of methyl esters of fatty acids (FAMEs) after a derivatisation step, as the compounds are thermally stable. The derivatives for the GC analysis are normally acetate, trimethylsilyl (TMS), or *t*-butyldimethylsilyl (BDMS) derivatives (Goh and Timms 1985; Christie 1989; Lechner *et al*. 1997).

On the other hand, HPLC can be useful in the determination of acylglycerols and their aliphatic ester derivatives, especially when the evaporative light scattering detector is used. The evaporative light scattering detector is only used for semi-quantitative determination however, due to the non-linear dependence of the signal on the analyte concentration (Heron *et al*. 1995; Amaral *et al*. 2004). Ultraviolet detection is always difficult as these compounds do not absorb strongly above 220 nm (Lin *et al*. 1997).

Another chromatographic detector that has been reported is mass spectrometry such as matrix-assisted laser desorption ionisation mass spectrometry (HPLC-MALDI-MS), although its sensitivity to these compounds was reported to be low (Hlongwane *et al*. 2001). Atmospheric pressure chemical ionisation mass spectrometry (APCI-MS) has however been reported to have more promising results, and allows the identification of both molecular ions as well as the fragments (Laakso and Voutilainen 1996; Holcapek *et al*. 1999; Byrdwell 2001; Careri *et al*. 2002; Cai and Syage 2006; Rezanka and Sigler 2007; Zehethofer and Pinto 2008). The method provides other advantages such as accuracy and selectivity, as well as quantitative measurements of the ions and fragments.

2.6   FOODS CONTAINING EMULSIFIERS

The food types in which emulsifiers are used include milk, cream, salad cream, mayonnaise, salad dressings, soups, sauces, butter, margarine, low-fat spreads, beverages, ice-cream and coffee whitener. The aqueous phase of many of these foods contains biopolymers that either enhance viscosity or cause gelation. The appearance and rheological properties of an aqueous phase are determined by the nature of the interactions between the biopolymer molecules (hydrogen bonding, hydrophobic interactions, van der Waals forces, electrostatic interactions and disulphide bond formation), as well as the kinetics of the aggregation process.

2.6.1 Mayonnaise

Mayonnaise is a food product prepared from a mixture of oil, egg yolk and either vinegar or lemon juice, with some other possible optional ingredients (David 1960; McGee 2004). The type of emulsion most commonly used in mayonnaise stabilisation is lecithin.

2.6.2 Margarine

Margarine is an emulsion of oil in water which also consists of a lipid fraction, water and some minor compositions of colouring agents, antimicrobials, lecithin emulsifiers and preservatives (Hui *et al.* 2007). To make margarine (which is semi-solid) with oils and water which exist as liquids at room temperature, hydrogenation under a nickel catalyst is always necessary. The compactness of margarine as a product after hydrogenation will mainly depend on the number of double bonds of the saturated fatty acid present (Lai *et al.* 1999).

2.6.3 Butter

Butter is a typical example of water-in-oil emulsion resulting from an inversion of cream, which is an oil-in-water emulsion. The stabilisation of butter ingredients is made possible by the milk proteins which play a very important role as emulsifiers.

The preparation of butter involves a number of ingredients including the fat-globules-rich components such as unhomogenised milk and cream. The fat globules in unhomogenised milk are known to be surrounded by phospholipid membranes and contain fatty acids; these play a significant role as emulsifiers to stabilise the mixture by preventing the fat in milk from pooling together to form a cluster of its own phase (Cook *et al.* 2005). During the process of emulsification, the cream is agitated to cause the disruption of phospholipid membranes, resulting in the partitioning of milk fats from the other cream ingredients. The fat then becomes available for the emulsification stabilisation role.

2.6.4 Ice-creams

Ice-cream is a foam as well as an emulsion, containing both ice crystals and an unfrozen aqueous phase with a polysaccharide component. The latter plays the role of stabiliser of the mixture (Newton *et al.* 2005). The inclusion of an emulsifying agent in ice-cream makes the ice whip (beaten into a froth or foam) more easily than otherwise. Other advantages include: the ice-cream becomes drier; has better resistance-to-melting properties; is smoother; and has a better texture (Arbuckle 1986).

The other role of emulsifiers in ice-cream is to improve the structure of the final ice-cream product by destabilising the milk protein present. Emulsifiers in ice-creams also tend to promote protein desorption from the surface of the fat droplets due to their relatively higher surface activity and the ability of the formation of liquid crystal metaphases. Emulsifiers act as nucleation centres for surface crystallisation of triglycerides, and also assist in the initial formation and stabilisation of the ice-cream foam prior to partial fat globule coalescence and freezing.

The main types of emulsifier used in making ice-creams are glycerol monostearate, polysorbates and glycerol monopalmitate which assist in the stabilisation of the mixture by displacing proteins from the fat droplet surface (Zhang and Goff 2005).

Monoglyceride emulsifiers in ice-creams tend to attain a complex liquid crystalline phase when present in an aqueous solution phase with water (Barfod *et al.* 1991) to form lamellar liquid crystal; when cooled enough, this transforms into a gel-like phase. This phase transition is said to be highly essential in the process of protein displacement to stabilise the mixture (Berger 1990).

The polysorbate types of emulsifiers in ice-cream are known to be more efficient in displacing proteins from the oil–water interface than the mono- and diglycerides; polysorbate types are therefore better emulsifiers for ice-creams (Keeney 1982). The mono- and diglycerides have better forming qualities however, thus greater ability in the formation of the initial foam before the agglomeration of the fat droplets at the air–water interface (Keeney 1982).

2.6.5   Dairy emulsifiers and emulsions

Milk is a mixture of fats and other liquids. If fresh milk is left to sit for a while, the fat will eventually settle out leaving a cream layer. To prevent this, milk is normally homogenised by forcing the milk at high pressure through a narrow opening. Cream globules are broken into very small droplets. These have larger surface areas, and are more spread out throughout the milk. This homogenisation process combined with the emulsifiers naturally found in milk (including casein and lecithin) allow the cream to stay evenly distributed throughout the liquid for longer (Lal *et al.* 2006).

REFERENCES

Aitzetmuller, K. (1977) High-performance liquid-chromatographic analysis of partial glycerides and other technical lipid mixtures. *Journal of Chromatography* 139 (1), 61–68.

Amaral, J. S., Cunha, S. C., Alves, M. R., Pereira, J. A., Seabra, R. M. & Oliveira, B. P. (2004) Triacylglycerol composition of walnut (*Juglans regia* L.) cultivars: Characterization by HPLC-ELSD and chemometrics. *Journal of Agriculture and Food Chemistry*, 52 (26), 7964–7969.

Arbuckle, W. S. (1986). *Ice Cream*, 4th edition. The Avi Pub. Co., New York, USA.

Baker, S. R. (2010) Maximizing the use of food emulsifiers. MSc thesis, Food science, Kansas State University, Manhattan, Kansas, USA.

Barfod, N. M., Krog, N., Larsen, G. and Buchheim, W. (1991) Effect of emulsifiers on protein-fat interaction in ice cream mix during aging I. Quantitative analyses. *Fat Science and Technology* 93, 24–29.

Belitz H. D., Grosch, W & Schienberle, P. (2004) *Food Chemistry*, Berlin, Springer, 331–332.

Berger, K. G. (1990) Ice cream. In *Food Emulsions*, 2nd edition, Larsson, K. and Friberg, S. (eds), Marcel Dekker Inc., New York, 367–444.

Bosworth, A. W. & Brown, B. (1933) Isolation and identification of some hitherto unreported fatty acids in butter fat. *Journal of Biological Chemistry* 103, 115–134.

Byrdwell, W. C. (2001) Atmospheric pressure chemical ionization mass spectrometry for analysis of lipids. *Lipids*, 36 (4), 327–347.

Cai, S. S. & Syage, J. A. (2006) Comparison of atmospheric pressure photoionization, atmospheric pressure chemical ionization, and electrospray ionization mass spectrometry for analysis of lipids. *Analytical Chemistry* 78 (4), 1191–1199.

Careri, M., Bianchi, F. & Corradini, C. (2002) Recent advances in the application of mass spectrometry in food-related analysis. *Journal of Chromatography A*, 970 (1–2), 3–64.

Christie, W. W. (1989) *Gas Chromatography and Lipids: A Practical Guide*. The Oily Press Ltd./P.J. Barnes and Assoc., Bridgwater, UK.

Cook, N. B., Bennett, T. B. & Nordlund, K. V. (2005) Monitoring indices of cow comfort in free-stall-housed dairy herds. *Journal of Dairy Science*, 88 (11), 3876–3885.

David, E. (1960) *French Provincial Cooking* (1999 edition). Penguin, UK.

Davies, J. T. (1957) A quantitative kinetic theory of emulsion type, I. Physical chemistry of the emulsifying agent gas/liquid and liquid/liquid interface. *Proceedings of the International Congress of Surface Activity*, Butterworths, London, pp. 426–438.

Dziezak, J. D. (1988) Emulsifiers: The interfacial key to emulsion stability. *Food Technology* 42 (10), 172–186.

EU Commission (2003) Official Methods of Analysis, 17th edition, 2nd revision, *Method 966.18*, AOAC International, Gaithersburg, MD.

FAO. (1963) Report of the fourth meeting of the FAO experts on livestock infertility, Rome, Italy, 14–19 January 1963.

FAO/WHO (1963) Report of the first session of the joint FAO/WHO Codex Alimentarius Commission, Rome, Italy, 25 June–3 July 1963.

Flack, E. (1987) The contribution of emulsifying agents to modem food production. *Food Science and Technology Today*, 1 (4), 240–243.

Friberg, S. E. & Larsson, K. (eds) (1990) Food Emulsions, 2nd edition. Marcel Dekker, New York.

Fu, H., Yang, L., Yuan, H., Xiao, F. & Lo, M. (2008) Production of low acid value edible oil with reduced TFAs by electrochemical hydrogenation in a diaphragm reactor. *Journal of American Oil Chemists' Society* 85 (11), 1087–1096.

Goh, E. M. & Timms, R. E. J. (1985) Determination of monoglycerides and diglycerides in palm oil, olein and stearin. *Journal of American Oil Chemists' Society* 62 (4), 730–734.

Gotoh, N. & Wada, S. (2006) The importance of peroxide value in assessing food quality and food safety. *Journal of American Oil Chemists' Society* 83 (5), 473–474.

Griffin, W. C. (1949) Classification of surface-active agents by HLB. *Journal of the Society of Cosmetic Chemists* 1, 311–326.

Griffin, W. C. (1954) Calculation of HLB values of non-ionic surfactants. *Journal of the Society of Cosmetic Chemists* 5, 249–256.

Guillard, V., Guillbert, S., Bonazzi, C. & Gontard, N. (2004) Edible acetylated monoglyceride films: Effect of film-forming technique on moisture barrier properties. *Journal of the American Oil Chemists' Society*, 81 (11), 1053–1058.

Hadeball, K, Kroll, J. & Franzke C. (1986) Synthesis and properties of succinylated monoglycerides. *Nahrung* 30 (2), 209–211.

Hasenhuettl, G. L. & Hartel, R. W. (eds) (2008) *Food Emulsifiers and Their Applications*. Chapman & Hall, New York, USA.

Heron, S., Lesellier, E. & Tchapla, A. (1995) Analysis of triacylglycerols of borage oil by RPLC identification by coinjection. *Journal of Liquid Chromatography* 18 (3), 599–611.

Hlongwane, C., Delves, I. G., Wan, L. W. & Ayorinde, F. O. (2001) Comparative quantitative fatty acid analysis of triacylglycerols using matrix-assisted laser desorption/ionization time-of-flight mass spectrometry and gas chromatography. *Rapid Communications in Mass Spectrometry* 15 (21), 2027–2034.

Holcapek, M., Jandera, P., Fischer, J. & Prokes, B. (1999) Analytical monitoring of the production of biodiesel by high-performance liquid chromatography with various detection methods. *Journal of Chromatography A*, 858 (1), 13–31.

Hui, Y. H., Chandan, R. C., Clark, S. & Cross, A. (2007) *Handbook of Food Products Manufacturing: Principles, Bakery, Beverages*. John Wiley & Sons, New Jersey, USA.

Jaynes, E. (1985) Application in the food industry, II: In *Encyclopedia of Emulsion Technology*, Becher, P. (ed.), Volume 2, Mercel Dekker, New York, pp. 367–385.

Keeney, P. G. (1982) Development of frozen emulsions. *Food Technology* 36 (11), 65–70.

Koseoglu, S. S. & Lucas, E. W. (1990) Hydrogenation of canola oil. In: *Canola and Rapeseed Production, Chemistry, Nutrition and Processing Technology*, Shahid. F. (ed.) Van Nostrand Reinhold, New York.

Krog, N. (1990) Food emulsifiers and their chemical and physical properties. In: *Food Emulsions*, Larsson, K. & Friberg, S. E. (eds), Marcel Dekker, Inc., New York, pp. 127–180.

Krog, N., Rilson, T. H. & Larsson, K. (1985) Application in the food industry, I. In: *Encyclopedia of Emulsion Technology*, Becher, P. (ed.), volume 2, Mercel Dekker, New York, pp 321–366.

Laakso, P. & Voutilainen, P. (1996) Analysis of triacylglycerols by silver-ion high-performance liquid chromatography: Atmospheric pressure chemical ionization mass spectrometry. *Lipids* 31 (12), 1311–1322.

Lai, O. M., Ghazali, H. M., Cho, F. & Chong, C. L. (1999) Flow properties of table margarine prepared from lipase-catalyzed trans esterified palm stearin: palm kernel olein feedstock. *Food Chemistry* 64 (2), 221–226.

Lal, S. N. D., O'Connor, C. J. & Eyres, L. (2006) Application of emulsifiers/stabilizers in dairy products of high rheology. *Advances Colloid and Interface Sciences* 123–126, 433–437.

Lechner, M., Bauer-Plank, C. & Lorbeer, E. (1997) Determination of acylglycerols in vegetable oil-methyl esters by on-line normal phase LC-GC. *Journal of High Resolution Chromatography* 20 (11), 581–585.

Lin, J. T., Woodruff, C. L. & McKeon, T. A. (1997) Non-aqueous reversed-phase high-performance liquid chromatography of synthetic triacylglycerols and diacylglycerols. *Journal of Chromatography A*, 782 (1), 41–48.

Mariani, C. & Fedeli, E. (1985) Partial glycerides of vegetable oils. *La Rivista Italiana Delle Sostanze Grasse* 57, 129–152.

Dieffenbacher, A. & Martin, E. (1987) Determination of emulsifiers in foods - separation of polar and nonpolar lipids by chromatography on silica-gel microcolumns. *Revue Francaise Des Corps GRAS* 34 (7–8), 323–326.

McGee, H. (2004) *Food and Cooking*. Scribner, New York.

Newton, J. M., Bazzigialuppi, M., Podczeck, F., Booth, S. & Clarke, A. (2005) The rheological properties of self-emulsifying systems, water and microcrystalline cellulose. *European Journal of Pharmaceutical Sciences* 26 (2), 176–183.

O'Brien, R. D. (2004) *Fats and Oils: Formulating and Processing for Application*, 2nd edition. CRC Press, Boca Raton.

Pyler, E. J. (1988) *Baking Science and Technology*, 3rd edition. Sosland Publishing Company, Kansas City.

Rezanka, T. & Sigler, K. (2007) The use of atmospheric pressure chemical ionization mass spectrometry with high performance liquid chromatography and other separation techniques for identification of triacylglycerols. *Current Analytical Chemistry* 3, 252–271.

Sax, N. I. & Lewis, R. J. Sr. (1989) Succinic anhydride. *Dangerous Properties of Industrial Materials*. Van Nostrand Reinhold Company, New York, pp. 3131–3132.

Schmidt, A. A. (1976a) Chromatographic analysis of succinylated and lactylated monoglycerides as food surfactants. *Khimicheskava Promyshlennost* 8, 598–600.

Schmidt, A. A. (1976b) Synthesis of lactylated monoglycerides. *Masolzhinonyaya Promyshlennost* 10, 19–20.

Schmidts, T., Dobler, D., Guldan, A.-C., Paulus, N. & Runkel, F. (2010) Multiple W/O/W emulsions: Using the required HLB for emulsifier evaluation. *Colloids and Surfaces A: Physicochemical and Engineering Aspects* 372 (1–3), 48–54.

Soe, J. B. (1983) Analyses of monoglycerides and other emulsifiers by gas-chromatography. *Fette Seifen Anstrichmittel*, 85 (2), 72–76.

Stampfli, L. & Nerden, B. (1995) Emulsifiers in bread making. *Food Chemistry* 52, 353–360.

Stauffer, C. E. (1999) *Emulsifiers*, 1st ed. Eagen Press, St Paul.

Suman, M., Silva, G., Catellani, D., Bersellini, U., Caffarra, V. & Careri, M. (2009) Determination of food emulsifiers in commercial additives and food products by liquid chromatography/atmospheric-pressure chemical ionization mass spectrometry. *Journal of Chromatography A*, 1216 (18), 3758–3766.

Szuhaj, B. F. (2005) Lecithins. Chapter 13 in *Bailey's Industrial Oil and Fat Products*, Shahidi, F. (ed.), John Wiley & Sons, Inc., Hoboken, pp. 361–456.

Zehethofer, N. & Pinto, D. M. (2008) Recent developments in tandem mass spectrometry for lipidomic analysis. *Analytica Chimica Acta* 627 (1), 62–70.

Zhang, Z. & Goff, H. D. (2005). On fat destabilization and composition of the air interface in ice cream containing saturated and unsaturated monoglyceride. *International Dairy Journal* 15, 495–500.

FURTHER READING

Ahlers, M., Muller, W., Reichert, A., Ringsdorf, H. & Venzmer, J. (1990) Specific interactions of proteins with functional lipid monolayers-ways of simulating biomembrane process. *Angewandte Chemie International Edition in English* 29, 1269–1285.

Aitzetmuller, K. (1975) The liquid chromatography of lipids. A critical review. *Journal of Chromatography* 113 (2), 231–266.

Almgren, M. & Rangelov, S. (2006) Polymorph dispersed particles from the bicontinuous cubic phase of glycerol monooleate stabilized by PEG-copolymers with lipid-mimetic hydrophobic anchors. *Journal of Dispersion Science and Technology* 27, 599–609.

Ananthapadmanabhan, K. P. (1993) Protein-surfactant interactions. In: *Interactions of Surfactants with Polymers and Proteins*, Ananthapadmanabhan K. P. & Goddard E. D. (eds), CRC Press, Boca Raton, Fl., pp. 319–366.

Andersson, S., Hyde, S. T., Larsson, K. & Lidin, S. (1988) Minimal surfaces and structures: From inorganic and metal crystals to cell membranes and biopolymers. *Chemical Reviews* 88, 221–242.

Angelova, A., Angelov, B., Papahadjopoulos-Sternberg, B., Ollivon, M., Bourgaux, C. (2005) Proteocubosomes: Nanoporous vehicles with tertiary organized fluid interfaces. *Langmuir* 21, 4138–4143.

Angelov, B., Angelova, A., Papahadjopoulos-Sternberg, B., Lesieur, S., Sadoc, J. F., Ollivon, M. & Couvreur, P. (2006) Detailed structure df diamond-type lipid cubic nanopartcles. *Journal of American Chemical Society* 128, 5813–5817.

Aynié, S., Le Meste, M., Colas, B. & Lorient, D. (2006) Interactions between lipids and milk proteins in emulsion. *Journal of Food Science* 57, 883–887.

Backstrom, K., Lindman, B. & Engstrom, S. (1988) Removal of tryglicerides from polymer surface in relation to surfactant packing – ellipsometer studies. *Langmuir* 4, 872–878.

Barauskas, J., Razumas, V. & Nylander, T. (1999) Solubilization of ubiqinone-10 in the lamellar and bicontinous cubic phases of aqueous monoolein. *Chemical Physics* 97 (2), 167–179.

Barauskas, J., Razumas, V. & Nylander, T. (2000) Entrapment of glucose oxidase into the cubic Q230 and Q224 phases of aqueous monoolein. *Progress in Colloidal and Polymer Science* 116, 16–20.

Barauskas, J., Johnsson, M. & Tiberg, F. (2005) Self- assembled lipid superstructures: Beyond vesicles and liposomes. *Nano Letters* 5, 1615–1619.

Barauskas, J., Johnsson, M., Johnson, F. & Tiberg, F. (2005) Cubic phase nanoparticles (cubosome): principles for controlling size, structure and stability. *Langmuir* 21, 569–577.

Barauskas, J., Johnsson, M., Nylander, T., Tiberg, F. (2006) Hexagonal liquid-crystalline nanoparticles in aqueous mixtures of glyceryl monooleyl ether and pluronic F127. *Chemistry Letters* 35, 830–831.

Barauskas, J., Misiunas, A., Gunnarsson, T., Tiberg, F. & Johnsson, M. (2006) "Sponge" nanoparticle dispersions in aqueous mixtures of diglycerol monooley ether and polysorbate 80. *Langmuir* 22, 6328–6334.

Benichou, A., Aserin, A. & Garti, N. (2002) Protein-polysaccharide interactions for stabilization of food Emulsions. *Journal of Dispersion Science and Technology* 23, 93–123.

Better, A. (2005) Stick of Butter? *Cook's Illustrated*, No. 72, p. 3.

Biswas, S. C. & Marion, D. (2006) Interaction between puroindolines and the major polar lipids of wheat seed endosperm at the air-water interface. *Colloids and Surfaces B: Biointerfaces* 53, 167–174.

Blomqvist, B. R., Ridoul, M. J., Mackie, A. R., Warnheim, T., Claesson, P. M. & Wilde, P. (2004) Disruption of viscoclastic beta-lactoglobulin surface layers at the air-water interface by nonionic polymeric surfactants. *Langmuir* 20, 10150–10158.

Blomqvist, B. R., Wilde, P. & Claesson, P. M. (2006) Competitive destabilization/stabilization of betalactoglobulin foam by PEO-PPO-PEO polymetric surfactants. *Journal of Dispersion Science and Technology* 27 (5), 27–536.

Bohnert, J. L. & Horbett, T. A. (1986) Changer in adsorbed fibrinogen and album ininteraction with polyers indicated by decrease in detergent elutability. *Journal of Colloids and Interface Science* 111, 363–378.

Borné, J. & Nylander, T. (2001) Phase behavior and aggregate formation for aqueous monoolein system mixed with sodium oeate and oleic acid. *Langmuir* 17, 742–751.

Borné, J., Nylander, T. & Ehan, A. (2002a) Effect of lipase on different lipid liquid crystalline phases formed by oleic acid based acylglycerols in aqueous systems. *Langmuir* 18 (23), 8972–8981.

Borné, J., Nylander, T. & Ehan, A. (2002b) Effect of lipase on monoolein-based cubic phase dispersion (cubosomes) and vesicles. *Journal of Physical Chemistry* 106 (40), 10492–10500.

Borné, J., Nylander, T. & Khan, A. (2000) Effect of lipase on monoolein-based cubic phase dispersion (cubosomes) and vesicles. *Journal of Physical Chemistry B*, 106 (40), 10492–10500.

Bos, M. & Nylander, T. (1996) The interaction between β-lactoglobulin and phospholipid sat the air/water interface. *Langmuir* 12, 2791–2707.

Bos, M., Nylander, T., Arnebrant, T. & Clark, D. C. (1997) Protein/emulsifier interactions. In: *Food Emulsifiersand Their Applications*, Hasenhuettl, G. L. & Hartel, R.W. (eds), Chapman and Hall, New York, pp. 95–146.

Cabras, P. & Martelli A. (eds) (2004) The name margarine. In: *Food Chemistry*, Piccin, Padova.

Fu, H., Yang, L., Yuan, H., Xiao, F. & Lo, M. (2008) Production of low acid value edible oil with reduced TFAs by electrochemical hydrogenation in a diaphragm reactor. *Journal of American Oil Chemists' Society* 85 (11), 1087–1096.

Hui, Y. H., Chandan, R. C., Clark, S. & Cross, A. (2007) *Handbook of Food Products Manufacturing: Principles, Bakery, Beverages*. John Wiley & Sons, New Jersey, USA.

List, G. R., Byrdwell, W. C., Steidley, K. R., Adlof, R. O. & Neff, W. E. (2005) Triacylglycerol structure and composition of hydrogenated soybean oil margarine and shortening basestocks. *Journal of Agriculture and Food Chemistry* 53 (12), 4692–4695.

Phillips, J. C., Gaunt, I. F. & Gangolli, S. D. (1975) Studies on the metabolic fate of ^{32}P-labelled emulsifier YN in the mouse, guinea-pig and ferret. *Food and Cosmetics Toxicology* 13 (1), 23–30.

3 Stabilisers, Gums, Thickeners and Gelling Agents as Food Additives

Abstract: Stabilisers are additives that are incorporated into food items to give them a firmer texture. A number of hydrocolloids and many other polymeric substances are employed in the food industry to play roles as gelling, stabilisers or thickening agents. The molecules used as stabilisers, thickeners or gelling agents are macromolecules which, when dissolved or dispersed in aqueous media, are capable of causing an increase in the viscosity or gel formation. Among the most widely used stabilisers in the food industry are alginates, carrageenans, agar, guar gum, arabic gum, methylcellulose and carboxymethylcellulose. The same molecules are also used in the food industry to perform a variety of roles, for example as emulsifiers or preservatives.

Keywords: anionic polysaccharides; cationic polysaccharides; food stabilisers; gelling agents; gums; neutral polysaccharides; proteins; thickeners

3.1 INTRODUCTION TO STABILISERS, THICKENERS AND GELLING AGENTS

Further Thinking

Food stabilisers, thickeners or gelling agents are compounds of high molecular weight such as proteins (e.g. gelatin) and carbohydrates (e.g. pectins, starches, alginates and gums). They are characterised by long chains of respective monomer molecules.

Like all other food additives, stabilisers, thickeners and gelling agents are a group of compounds which play a very important role in the food industry to improve stability, texture, viscosity and make food products firmer (Feddersen and Thorp 1993). They are a diverse group of compounds with different chemistries and functional groups including amino acids, polyalcohols, carboxylic acid salts, carbohydrates and some proteins (Rambourg *et al.* 1982; Reis *et al.* 1993; Tarelli *et al.* 1998; Maclean *et al.* 2002). Most of these compounds are natural from either plant or animal origin (Gómez-Díaz and Navaza 2004).

Stabilisers in foodstuffs prevent food components from separating; they also give food a consistent texture. Gelling agents are useful when the consistency of a particular food has

Chemistry of Food Additives and Preservatives, First Edition. Titus A. M. Msagati.
© 2013 John Wiley & Sons, Ltd. Published 2013 by John Wiley & Sons, Ltd.

to be altered to meet consumer demand. On the other hand, thickeners play a useful role in giving food body.

Stabilisers and thickeners may be divided into two main classes: polysaccharides (Section 3.2) and proteins (Section 3.3). Polysaccharides comprise three main subclasses: (1) non-ionic (neutral), for example hydroxyethylcellulose and dextran; (2) anionic, for example xanthan, carrageenan, guar gum, alginate and carboxymethylcellulose (Cao *et al.* 1990; Dickinson and Galazka 1991; Samant *et al.* 1993; Ward-Smith *et al.* 1994; Dickinson 1996); and (3) cationic such as arginine hydrochloride and chitosan (a linear copolymer of D-glucosamine and N-acetyl-D-glucosamine units connected through β-(1–4) linkages).

3.2 POLYSACCHARIDES

3.2.1 Non-ionic (neutral) polysaccharide stabilisers

Non-ionic (neutral) polysaccharides such hydroxyethyl cellulose and hydroxypropyl cellulose are widely used as food stabilisers (Figure 3.1a–c). Their mechanism of adsorption involves the formation of steric barriers which are produced when the adsorbed polymer extends its chain to the water phase (Ain-Ai and Gupta 2008).

Hydroxypropyl cellulose is a cellulosic ether polymer containing OH groups attached to the chains of hydroxypropylated glucose molecules. The hydroxypropylation of glucose units is carried out chemically using the reaction between a base and propylene oxide. Because of the hydroxypropylation, hydroxypropyl cellulose contains both hydrophobicity and hydrophilicity characters.

3.2.2 Anionic polysaccharide stabilisers

Anionic polysaccharides (e.g. carboxymethyl cellulose, pectin, alginates, carrageenans, guar gums and gelatins) are employed as stabilisers in dairy products in particular (Marshall 1995). Their mechanism involves interaction with the positive charges on the surface of foodstuffs such as milk (casein micelles) to stabilise the casein network and thus lower syneresis.

Neutral hydrocolloids (e.g. xanthan gum, guar gum) form steric barriers using a different mechanism to that described in Section 3.2.1.1, which involves enhancement of the viscosity of the continuous phase, making them more of non-adsorbing polysaccharides (Hansen 1993).

3.2.2.1 Xanthan gum

Xanthan gum is an anionic polysaccharide employed in food industries as an additive as well as an agent to modify rheological properties of food products. It is obtained by precipitating the fermentation products of saccharides (glucose, sucrose or lactose) by bacterial species *Xanthomonas campestris*. The precipitation of the fermentation products is normally performed using isopropanol (Davidson 1980).

Structure-wise, xanthan gum is a polymer consisting of β-D-glucose repeating units such that trisaccharide molecules of mannose and glucuronic acid are attached at every second glucose of the polymer chain (Figure 3.2). The presence of carboxyl groups in the structure of xanthan gum, which may ionise to create negative charges, are known to assist in increasing viscosity of the solution in water (Davidson 1980).

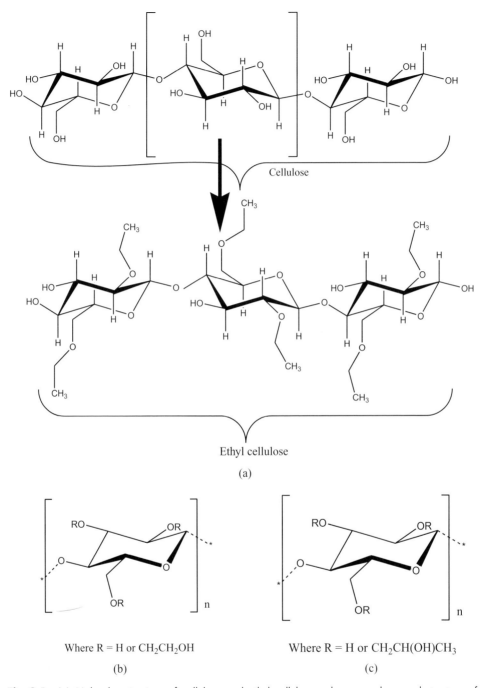

Cellulose

Ethyl cellulose

(a)

Where R = H or CH$_2$CH$_2$OH

(b)

Where R = H or CH$_2$CH(OH)CH$_3$

(c)

Fig. 3.1 (a) Molecular structure of cellulose and ethyl cellulose polymers and general structure of (b) hydroxyethyl cellulose polymers and (c) hydroxypropyl cellulose polymer (Hansen 1993; Ain-Ai and Gupta 2008).

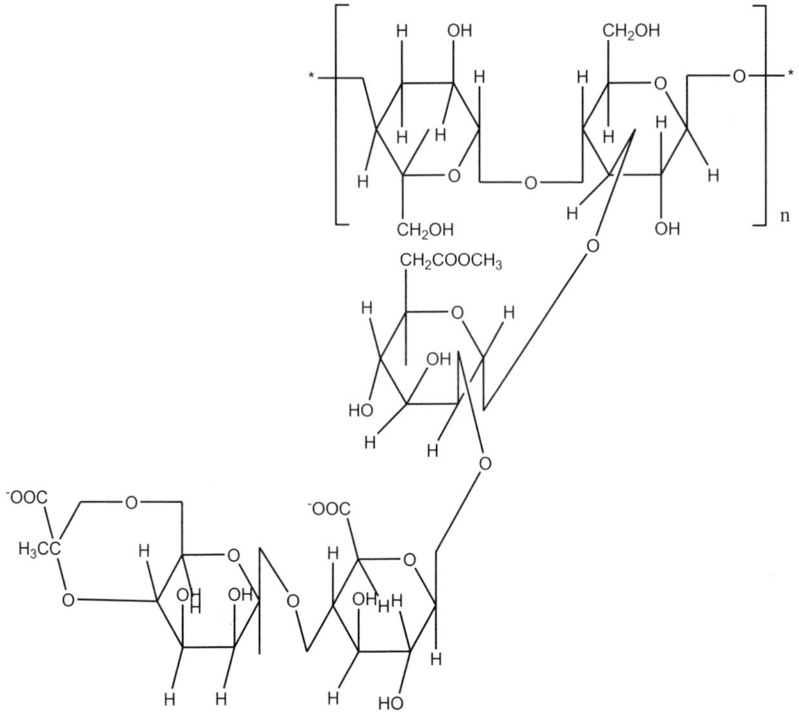

Fig. 3.2   Chemical structure of xanthan gum (Davidson 1980).

3.2.2.2 *Alginates*

Alginates (Figures 3.3 and 3.4) are polysaccharide compounds naturally found in the brown seaweed phaeophyceae where they occur as mixed salts of sodium, potassium, calcium and magnesium (Lal *et al.* 2006). The chemical structures of alginate are composed of random sequences of chains of β-D-mannuronic and α-L-guluronic acids attached with 1→4 linkages (Figure 3.3). Alginates are insoluble in water, although they do absorb water readily; they therefore find application in the food industry as gelling and thickening agents.

Sodium alginates (Figure 3.3) are attractive as they play a number of roles in the food industry including as emulsifying agents since they have the ability to develop a wide range of textures, are soluble in cold water and are also suitable as thickening agents (Saltmarsh 2000). However, sodium alginates are known to have one limitation, especially when used in calcium-rich foodstuffs such as milk: they are insoluble in a high concentrated calcium media (Lal *et al.* 2006).

To overcome such limitations, the addition of sequestration agents (e.g. citric acid or its salt) has been successfully demonstrated (Lal *et al.* 2006). The availability of hydrogen or carboxyl radicals in sequestering agents, or the presence of the corresponding salt, plays an important role in sequestering calcium ions (Tamine and Robinson 1999). This enables polysaccharide molecules used as food stabilisers to form a network of linkages between themselves and the chemical components of the food items (e.g. milk or yogurt). In this process, the negative groups in the polysaccharides are positioned at the interface, such that the binding of water molecules in the foodstuff becomes possible through the action of the

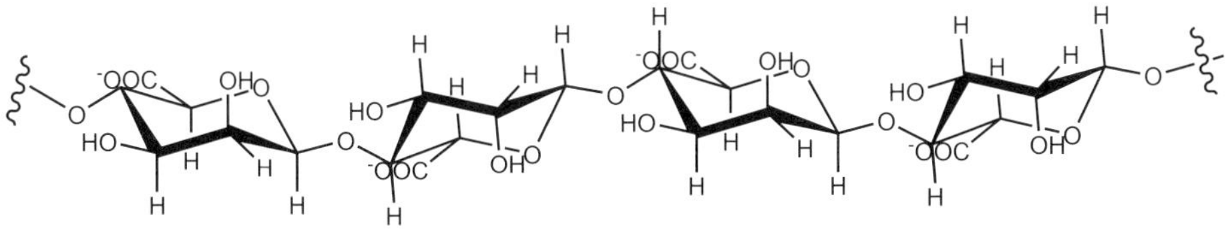

Alginate polymer made up of poly D-manauronic acid monomers

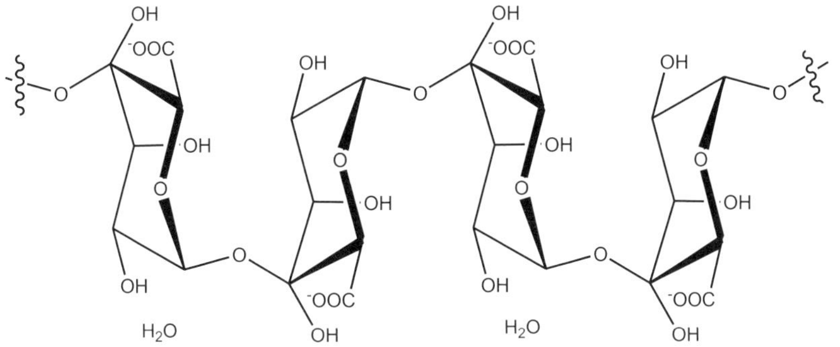

Alginate polymer made up of poly L-guluronic acid monomers

Fig. 3.3 Chemical structure of alginates showing sugar molecules D-manunuronic acid and L-guluronic acids (Lal *et al.* 2006).

polysaccharide stabiliser in a series of processes. Firstly, the water molecules hydrate to the stabiliser's surface (that is, as water of hydration). Water molecules then bind to the hydrophobic/protein part of the food item (milk) with the assistance of the polysaccharide stabiliser, enhancing their level of water absorption and thus stabilising the protein molecules. This enables them to form a chain which discourages the free movement of water (Ingenpass 1980).

3.2.2.3 *Propylene glycol alginate*

Propylene glycol alginate (PGA) is an ester of alginic acid (Figure 3.4) and propylene glycol. It is an emulsifier, stabiliser and thickener used in food products such as milk, where it is

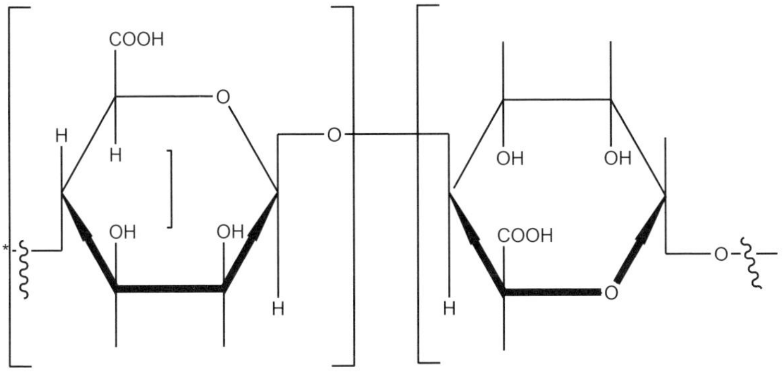

Fig. 3.4 Chemical structure of alginic acid (Lal *et al.* 2006).

utilised in the stabilisation of milk proteins under acidic conditions. The esterified alginate is far more effective, particularly as a thickener and stabiliser in several acidic food applications, than the standard alginate. This is because PGA contains both hydrophilic and lipophilic groups in its molecular structure, thus possessing important properties of emulsification and thickening.

3.2.2.4 Carrageenans

Carrageenan is the name given to a class of galactan polysaccharides that are found in agars and in the cell wall compositions of marine algae of the class Rhodophyta. Carrageenans exist in three main types as kappa, iota and lambda carrageenan (Figure 3.5) and are found naturally in red seaweeds (Lal et al. 2006). The chemical structure of carrageenan is composed of sugar units of galactose and anhydrogalactose joined with sulphate linkages to form a polymer chain.

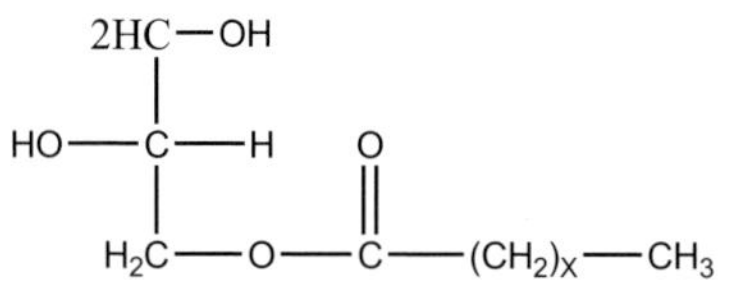

Monoglyceride

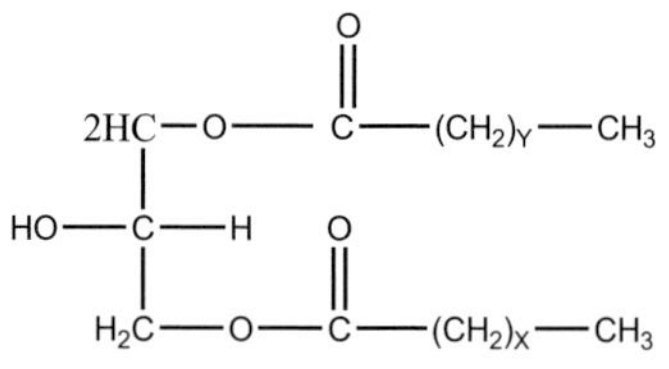

Diglyceride

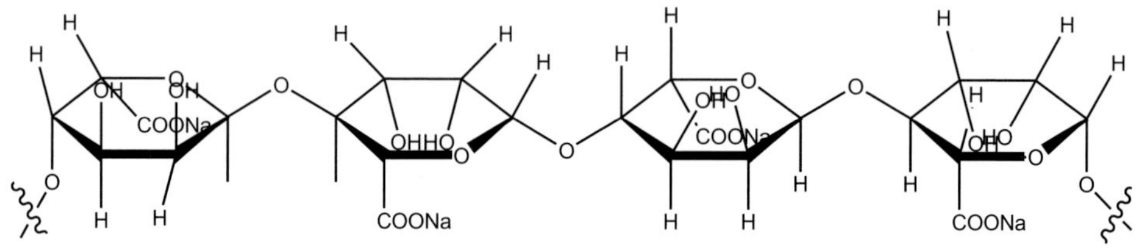

Sodium alginate

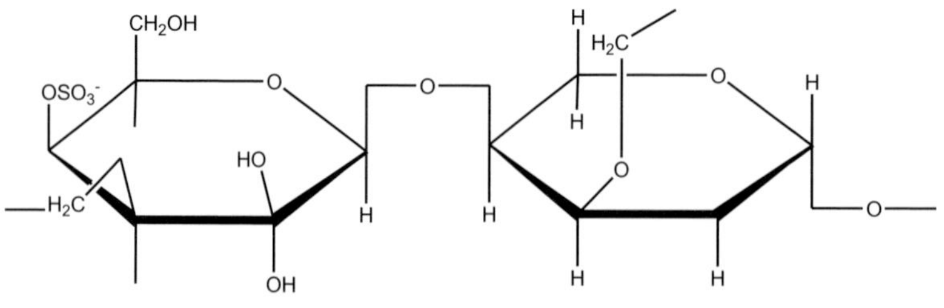

Kappa Carrageenan   (a)

Fig. 3.5 Chemical structures of selected thickeners/stabilisers (Lal et al. 2006).

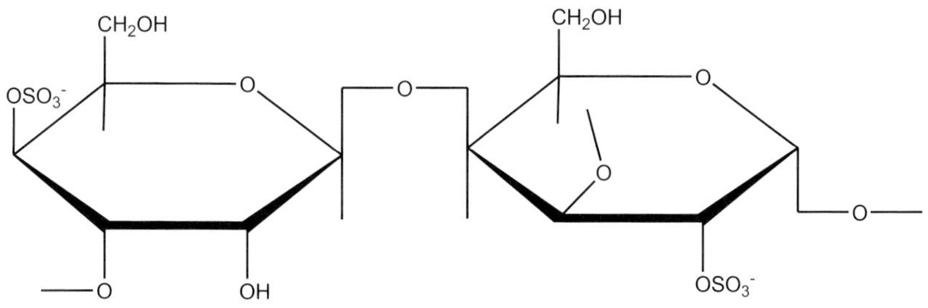

Iota Carrageenan

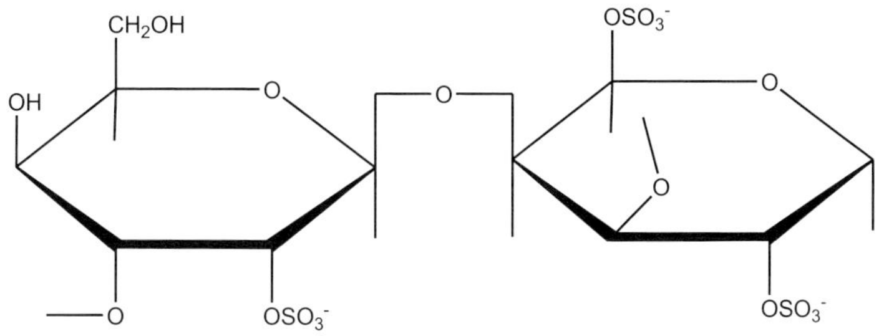

Lamba Carrageenan

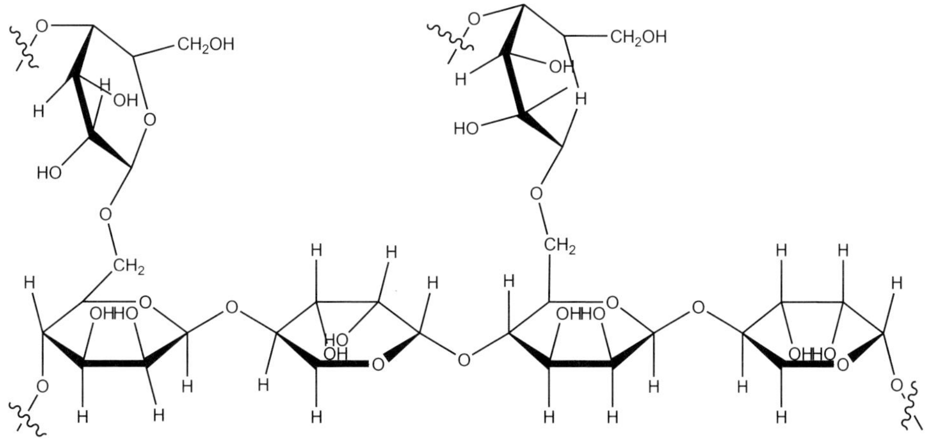

Guar gum   (b)

Fig. 3.5 *(Continued)*

In terms of their chemical properties, carrageenans are hydrophilic and hence water soluble. They are also sulphated galactans built on the repetition of 1,3-linked β-δ-galactopyranose alternating with 1,4-linked α-galactopyranose (Rees 1969; Figure 3.5), with the latter sugar occurring mostly as 3,6-anhydro-D-galactose. Note that the sulphation pattern of carrabiose is actually the basis of the classification of carrageenans (lambda, kappa and iota; Rees 1969). Other carrageenan variants with methyl ether or pyruvate acetal substitutions are known of (Craigie 1990); however, only the kappa- and iota-carrageenans are applied as food stabilisers or gelling agents.

Lambda carrageenan differs from iota- and kappa-carrageenan by the level of sulphation which is higher in the lambda form per di-saccharide residue on average. In the lambda form, the $\beta(1$—$4)$-D-galactose ring is not conformationally locked in the -3, 6-anhydro form, and it carries the sulphate groups at C2 and C6. Sulphation in the $\alpha(1$—$3)$-D-galactose-1 residue is at C2. Moreover, the lambda-carrageenan does not undergo conformational ordering and hence does not form gels (Glicksman 1983; Piculell 1995; Whistler and BeMiller 1997); for this reason, it is not used as a gelling agent in foods.

For long-time storage of carrageenan or alginates, a stable and homogeneous suspension is always desired. To attain this, a system that is capable of reforming its network structure to ensure stability (a thixotropic system) is introduced (Tziboula and Horne 2000). This is made possible due to the presence of reactive sites at the surface of the kappa carrageenan molecules which undergo reaction at elevated temperatures with the charged surface of the foodstuffs (e.g. casein micelles in milk). This will then bring together the carrageenan molecules, forming a stable homogenous suspension and thus creating a stable thixotropic gel. This is important in slowing down the coalescence and separation of the fat globules (Lal *et al.* 2006).

3.2.2.5 Pectin

Anionic polysaccharide pectin (Figure 3.6) has found application in the food industry as a gelling agent, thickener, texturiser, emulsifier and stabiliser. This compound is obtained naturally as sugar-beet pulp, which is the residue left from sugar extraction (Mesbahia *et al.* 2005).

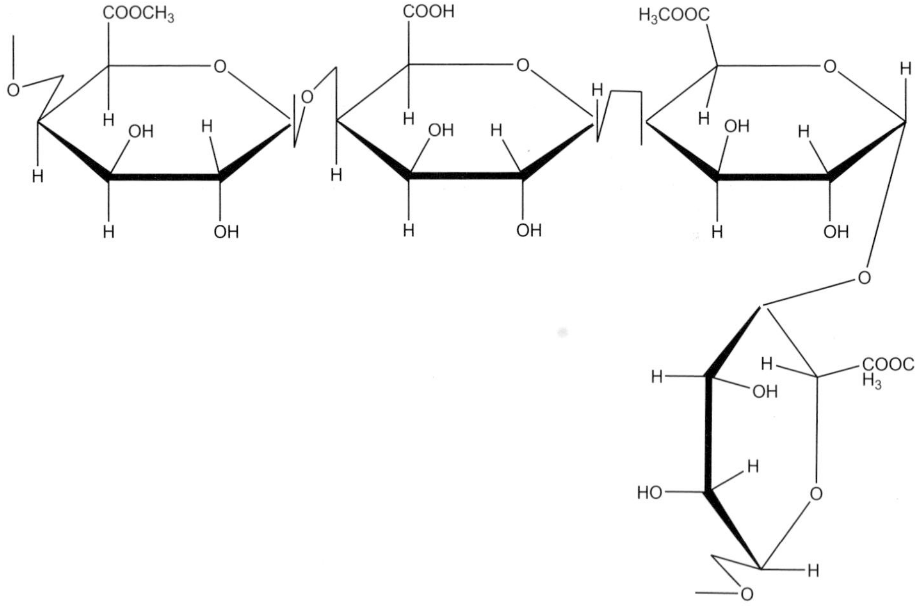

Fig. 3.6 Chemical structure of pectin (Baker 2010).

Chemically, pectin is a polysaccharide polymer comprising D-galacturonic acid units which are linked through α-1, 4-glycosidic linkages (Van Buren 1991; Thakur 1997). Its structure (Figure 3.6) shows that some of the carboxylic groups of galacturonic acid molecules are methylated (methyl esterified). The extent of pectin esterification is important, such that those with a degree of esterification greater than 50% are classified as high methoxyl pectin while those with a degree of esterification less than 50% are classified as low methoxyl pectin. (Thibault 1991; Thakur 1997). Differences in the side chain composition of pectins are usual, and they commonly involve side chains of arabinose, galactan, arabinogalactan, glucose, mannose and xylose (Ryden and Selvendram 1990).

3.2.2.6 Guar gum food stabilisers

Guar gum is another polysaccharide soluble in cold water which is used as a food additive (thickener). It plays the important role of increasing viscosity while maintaining the flavour and all the other qualities of the food. It also delays the sedimentation of solids or the creaming of fats (Lo et al. 1996). Guar gum is composed of molecules of galactose and mannose (galactomannan) and is obtained naturally from plant species (*Cyamopsis tetragonoloba* (L.) Taub.) (Barceloux 2008). Since this polysaccharide is hydrophilic, when subjected to water it forms a thick viscous gel material which ferments in the large intestines to form short chain fatty acids.

3.2.2.7 Carboxymethyl cellulose/cellulose gum

Carboxymethyl cellulose or cellulose gum (Figure 3.7), a derivative cellulose, finds application in the food industry as a food stabiliser and thickener. It contains carboxymethyl groups (-CH$_2$-COOH) attached to OH groups within the glucopyranose monomers forming the caroboxymethyl gum backbone.

This anionic polysaccharide is often used as a food additive in its sodium salt (sodium carboxymethyl cellulose) (Figure 3.8), where some of the hydroxyl groups in the structure of carboxymethyl gums are substituted with sodium atoms. There are three factors that determine the functional properties of carboxymethyl cellulose: the degree of substitution of carboxymethyl gum by sodium atom; the chain length of the cellulose backbone structure; and the degree of clustering of the carboxymethyl substituents.

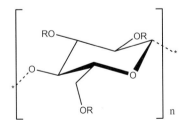

Where R = H or CH$_2$CO$_2$H

Fig. 3.7 General structure of carboxymethyl cellulose (cellulose gum) polymer.

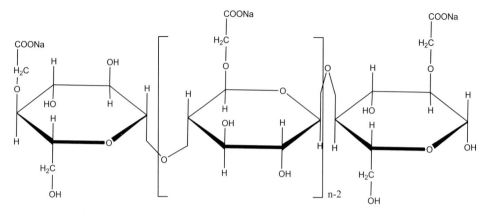

Fig. 3.8   Chemical structure of sodium salt of carboxymethyl cellulose.

In Figure 3.8 (which depicts an idealised unit structure of the carboxymethyl cellulose shown in Figure 3.7), this carboxymethyl cellulose has a degree of substitution of 1.0. If the remaining two hydroxyl groups in this unit are substituted, then the degree of substitution will be 3.0; this is the theoretical maximum that can be attained.

3.2.2.8   Agar (agar agar)

This is a natural compound extracted from seaweed which has numerous applications in the food industries, mainly as a gelling agent. Chemically, agar is made up of agarobiose polymers which are themselveshaving repeating units of disaccharides composed of D-galactose and 3,6-anhydro-L-galactose (Figure 3.9).

3.2.3   Cationic polysaccharide food stabilisers

3.2.3.1   Chitosan food stabilisers

Chitosan is a linear copolymer composed of D-glucosamine and N-acetyl-D-glucosamine units linked by β-(1–4) linkages. It is attractive for use as a food additive (stabiliser) due to its abundance and chemical properties including its hypocholesterolaemic and hypolipidaemic characteristics (Nauss *et al.* 1983; Winterowd and Sandford 1995; Muzzarelli 1996, 1997). Chitosan is also known for its good co-adsorption properties on lipid droplets covered with

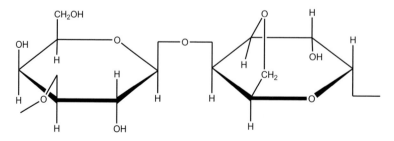

Fig. 3.9   Chemical structure of agarobiose showing the repeating disaccharide unit in agar.

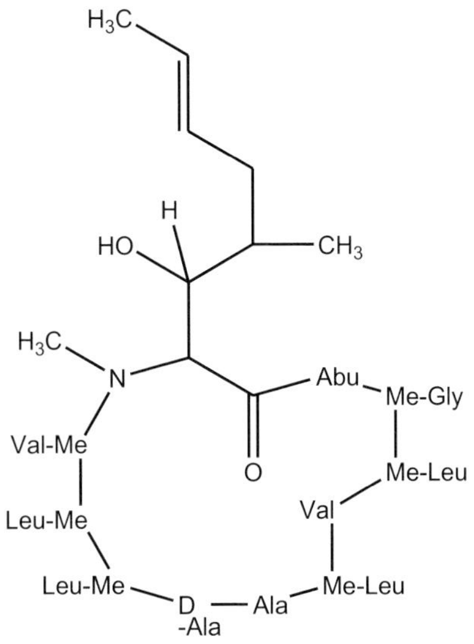

Fig. 3.10 Chemical structure of gelatin (Pouradier and Aribat 1950; Pouradier and Venet 1950; Veis 1964; Kozlov and Burdygina 1983) where Me: methionine, Leu: leucine, Ala: alanine, Gly: glycine and Val: valine.

anionic lipid emulsifiers (Alamelu and Panduranga Rao 1990; Magdassi *et al.* 1997) as well as its emulsifying properties (Schulz *et al.* 1998; Del Blanco *et al.* 1999).

Arginine hydrochloride is another example of a cationic polysaccharide.

3.3 PROTEIN-BASED FOOD STABILISERS

Protein-based stabilisers such as soya proteins are normally used in foods such as burgers or sausages for vegetarians. Soya proteins in these vegetarian food preparations prevent the food from disintegration during cooking. The soya protein will tend to gel the whole product, thus keeping it together. Food stabilisers play an important role in maintaining the integrity of foods in terms of physical and textural attributes, especially during heat treatment of foods, transport or storage. Thickeners are responsible for altering the texture of foods while gelling agents impart the desired shape and structure.

Gelatin (Figure 3.10) is a product of the structural and chemical degradation of collagen derived from animal skin and bones (Pouradier and Aribat 1950; Pouradier and Venet 1950; Veis 1964; Kozlov and Burdygina 1983).

Further Thinking

Stabilisers, thickeners, gums and gelling agents all play an important role in the processing of fruits such as strawberry and other industrially processed foods, especially the types of foods which do not contain gelatinous compounds or any other stabilising ingredients (e.g. pectin). These have to be added to maintain their structure as well as their physical properties during the processing period.

3.4 QUALITY CONTROL OF FOOD STABILISERS AND THICKENERS

Quality control of foods stabilisers is normally carried out by assessing rheological measurements (i.e. investigating the flow tendencies of matter in their liquid state). The flow behaviour can also be studied in non-liquid state (e.g. solid state or semi-solid state), where plasticity rather than deformation properties of the stabilisers in response to the force being applied can be assessed and characterised (Schowalter 1978).

The rheological qualities of hydrocolloids in solution are governed by a number of factors such as the amount of the active compound, dissolution, temperature and extent of dispersion (Rao and Anantheswaran 1982; Marcotte *et al.* 2001a, b).

Scientific research has linked beer-drinking patients to congestive heart failure and pericardial effusion, which could be attributed to the addition of cobalt as a foam stabiliser in the brewing process. This addition is also thought to have contributed to various cardio toxicity problems (Seghizzi *et al.* 1994). The use of guar gum has also been linked to cases of type-2 diabetes and hypercholesterolemia (Todd *et al.* 1990). Other reports have associated the use of guar gum with esophageal obstruction, this being more pronounced in the case of individuals who are already suffering from esophageal disease, for example, strictures (Opper *et al.* 1990; Seidner *et al.* 1990).

3.5 ANALYTICAL METHODS

Analysis of most additives incorporated in foods as stabilisers, thickeners or gelling agents is always a challenge. This is due to a number of factors, including the fact that most of these agents (e.g. the polysaccharide hydrocolloids) are applied in food matrices at minute quantities (0.001 to 1%). In addition to this, the food matrices also contain other polysaccharide molecules which are quite similar in terms of their chemistry to the stabiliser polysaccharide hydrocolloids. The analytical regimes for stabilisers and thickeners in foodstuffs is also complicated by the tendency to use a combination of more than one type/class of stabiliser (synergy). This creates an environment with a number of residues of stabilisers from different classes. Some natural biomolecules found as part of natural food constituents such as carbohydrates (sugars), proteins or lipids may also present interference problems in the sense that they may either extract the stabiliser molecules or mask the signal of stabilisers as they arrive at the detection system of the analytical device.

3.5.1 Acid hydrolysis and methanolysis

Another problem is that some stabiliser food additives (e.g. polysaccharide hydrocolloids) are actually polydisperse biomolecules, and can only be determined by using a combination of a number of analytical techniques. These techniques might include acid hydrolysis, which allows the polysaccharide to depolymerise (Rourke *et al.* 1993), or methanolysis followed by chromatographic separation of components of sugars with subsequent chromatographic analysis of individual sugars released.

Since there are different categories of stabilisers – non-ionic (neutral), anionic and cationic – the choice of conditions for aqueous hydrolysis is crucial, especially for anionic stabilisers which includes alginates and carrageenans. The difficultly in selecting analysis conditions is a result of the fact that the monosaccharides that will be released to the solution after the acidic hydrolysis procedure will exert different stabilities (Saeman *et al.* 1954; Albersheim *et al.* 1967). Contrary to this, methanolysis is preferable over acid hydrolysis as it results in stable products (stable methylglycosides) (Chambers and Clamp 1971; Quemener *et al.* 1995). Methanolysis is also more attractive compared to acid hydrolysis as it creates the possibility of simultaneously analysing both acidic and neutral monosaccharides using either gas chromatography (GC; Chaplin 1982) or high-performance liquid chromatography (HPLC; Cheetham and Sirimanne 1983; Quemener and Thibault 1990; Quemener *et al.* 1995).

3.5.2 Alternative techniques

Alternative techniques such as electrophoresis or immunoassays (e.g. enzyme linked immunoassay or ELISA) which do not need a depolymerisation step before the determination procedures may also be employed (Do *et al.* 1971; Pechanek *et al.* 1982; Patel and Hawes 1988; Arakawa *et al.* 1991; Roberts *et al.* 1998). One precaution which must be observed when using ELISA, which has been described as highly sensitive, is that it requires highly specific antisera in order to avoid cross-reactions.

Fluorophore-assisted carbohydrate electrophoresis (FACE) is another technique that has been introduced for the analysis of sugar containing polymers (Marrs *et al.* 1989). This separation is based on the fluorimetric labelling of the sugar molecules (saccharide units) by reductive amination, followed by electrophoretic separation on a highly cross-linked acrylamide gel.

Polarimetry and refractometry are methods which take advantage of the ability of sugars to refract light or rotate polarised light. These techniques are very popular in the determination of stabilisers and thickeners in foodstuffs (Zhang *et al.* 1994). For example, the angle of rotation of polarised light passing through a sucrose solution is used to determine its concentration. Sucrose comprises two monosaccharides: glucose (dextrorotatory) and fructose (levorotatory). Since the fructose fraction has a relatively stronger molar rotation than the glucose fraction, the sucrose solution will therefore tend to be levorotatory. The difference between the initial angle of rotation (when the reaction starts) and the magnitude of the rotation when the reaction comes to a completion gives a measure of the concentration of the sucrose.

3.5.3 Protein-binding assays

Protein-binding assays exploit the binding specificity tendencies of antibodies and lectins towards the sugar moiety or functionalities of polysaccharides. These assays make use of protein biomolecules such as antibodies or lectins which are capable of reacting specifically with some known and well-defined portion (targets) of sugar molecules within the polymeric structure of polysaccharides. The binding interaction of the protein molecules with these specific targets is always highly stereospecific. Factors that may influence this stereo-specificity include the presence of divalent cations such as calcium (Ca^{2+}) or magnesium (Mg^{2+}) as well as the degree of accessibility of the target sites.

REFERENCES

Ain-Ai, A. & Gupta, P. K. (2008) Effect of arginine hydrochloride and hydroxypropyl cellulose, as stabilizers on the physical stability of high drug loading, nanosuspensions of a poorly soluble compound. *International Journal of Pharmaceutics* 351, 282–288.

Alamelu, S. & Panduranga Rao, K. (1990) Effect of surfactants on the stability of modified egg-yolk phosphatidylcholine liposomes. *Journal of Microencapsulation* 7, 541–551.

Albersheim, P., Nevins, D. J., English, P. D. & Karr, A. (1967) A method for the analysis of sugars in plant cell-wall polysaccharides by gas-liquid chromatography. *Carbohydrate Research* 5, 340–345.

Arakawa, S., Ishihara, H., Nishio, O. & Isomura, S. (1991) A sandwich enzyme-linked immunosorbent assay for kappa-carra-geenan determination. *Journal of the Science of Food and Agriculture* 57, 135–140.

Barceloux, D. G. (2008) *Medical Toxicology of Natural Substances: Foods, Fungi, Medicinal Herbs, Toxic Plants, and Venomous Animals. Hoboken.* John Wiley & Sons, Inc., New York, pp. 22–33.

Cao, Y., Dickinson, E. & Wedlock, D. J. (1990) Creaming and flocculation in emulsions containing polysaccharide. *Food Hydrocolloids*, 4, 185–195.

Chambers, R. E & Clamp J. R. (1971) An assessment of methanolysis and other factors used in the analysis of carbohydrate-containing materials. *Biochemical Journal* 125 (4), 1009–1018.

Chaplin, M. F. (1982) A rapid and sensitive method for the analysis of carbohydrate components in glycoproteins using gas-liquid-chromatography. *Analytical Biochemistry* 123, 336–341.

Cheetham, N. W. H. & Sirimanne, P. (1983) Methanolysis studies of carbohydrates, using HPLC. *Carbohydrate Research* 112 (1), 1–10.

Craigie, J. S. (1990) Cell walls. In: *Biology of the Red Algae*, Cole, K. M. & Sheath, R. G. (eds), Cambridge University Press, Cambridge. pp. 221–257.

Davidson, R. L. (1980) *Handbook of Water-soluble Gums and Resins.* McGraw Hill, New York.

Del Blanco, L. F., Rodriguez, M. S., Schulz, P. C. & Agullo, E. (1999) Influence of the deacetylation degree on chitosan emulsification properties. *Colloid and Polymer Science* 277, 1087–1092.

Dickinson, E. and Galazka, V. B. (1991) Emulsion stabilization by ionic and covalent complexes of -lactoglobulin with poly-saccharides. *Food Hydrocolloids* 5, 281–296.

Dickson E. (1996) Biopolymer interactions in emulsion system: Influnces on creaming, flocculation and rheology. In: *Macromolecular Interactions in Food Technology*, Parris, N (ed.) American Chemical Society, Washington, DC, pp. 197–207.

Do, J. Y., Ioannou, J. & Haard, N. F. (1971) A research note. Polyacrylamide gel electrophoresis of pectic substances. *Journal of Food Science* 36, 1137–1138.

Feddersen, R. L. & Thorp, S. N. (1993) Sodium carboxymethylcellulose. In: *Industrial Gums: Polysaccharides and derivatives*, 3rd edition. Whistler R. L. & BeMiller J. N. (eds.) Academic Press, San Diego.

Glicksman, M. (ed.) (1983) Red seaweed extracts (agars, carrageenans, furcellaran). In: *Food Hydrocolloids*, CRC Press, Boca Raton, Florida, Vol. 2, pp. 73–113.

Gómez-Díaz, D. & Navaza, J. M. (2004) Rheology of food stabilizers blends. *Journal of Food Engineering* 64, 143–149.

Hansen, P. M. T. (1993) Food hydrocolloids in the dairy industry. In: *Food Hydrocolloids: Structures, Properties and Functions*, Nishinari, K. & Doi E. (eds), Plenum Press, New York, pp. 211–224.

Ingenpass, P. (1980) Food flavouring. *Packaging and Processing Journal*, 2 (1), 16–17.

Kozlov, P. V. & Burdygina, G. I. (1983) The structure and properties of solid gelatin and the principles of their modification. *Polymer*, 24, 651–666.

Lal, S. N. D., O'Connor, C. J. & Eyres, L. (2006) Application of emulsifiers/stabilizers in dairy products of high rheology. *Advances in Colloid and Interface Science*, 123–126, 433–437.

Lo, C. G., Lee, K. D., Richter, R. L., Dill, C. W. (1996) Influence of guar gum on the distribution of some flavor compounds in acidified milk products. *Journal of Dairy Science* 79 (12), 2081–2090.

Maclean, D. S., Qian, Q. & Middaugh, C. R. (2002) Stabilization of proteins by low molecular weight multi-ions. *Journal of Pharmaceutical Sciences* 91 (10), 2220–2229.

Magdassi, S., Bach, U. & Mumcuoglu, K. Y. (1997) Formation of positively charged microcapsules based on chitosan-lethicin interaction. *Journal of Microencapsulation* 14, 189–195.

Marcotte, M., Taherian, A. R. & Ramaswamy, H. S. (2001a) Rheological properties of selected hydrocolloids as a function of concentration and temperature. *Food Research International* 34, 695–704.

Marcotte, M., Taherian, A. R. & Ramaswamy, H. S. (2001b) Evaluation of rheological properties of selected salt enriched food hydrocolloids. *Journal of Food Engineering* 48, 157–167.

Marrs, W. M., Sworn, G. & Hart, R. J. (1989) Improved Method for Electrophoresis of Polysaccharides. Leatherhead Food RA Research Report No. 635, Leatherhead Food RA, Leatherhead, UK.

Marshall, S. (1995) Food ingredients – the role for dairy-products. *Food Australia* 47 (3), 105–107.

Mesbahia, G., Jamaliana, J. & Farahnaky, A. (2005) A comparative study on functional properties of beet, and citrus pectins in food systems. *Food Hydrocolloids* 19, 731–738.

Muzzarelli, R. A. A. (1996) Chitosan-based dietary foods. *Carbohydrate Polymers*, 29, 309–316.

Muzzarelli, R. A. A. (1997) Human enzymatic activities related to the therapeutic administration of chitin derivatives. *CMLS Cellular and Molecular Life Science*, 53 (2), 131–140.

Nauss, J. L., Thompson, J. L. & Nagyvary, J. (1983) The binding of micellar lipids to chitosan. *Lipids* 18 (10), 714–719.

Opper, F. H., Isaacs, K. L. & Warshauer, D. M. (1990) Esophageal obstruction with a dietary fiber product designed for weight reduction. *Journal of Clinical Gastroenterology* 12, 667–669.

Patel, P. D. & Hawes, G. B. (1988) Estimation of food-grade galactomannans by enzyme-linked lectin assay. *Food Hydrocolloids* 2 (2), 107–118.

Pechanek, V., Blaicher, G., Pfannhauser, H. & Woidich, H. (1982) Electrophoretic methods for qualitative and quantitative analysis of gelling and thickening agents. *Journal of the Associate Office of Analytical Chemistry* 65, 745–752.

Piculell, L. (1995) Gelling carrageenans. In: *Food Polysaccharides and their Applications*, Stephen, A. M. (ed.), Marcel Dekker, New York, pp. 205–244.

Pouradier, J. & Abribat, M. (1950) Dissolution of gelatin in cold water. *Bulletin de la Societe de Chimie Biologique* 32 (11–12), 947–951.

Pouradier, J. & Venet, A. M. (1950) Contribution a letude de la structure des gelatins. *Journal de Chimie Physique et de Physico-Chimie Biologique* 47, 11–14.

Quemener, B. & Thibault, J.-F. (1990) Assessment of methanolysis for the determination of sugars in pectins. *Carbohydrate Research* 206 (2), 277–287.

Quemener, B., Lahaye, M. & Metro, F. (1995) Assessment of methanolysis for the determination of composite sugars of gelling carrageenans and agarose by HPLC. *Carbohydrate Research* 266, 53–64.

Rambourg, P., Le Graet, D., Severac, P., Doucet, D., Larcher, D. & Labrude, P. (1982) The n-acetyltryptophan as a protector of hemoglobin during the freeze-drying process. *Annales Pharmaceutiques Françaises* 40, 535–544.

Rao, M. A. & Anantheswaran, R. C. (1982) Rheology of fluids in food processing. *Food Technology* 36, 116–126.

Rees, D. A. (1969) Structure, conformation and mechanism in the formation of polysaccharide gels and networks. *Advances in Carbohydrate Chemistry and Biochemistry* 24, 267–332.

Reis, F. M., de-Koning, B., Das, P. C., Smit-Sibinga, C. T. (1993) Recovery of fibrinogen in cryoprecipitate pasteurized in the presence of sucrose and glycine. *Brazilian Journal of Medicine and Biology Research* 26 (5), 473–476.

Roberts, M. A., Zhong, H. J., Prodolliet, L. & Goodall, D. M. (1998) Separation of high molecular weight carrageenan polysaccharides by capillary electrophoresis with laser induced fluorescence detection. *Journal of Chromatography A*, 817, 353–366.

Rourke, T. J., Clarke, A. D., Bailey, M. E. & Hedrick, H. B. (1993) HPLC quantification of alginate or pectin added to lean ground pork. *Journal of Food Science* 58, 973–977.

Ryden, P. & Selvendram, R. R. (1990) Structural features of cell wall polysaccharides of potato. *Carbohydrate Research* 195, 257–272.

Saeman, J. F., Moore, W. E., Mitchell, R. L. & Millet, M. A. (1954) Techniques for the determination of pulp constituents by quantitative paper chromatography. *Tappi* 37, 336–343.

Saltmarsh, M. (ed.) (2000) *Essential Guide to Food Additives*. Leatherhead Publishing, Surrey, UK.

Samant, S. K., Singhal, R. S., Kulkarni, P. R. & Rege, D. D. (1993) Protein-polysacchride interactions: a new approach in food formulations. *International Journal of Food Science and Technology* 28, 547–562.

Schowalter, W. R. (1978) *Mechanics of Non-Newtonian Fluids*. Pergamon Press, Oxford, Frankfurt.

Schulz, P. C., Rodriguez, M. S., Del Blanco, L. F., Pistonesi, M. & Agullo, E. (1998) Emulsification properties of chitosan. *Coll. & Polymer Sci.* 276 (12), 1159–1165.

Seghizzi, P., D'Adda, F. & Borleri, D. (1994) Cobalt myocardiopathy. A critical review of literature. *Science of the Total Environment* 150, 105–109.

Seidner, D. L., Roberts, I. M. & Smith, M. S. (1990) Esophageal obstruction after ingestion of a fiber-containing diet pill. *Gastroenterology*, 99, 1820–1822.

Tamine, A. Y. & Robinson, R. K. (1999) *Yoghurt Science and Technology*. Woodhead Publishing Limited, Cambridge.

Tarelli, E., Mire-Sluis, A., Tivnann, H. A., Bolgiano, B., Crane, D. T., Gee, C., Lemercinier, X., Athayde, M. L., Sutcliffe, N., Corran, P. H. & Rafferty, B. (1998) Recombinant human albumin as a stabilizer for biological materials and for the preparation of international reference reagents. *Biologicals* 26 (4), 331–346.

Thakur, B. R. (1997) Chemistry and uses of pectin. *Critical Reviews in Food Science and Nutrition* 37 (1), 47–73.

Thibault, J. F. (1991) Gelation of sugar beet pectin by oxidative coupling. In: *The Chemistry and Technology of Pectin*, Reginald H. W. (ed.), Academic Press, New York, pp. 119–133.

Todd, P. A., Benfield, P. & Goa, K. L. (1990) Guar gum a review of its pharmacological properties, and use as a dietary adjunct in hypercholesterolaemia. *Drugs* 39, 917–928.

Tziboula, A. & Horne, D. S. (2000) Influence of milk proteins on the gel transition temperature and the mechanical properties of weak-carrageenan gels. In: *Gums and Stabilisers for the Food Industry*, Williams, P. A. & Phillips, G. O. (eds), The Royal Society of Chemistry, Cambridge.

Van Buren, J. P. (1991) Function of pectin in plant tissue structure and firmness. In: *The Chemistry and Technology of Pectin*, Reginald, H. W. (ed.), Academic Press, New York.

Veis, A. (1964) *The Macromolecular Chemistry of Gelatin*. Academic Press, New York.

Ward-Smith, R. S., Hey, M. J. & Mitchell, J. R. (1994) Protein-polysaccharide interaction at oil-water interface. *Food Hydrocolloids* 8, 309–315.

Whistler, R. L. & BeMiller, J. N. (1997) *Carbohydrate Chemistry for Food Scientists*. American Association of Cereal Chemists, St Paul, Minnesota, USA.

Winterowd, J. G. & Sandford, P. A. (1995) Chitin and chitosan. In: *Food Polysaccharides and their Applications*, Stephen A. M. (ed.), Marcel Dekker, New York, pp. 441–462.

Zhang, W., Piculel, L., Nilsson, S. & Knutsen, S. H. (1994) Cation specificity and cation binding to low sulfated carrageenans. *Carbohydrate Polymers* 23 (2), 105–110.

FURTHER READING

Hasenhuettl, G. L. (2008) Synthesis and commercial preparation of food emulsifiers. In: *Food Emulsifiers and their Applications*, Hasenhuettl, G. L. & Hartel, R. W. (eds), Springer Science, 11–37.

4 Sweeteners

Abstract: Sweeteners are additives that are added to foods mainly for flavouring purposes and also as supplements. Among the well-known nutritive sweeteners that have been used in the food industry are sugar alcohols such as glycol, glycerol, erythritol, threitol, arabitol, xylitol and ribitol. Natural sweeteners such as honey, which is naturally manufactured by bees using flower nectar and other ingredients, and syrups (thick and viscous liquid products containing sugars) are also commonly used.

Keywords: bulk sweeteners; intense sweeteners; sweeteners; sweetness sensation

4.1 INTRODUCTION TO SWEETENERS

Sweeteners are added to foodstuffs as a complement to caloric sugar, to bring about a precise taste sensation and to optimally support, reinforce and distribute the taste in general (Wasik *et al.* 2007). The active functional structural feature in these sweet taste compounds responsible for the sweetness sensation is a glycophore (Shallenberger 1996). All sweeteners and other sweet compounds contain this common structure . The chemistry of a glycophore is that of different configurations of atoms as well as groups of atoms, with a hydrogen atom such that two vicinal atoms can attract extra electrons close enough to exert negative charges. This flexibility allows glycophores to exist in various sizes and shapes, for example, in large sweetener molecules the glycophore may exist in a branched configuration or form while in smaller molecules it will exist in a simple linear form.

As the name suggests, sweeteners are responsible for a sweet taste. Sweetness as a sensation refers to one of the fundamental components of taste and is contributed to by a variety of compounds such as sugars, amino acids, proteins, cyclic and aromatic organic derivatives as well as inorganic salts. This sensation of sweetness is elicited through a bilaterally symmetrical and concerted dipolar interaction between a glycophore and the receptors located mainly in the tongue (Shallenberger 1996).

Further Thinking

Sweeteners (sugars) are added to foodstuffs to accomplish certain purposes such as adding flavour to foods, imparting a sensation of sweetness and providing energy sources. They also provide bulk and texture (especially in bakery products) and prevent foodstuffs such as cakes from drying out, hence acting as humectants. Sugars play the

Chemistry of Food Additives and Preservatives, First Edition. Titus A. M. Msagati.
© 2013 John Wiley & Sons, Ltd. Published 2013 by John Wiley & Sons, Ltd.

role of a preservative in jams and are also an important additive in ice-creams, where they lower the freezing point. Unfortunately, too much sugar is harmful to the body system, causing health problems such as obesity and diabetes. To avoid these problems, a number of other alternatives sweeteners (e.g. saccharin) have been synthesised which are several times sweeter than sucrose; a small amount of saccharin can therefore replace a large amount of sucrose, thus reducing the energy content several times over.

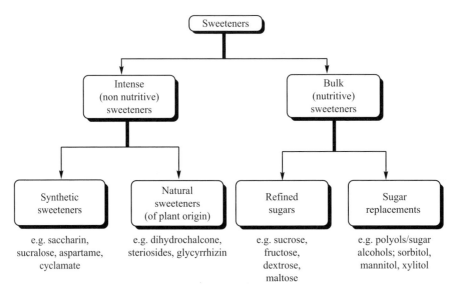

Scheme 4.1   Classification of sweeteners based on their nutritive status (Rogers *et al.* 1988; Ager *et al.* 1998).

Molecules referred to as sweeteners are numerous. A common method of classifying them is according to their nutritive status (Scheme 4.1): (1) intense (non-nutritive) sweeteners (see Section 4.3) and (2) bulk (nutritive) sweeteners (see Section 4.4).

4.2   PROPERTIES OF SWEETENERS

4.2.1   Structure–activity relationship

Sweeteners are known to be dipolar compounds. This configuration is important as it enables the concerted intermolecular hydrogen bonding interactions which involve a proton donor (AH) and a proton acceptor (B) to take place effectively at the receptor site in the taste buds (Shallenberg and Acree 1967). This process is further discussed under the section dealing with the mechanisms of action (Section 4.2.3).

The dipole configuration in sweeteners is such that the dipole functions are equal in magnitude but of opposite charges, meaning the molecule is symmetrical. If the molecule was asymmetrical, it would result in a charge imbalance and hence the possibility of an admixture of sweetness and bitterness or even only bitterness.

4.2.2 Structure–taste relationship

There are different classes of sulphamate sweetener including: carbinyl sulpha-mates (ureasulphonates) ($RNHCONHSO_3^-Na^+$); sulphonyl sulphamates (sulphonamide sulphonates) ($RSO_2NHSO_3^-Na^+$); sulphamidyl sulphamates (sulphamide sulphonates) ($RNHSO_2NHSO_3^-Na^+$); and iminyl sulphamates (hydrazonyl sulphonates) (Ar-C(R) =N-$NHSO_3^-Na^+$). The general structure of sulphamate sweeteners comprises three parts: the R group, the middle sulphamate functionality and the cationic part, i.e. R - $NHSO_3^-$ - M^+.

The sulphamate moiety $NHSO_3^-$ is considered to contain the Shallenberger AH/B centres required for the generation of sweetness taste. This implies that the NH group represents the AH entity while the SO_3^- moiety can be regarded as the B centre of the sulphamate function (Benson and Spillane 1976). The influence of the cation M^+ (which can either be silver, sodium, ammonium or cyclohexyl ammonium salts of cyclamate) on the taste of cyclamate is that of 'marked sweetness' (Audrieth and Sveda 1944).

4.2.3 Mechanism of sweetness

The sweet taste sensation is made possible by the intermolecular hydrogen bonding that is formed between a glycol unit in sweeteners and the taste bud receptor sites in the tongue. The structural unit of a glycophore is composed of AH and BB species. A and BB units are electronegative atoms separated by a distance magnitude of 22.55–44.0 Å (Shallenberger 1996), while H is the hydrogen atom attached covalently to one of the electronegative atoms (Figure 4.1).

Figure 4.1 illustrates how a sweet sensation is perceived in the taste buds, whereby sweet molecule hydrogen bonds in aqueous media and the hydrogen atoms in the sweet molecule are concurrently attracted to its own negative atoms and to negative atoms of vicinal molecules. Chemical bonds are therefore formed with water molecules and then a receptor site in the tongue (taste bud). This chemical bonding triggers a chemical stimuli or electrical relay that sends a taste message to the appropriate part of the brain for interpretation (Shallenberger 1996).

In this mechanism, water (aqueous media) plays a very important role because it creates a state of hydration which influences the movements and alignments of molecules so that it can properly catalyse the mechanism (Shallenberger 1996). This means that a sweetener must be hydrophillic and it must have the capability of specific binding to appropriate receptors in the taste buds at the surface of the tongue. The receptor site is linked to a glucogene unit which dissociates when the hydrated sweetener binds to the receptor site in the taste

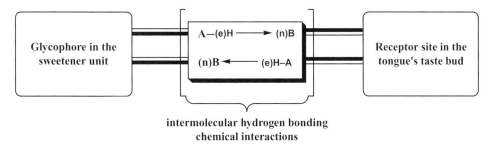

Fig. 4.1 Illustration of the mechanism of sweetness and sweet sensation (Shallenberger 1996).

buds. This stimulates a specific enzyme and sets in motion a series of events which will eventually generate signals to the specific part of the brain for interpretation. The nature of the sweetness signal will be determined by this interaction between receptor and sweetener. Water also influences the activity of the sweet molecule when it interacts with the receptor in the taste buds of the tongue.

The influence of hydration differs between sweeteners when the dissolution factor is taken into account, and this is used to evaluate the apparent molar volume of sweeteners. A sweetener's dissolution is normally determined by the change in water volume which occurs when a certain amount of sweetener (in moles) dissolves in a particular volume of water. This ratio to the molecular weight of that particular sweetener is what is known as the apparent molar volume, used to evaluate how deep a molecule moves into the cell membrane (taste buds). The smaller the apparent molar volume, the deeper the penetration into the taste buds.

As discussed in Section 4.2.1, the shape of the sweetener molecule (its stereo chemistry) also matters as this determines whether the molecule has a bitter or a sweet taste.

Sweeteners also differ from one another in that some give a sweet taste as soon as the molecule touches the taste buds which then disappears instantly, while for other sweeteners the sweetness is sensed gradually. This can best be explained by the rate and order at which the sweetener molecule reaches and interacts with the receptor as well as the length of time taken by the interaction with the receptors.

4.2.4   Sensory properties

These qualitative measurements have three main components: identification; analysis; and interpretation of the investigated properties of the product item (Carpenter *et al*. 1999). The bases of these measurements are the natural human senses which include taste (tongue), sight, smell, hearing and touching which are used to describe the quality of the product under three main criteria: discrimination, description and preference.

The discrimination criterion is intended to obtain an opinion on the possible differences that may exist between products and its qualitative magnitude (large, small or no difference). The description criterion is for the purpose of measuring whatever difference may exist between products, while the preference criterion is intended to identify the acceptability or rejection of one product over the other.

Panellists who are chosen for sensory properties testing normally undergo extensive training in the whole practice of the procedures for the test. Participants measuring 'taste' will be trained in methods such as sip-and-spit and how cleanse their palates, and will be instructed on the important of leaving a certain period of time (e.g. 2 minutes) before performing the next 'taste' test. Data from these types of analyses are processed by digital systems as well as by statistical tools (Portmann and Kilcast 1996).

4.3   INTENSE SWEETENERS IN FOODS

This group is comprised of very sweet molecules with a variety of chemical structures and functional groups, and they generally find application in products which are designed to reduce calories or in tooth-friendly types of confectionary (Rogers *et al*. 1988). Intense sweeteners are used as an alternative to sugar by people who, for health reasons, are trying to lose or control their weight. Since intense sweeteners do not promote tooth decay, they also

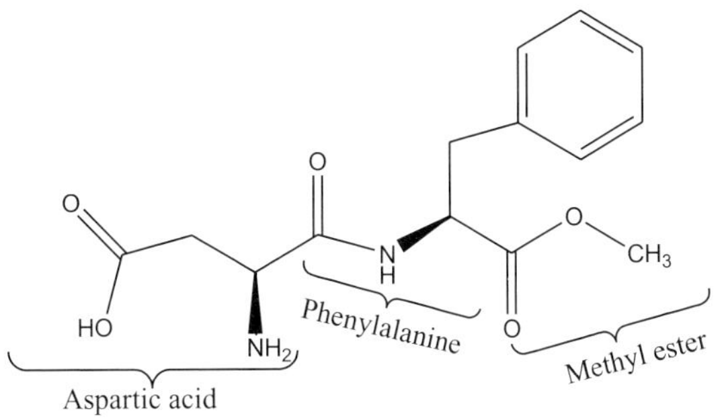

Fig. 4.2 Chemical structure of aspartame (Ager *et al.* 1998).

have applications in items such as toothpaste and mouthwash solutions. These sweeteners contribute positively to the status of health diet without compromising the pleasure of sweetness in foods. The class is further subdivided into synthetic or artificial sweeteners and natural sweeteners (Scheme 4.1).

4.3.1 Synthetic intense sweeteners

4.3.1.1 *Aspartame*

The synthetic subclass includes compounds such as aspartame, which is a methyl ester of the dipeptide compound of L-aspartic acid and L-phenylalanine amino acids (Figure 4.2). Aspartame is 200 times sweeter than table sugar (sucrose) (Ager *et al.* 1998). However, the compound is not stable under extreme pH conditions as it hydrolyses to yield methanol. If the pH conditions or even temperature are more stringent, the peptide bonds undergo hydrolysis to form free amino acids (Ager *et al.* 1998). For this reason, the application of aspartame as sweeteners in baked products is unrealistic and other ways to improve its stability under heating conditions have been devised. These techniques include the encasing of aspartame in fats or maltodextrin (Prodolliet and Bruelhart 1993). Alternatively, aspartame has been blended with other more stable sweeteners to improve its heat stability (Prodolliet and Bruelhart 1993).

4.3.1.2 *Sucralose*

Another type of artificial intense sweetener is a chlorinated sugar known as sucralose, produced by substituting the three hydroxyl groups in sucrose with chlorine atoms (Wasik *et al.* 2007; Figure 4.3). Sucralose is 600 times sweeter than sucrose and is known to be heat stable, an attribute which qualifies it to be used in fried and baked products in addition to other uses in beverages, chewing gums and frozen desserts (Mazurkiewicz *et al.* 2006).

4.3.1.3 *Saccharin*

Saccharin, a chemical compound containing a benzoic sulphimide base structure (Figure 4.4), is another artificial sweetener that finds application in drinks, cakes and biscuits (Nofre and Tint 2000), among other products. It is also unstable at higher temperatures and is therefore

Fig. 4.3   Chemical structure of sucralose (Wasik *et al.* 2007).

Fig. 4.4   Chemical structure of saccharin (Nofre and Tint 2000)

used in synergy with other stable sweeteners. It is mostly used as a sodium salt which is very soluble, unlike its acidic form which is insoluble.

4.3.1.4   Cyclamate

The artificial sweetener which has been in existence longer than many others currently in use is cyclamate (Figure 4.5), which is 30 times sweeter than sucrose (Nofre and Tint 2000). Cyclamate has a wide application because of its safety record, its stability in heat and its good qualities in the context of combined sweetening (Hellekant and Danilova 1996).

4.3.2   Natural intense sweeteners

Plants such as sugarcane, sugar beets, maple trees and corn produce sugars (sweeteners) via the process of photosynthesis. The sweeteners from natural or plant origin include perillalde-hyde, stevioside, rabaudioside, glycyrrhizin, osladin, thaumatins, monellin, dihydrochalcones and miraculin. Although the former is not sweet, it has the property of modifying the taste of sour food into a delightfully sweet taste (Sardesai and Waldshan 2005).

Fig. 4.5   Chemical structure of cyclamate (Nofre and Tint 2000)

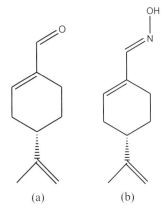

(a)   (b)

Fig. 4.6 (a) Perillaldehyde structure and (b) perillartine structure (Hellekant and Danilova 1996).

4.3.2.1 Perillaldehyde

Perillaldehyde, also known as perilla aldehyde, is a monoterpenoid natural organic compound found most abundantly in the perennial herb perilla (Hellekant and Danilova 1996). Perillartine, the oxime of perillaldehyde which is also known as perilla sugar, is about 2000 times as sweet as sucrose and is used as an intense natural sweetener (Hellekant and Danilova 1996; Figure 4.6a, b).

4.3.2.2 Glycyrrhizin

Glycyrrhizin is another natural intense sweetener that exists in liquorice and is about 170 times sweeter than sucrose (Hellekant and Danilova 1996). One of the derivatives of this compound, the glycyrrhizic acid monoglucuronide, is also an intense natural sweetener which is 1000 times sweeter than sucrose. Structurally, glycyrrhizin is a triterpenoid saponin glycoside found in the roots of liquorice plants of the species *Glycyrrhiza glabra*. It is used as sweetener in baked products, frozen dairy, non-alcoholic beverages and vinegars (Hellekant and Danilova 1996).

4.3.2.3 Steviol glycosides

Steviol glycosides are natural sweeteners extracted from the leaves of *Stevia rebaudiana* (Bertoni) Bertoni (Geuns 2003). Stevioside compounds are prepared by connecting glucose molecules to the steviol molecule. This diterpene exists as an aglycone and is synthesised by replacing the carboxyl hydrogen atom of steviol (Figure 4.7) with glucose to generate an ester, and replacing the hydroxyl hydrogen with mixtures of glucose and rhamnose sugars. When only glucose is used, the resulting compounds are stevioside which contains two glucose units and rebaudioside which is made up of three glucose units (Brandle 2004). Stevioside and rebaudioside are the two main compounds that are responsible for the sweet taste of stevia species (Bridel and Lavielle 1931).

Previous reports described serious concerns with regards to steviol, as it was linked to the possibility of triggering carcinogenic and mutagenesis activities (Pezzuto *et al.* 1985). This report was later disapproved by the World Health Organisation (WHO) (Benford *et al.* 2006)

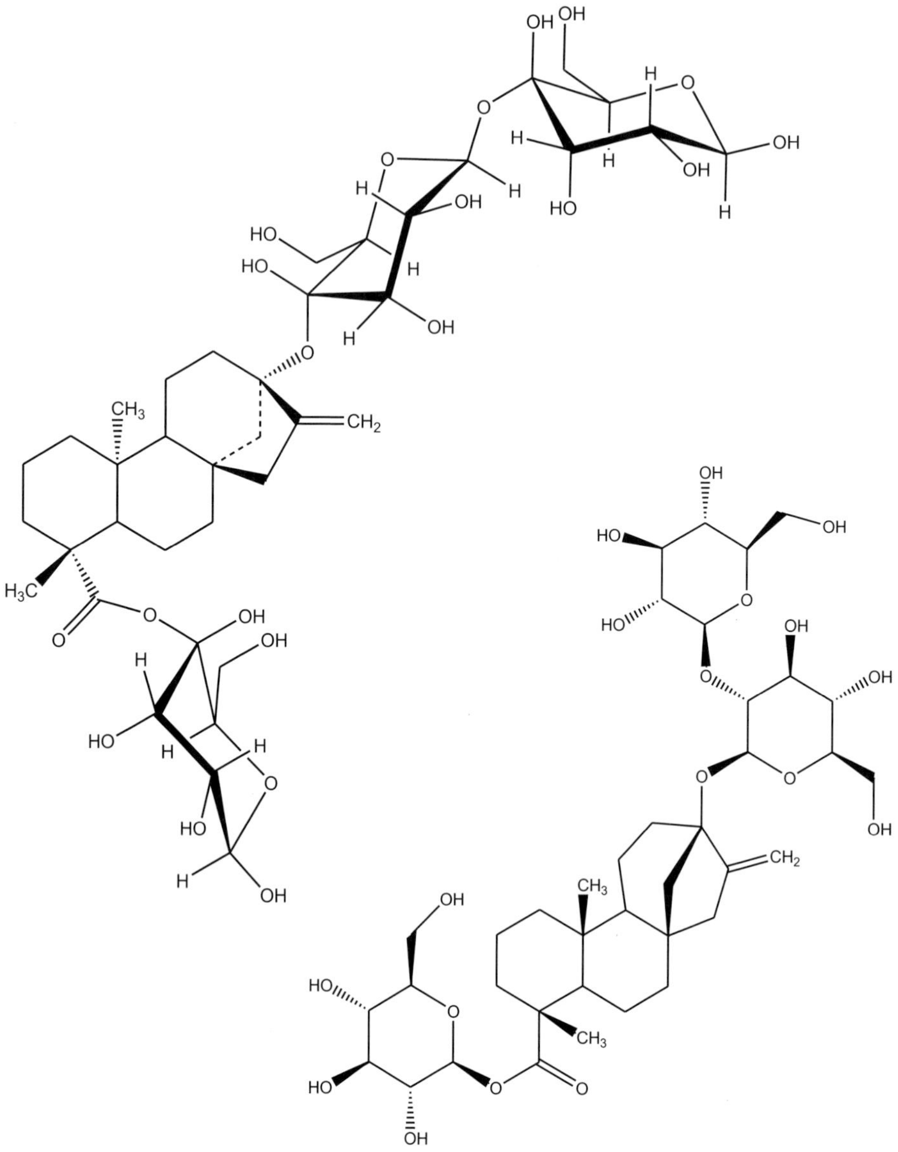

Fig. 4.7   Chemical structures of stevioside and rebaudioside (Bridel and Lavielle 1931; Geuns 2003).

as well as other researchers (Procinska *et al.* 1991). On the contrary, WHO has suggested that steviol has health benefits, especially to people suffering from hypertension and diabetes type-2 (Benford *et al.* 2006).

4.3.2.4   Naringin dihydrochalcone

From the dihydrochalcone class of natural intense sweetener, there is a compound known as naringin dihydrochalcone (Ikan 1991) which is a phloretin glycoside (Figure 4.8). The

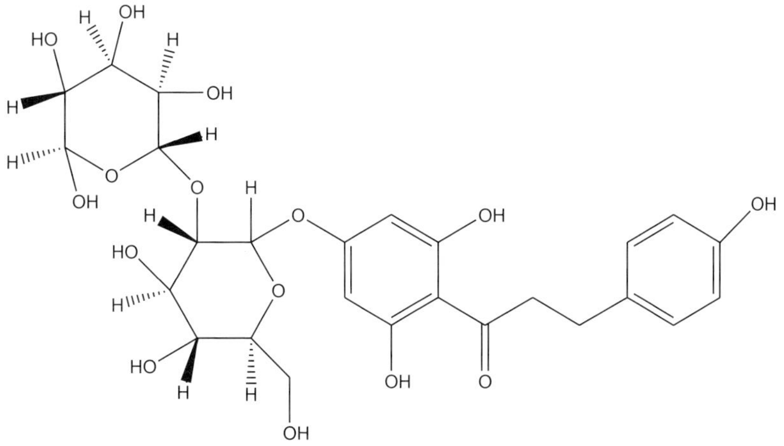

Fig. 4.8 Chemical structure of naringin hydrochalcone (Tomasik 2003).

reaction of naringin dihydrochalcone with strong bases such as potassium hydroxide with a further catalytic hydrogenation results in dihydrochalcone, a compound which is about 300–1800 times sweeter than sucrose at threshold concentrations (Tomasik 2003).

4.3.2.5 Neohesperidin dihydrochalcone

Another natural intense non-caloric sweetener derived from citrus is a glycosilated compound known as the neohesperidin dihydrochalcone (Neo-DHC; Figure 4.9), which is about 1000 times as sweet as sucrose and very stable to heat (Tomasik 2003). This sweetener has limited use however due to a number of drawbacks. It has an intense cooling effect on the tongue, has a liquorice-like bitter off-taste, tends to be slow in its onset and has a lingering taste that differentiates it from the taste of sucrose; it is not popular with consumers. Another limitation of neohesperidin dihydrochalcone is that it has very limited solubility in water. Neo-DHC is synthesised by chemical treatment of neohesperidin (Scheme 4.2), a bitter component of bitter orange, grapefruit and other citrus fruit peel and pulp. It is used to mask the bitter taste

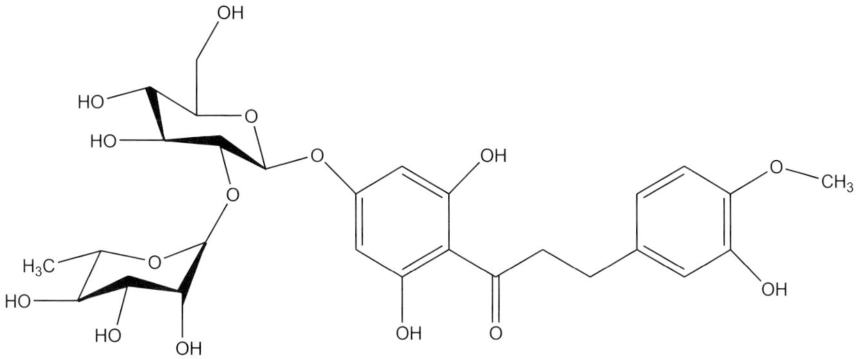

Fig. 4.9 Chemical structure of neohesperidin dihydrochalcone (Tomasik 2003).

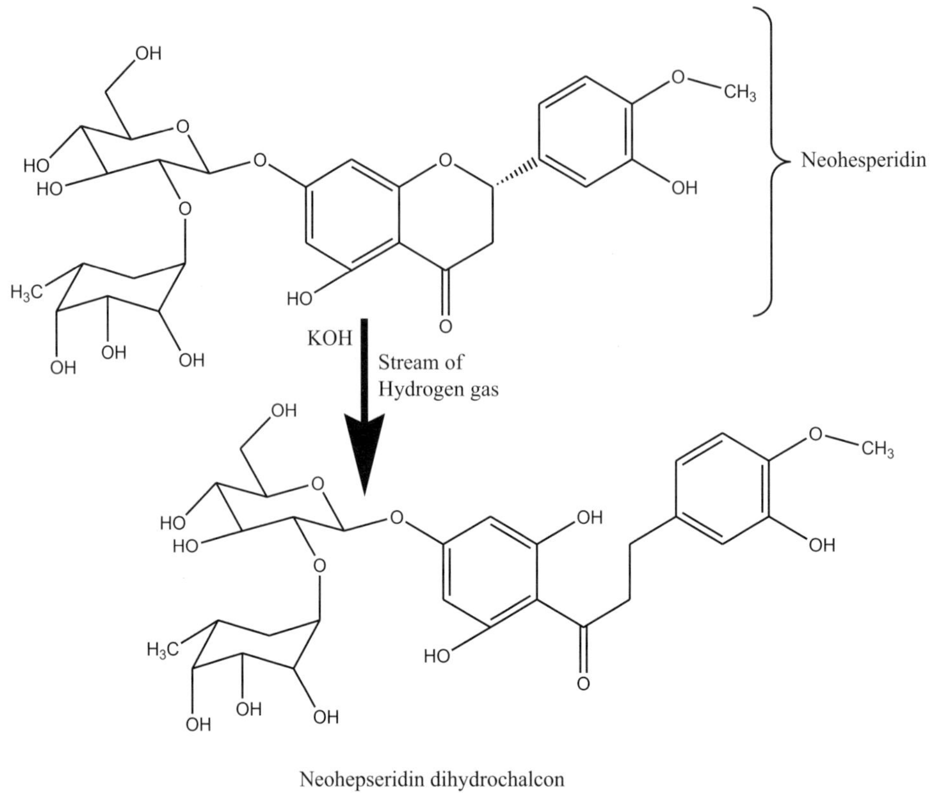

Neohesperidin

KOH

Stream of
Hydrogen gas

Neohepseridin dihydrochalcon

Scheme 4.2   Synthesis of neo-dihydrochalcone (Tomasik 2003).

in citrus, in the enhancement of the sweetness of other sweeteners and to mask bitterness in pharmaceuticals (Tomasik 2003).

4.4   BULK FOOD SWEETENERS

Bulk food sweeteners are bulking agents such as starch that increase the bulkiness of food products without compromising their nutritional integrity. This class of nutritive bulk sweeteners is composed of two subgroups: (1) refined sugars (e.g. dextrose, fructose, sucrose, maltose, etc.) and (2) sugar replacements which are comprised of mainly sugar alcohols or polyols (e.g. sorbitol, lactitol, erythritol, isomalt, xylitol and mannitol).

4.4.1   Refined sugars

Refined sugars (sweeteners) such as dextrose (Figure 4.10a–i) are obtained naturally in foodstuffs, and have the attributes of a moderately sweet saccharide. Saccharide is a monosaccharide which forms the basic building block unit of carbohydrates ($C_6H_{12}O_6$) and has a high glycemic index (GI), a parameter reflecting the ability of digested carbohydrates to raise blood glucose. Another useful parameter is the glycemic load (GL), which gives a measure of blood glucose of any food product. This parameter is important as it provides a measure

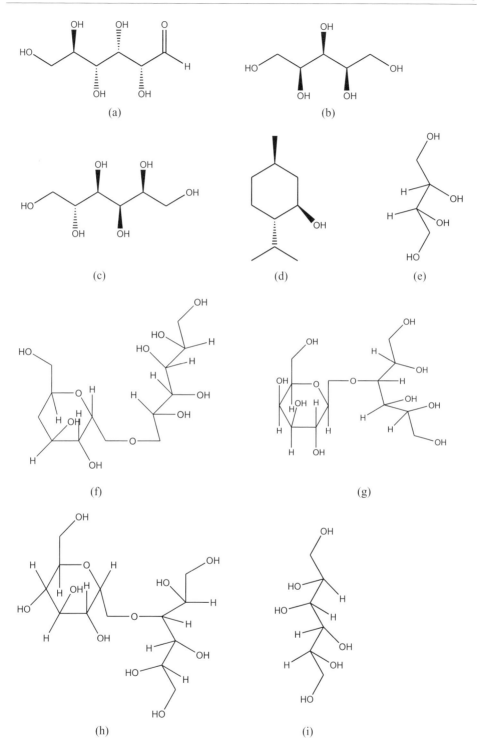

Fig. 4.10 Chemical structures of bulk sweeteners: (a) D-dextrose; (b) xylitol; (c) sorbitol; (d) 1-menthol; (e) erythritol; (f) isomalt; (g) lactitol; (h) maltitol; and (i) mannitol.

of the total glycemic response of food items. Mathematically, the glycemic load is calculated by multiplying the amount of carbohydrate (weight in grams or volume in millilitres) by the GI value of that particular food item, divided by 100. The glycemic response for a particular food item is therefore a measure of the impact of a food on the level of blood glucose.

Food items which hydrolyse easily and rapidly or which are absorbed easily also have a faster and greater impact on blood glucose (i.e. have high impact) and thus have high glycemic indices. On the other hand, food products with either slow or incomplete digested carbohydrates also lower glycemic indices because glucose is released gradually into the blood and the corresponding blood glucose response is lower and steady (Jenkins *et al.* 1981).

It should however be noted that the glucose response to foods tends to vary between individuals and on the way the food is being consumed.

4.4.2 Sugar replacements

4.4.2.1 Polyols

Polyols have a cooling effect and play the role of distribution of taste over time where they have a high molecular weight (Jenkins *et al.* 1981). Xylitol (a sugar alcohol) and sorbitol (known as glucitol, also a sugar alcohol) are the most cooling (Figure 4.10b and c). Their sweetness is 1.0 and 0.62 that of sucrose, respectively. Contributions from, for example 1-mentol (Figure 4.10d), may be incorporated in the total perception of coolness.

For consumers with effective metabolism, polyols are slowly and incompletely absorbed in the intestines. They normally require no or very little insulin and do not cause spikes in blood sugar (Jenkins *et al.* 1981). For this reason, polyols are suitable for use by diabetics because they play a key role in reducing glycemic index as well as reducing the risk of tooth decay. However, some of the polyols have drawbacks in that they are not absorbed by the blood and instead just pass through the small intestines (Jenkins *et al.* 1981). Also, large consumption of polyols can result in intestinal gas or diarrhoea.

4.4.2.2 Sugar alcohols

The compounds here referred to as sugar alcohols are neither sugar by their nature nor alcohols; however, their chemical structures partially resemble that of sugars and of alcohols. As food additives (sweeteners), they have a sweet taste which can mask the aftertaste of other sweeteners and can also add bulk and texture. They have the property of providing the cooling effect or taste, they inhibit browning during heating and retain moisture in foods.

4.4.3 Alternative classification

Sweeteners can also be classified based on their origin (natural and synthetic) into three main groups namely; natural sweeteners, nutritious sweeteners (derived from natural products) and intense sweeteners (which can be either of artificial or vegetable origin) (Scheme 4.3). The class of neo-azucares (a nutritious sweetener) in Scheme 4.3 comprises mainly a fructo-oligosaccharides mixture of actilight and raftilose. The neo-azucares are neo-sugars that are obtained by the fixation of fructo-oligosaccharides.

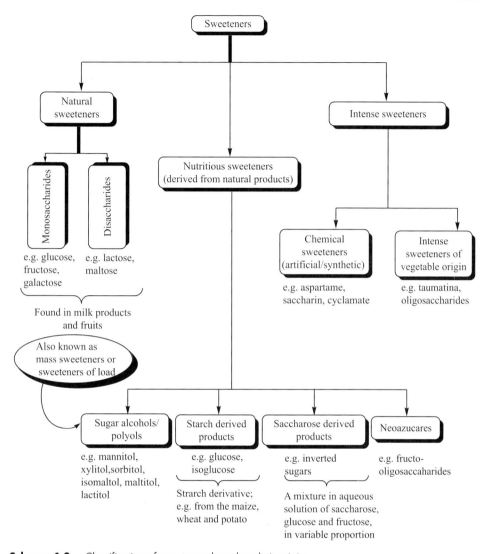

Scheme 4.3   Classification of sweeteners based on their origin.

4.5 QUALITY ASSURANCE AND QUALITY CONTROL

4.5.1 Methods of quality assessment

There are a number of tests and analyses that are normally used to ensure the quality of sweeteners, including the degrees Brix value, spectrometry, instrumental texture value as well as pH, sensory properties and the assessment of microbiological quality.

4.5.1.1 Degrees Brix

Degrees Brix (°Bx) gives the weight-by-weight sucrose percentage (the fraction of sugar per hundred parts) in aqueous solution by mass, that is, one degree Brix (1 °Brix) = 1 gram of sucrose in 100 grams of solution. Degrees Brix measurements may also give a measure of the total soluble solids (TSS) in the solution and, in this particular case, the total soluble solids in reference is strictly sucrose (sugar). In cases where the solution or matrix (e.g. fruits, vegetables, juices, wine, soft drinks, etc.) contains dissolved solids other than sucrose (e.g. other fructose or other sugars, minerals, acids, salts, etc.), then the measured °Bx value will represent an approximation of the dissolved solid content. From the definition of Brix measurements it follows that the higher the TSS or Brix value, the higher the sugar content and therefore the higher the quality of the sweetener. Brix values are measured using either refractometers or densitometers.

4.5.1.2   FT-IR spectrometry

Sugars also have chromophore functional groups that are capable of absorbing infrared light at near infrared region; it is therefore possible to measuring sugar levels using Fourier transform infrared spectrometry (FT-IR) techniques.

4.5.1.3   Instrumental texture

The instrumental texture value is another very important attribute that reflects both the quality of the sweeteners and the acceptance by consumers. There are a number of instrumental methods for determining the texture. These mainly mechanically based methods involve measuring the resistance of the sweetener product in relation to the force that was applied. There are a number of configurations with regard to the instrumentation required to measure texture and, for convenience, may be subdivided into four groups as follows.

1. A probing unit or device which makes physical contact with the product to be measured (sweetener food product). The probing device may be similar to a rod or cone structure that can pierce, it can also be like a plunger which has a flattened surface or it can have a shape like a cutting blade.
2. A driving mechanism to impart the required motion to the probing component unit. Depending on the design, the motion or rotation can be constant throughout or varied with time.
3. A detecting mechanism with the ability to sense or detect the resistance of the sample (sweetener food product on this case) in relation to the force that was applied (during pressing, piercing, cutting etc.).
4. An intelligent device with a read-out-mechanism such as a recorder, an digital integrator or a computer (a device consisting of electronic technology).

There are a number of texture-measuring meters that have been fabricated to perform instrumental texture analysis and which contain all four configurations listed above. One of these measuring devices is known as a penetrometer which penetrates the sample material and measures either the force needed for a certain penetration depth or the penetration depth achieved for a fixed force. The higher the force reading or the smaller the depth penetrated, the higher the resistance of the sample being analysed. This measure gives the firmness or toughness of the sample material (sweetener food material).

Another type of texture-measuring instrument is a compressometer which measures the resistance of the sample material to compression. The measured parameter in this case will either be the force needed to produce a certain deformation or a deformation that has been achieved as a result of the applied force. In this case, the extent of compression under constant load is regarded as an index of softness and the force needed to incur a certain deformation is an index of firmness. Compression as a method to measure instrumental texture is more attractive than penetration, because the sample material being tested is not pierced or punctured and normally the yield point is never exceeded (Szczeniac 1973).

4.5.2 Toxicity

Due to an increase in the use of sweeteners for the purpose of reducing calorie content of food products without compromising their taste, there exists the possibility of excessive levels of sugars in beverages and other foodstuffs (Weihrauch and Diehl 2004; Nunes *et al.* 2007). There are two types of health risks to consumers associated with excessive sugars in food stuffs: (1) allergenic reaction conditions (e.g. asthma) and (2) metabolic disruption (Cohen *et al.* 2006; Anderson and Young 2008; Brown 2008).

Some synthetic sweeteners such as aspartame, which is a dipeptide product of the reaction of L-aspartic acid and L-phenylalanine methyl ester, are normally hydrolysed after ingestion to produce aspartyl phenylalanine and methanol. The aspartyl phenylalanine produced undergoes further hydrolysis in the digestive system to produce aspartic acid and phenylalanine, which is actually a health risk for patients who suffer from a genetic disorder in homozygous gene for phenylketonuria (PKU). Unlike healthy consumers who do not suffer from this genetic disorder and who are able to hydroxylate phenylalanine to tyrosine after absorption by digestive system, those with this genetic disorder lack phenylalanine hydroxylase; they are therefore unable to hydroxylate phenylalanine, which causes accumulation of phenylalanine in their blood (Erlandsen *et al.* 2003).This implies that aspartame should not be taken by such people or it will complicate their health status (Northam 2004).

Another synthetic sweetener, saccharin, has also been reported to be a possible human carcinogen. Experiments conducted on male rats demonstrated an elevated incidence of bladder cancer (Barceloux 2008).

4.5.3 Link to weight gain and obesity

Obesity, a condition characterised by having a high proportion of body fat measured as a body mass index (BMI) of 30 or higher, has been speculated to be associated with additives such as high-fructose corn syrup (HFCS) sweeteners (Kroger *et al.* 2006; Renwick 2006). To date however, there is no definitive scientific link between obesity and sweetener intake, although it has been confirmed in research studies that HFCS sweeteners are less satiating despite being rich in calories (Kroger *et al.* 2006; Renwick 2006). Some scientific reports have indicated that a decrease in physical activities contributes significantly to weight gain.

On the other hand, non-nutritive sweeteners (NNS) have been reported to have the potential to promote weight loss due to their low calorie content (Porikos and Koopmans 1988; Renwick 2006; Mattes and Popkin 2009). Although NNSs are known to increase appetite and food intake, there is no concrete research to date that has shown that they contribute to weight gain. NNSs also have the potential to regulate sugar and energy intakes without compromising diet palatability, therefore posing no positive or negative effects as far as weight loss or reduced weight gain is concerned (Kroger *et al.* 2006).

4.6   ANALYTICAL METHODS

Due to the evidence of harmful effects on health from the use of certain sweeteners, there has been a strategy to ensure proper implementation of the legislation which limits the use of or bans some sweetener food additives. The main chemical properties of sweeteners (sugars) that are of importance regarding the development of analytical procedures include: the refractive index, polarisation (chirality), ability to act as reducing agents, reaction with specific enzymes, ability to adsorb onto stationary phases (chromatography) and ionic strength (ion chromatography).

Polarimetry and refractometry methods, techniques which are possible due to the ability of sugars to refract light or rotate polarised light, are very sensitive and popular for the analysis of food components containing sugars (Zhang *et al.* 1994). However, these techniques are mostly suitable for measuring sugar contents in foodstuffs with high sugar content; they are not intended for foodstuffs with lower sugar contents, which require more sensitive techniques.

A number of analytical methods have been developed for individual sweeteners as well as for combinations of them. For example, saccharin in food products has been determined singly using different techniques including spectrometry (Viannnasoares and Martins 1995); differential pulse polarography (AOAC 2000a); potentiometry (Dossantos 1993; Fatibello-Filho 1994; Negash *et al.* 1997); micellar electrokinetic capillary chromatography (Boyce 1999); high-performance liquid chromatography (HPLC; Sjoeberg and Alanko 1987; Sjoeberg 1988; Willets *et al.* 1996; AOAC 2000a); high-performance anion exchange chromatography (HPAEC; Guo 1995) and gas-liquid chromatography (GLC; Kauschus and Their 1985).

Electrically separation-driven process techniques such as capillary electrophoresis have been reported in the determination of a number of artificial sweeteners (Thompson *et al.* 1995; Pesek and Matyska 1997; Walker *et al.* 1997). Since most sweeteners lack the presence of strong chromophores (allowing them to be detected using HPLC traditional detectors such as ultra-violet visible or UV-Vis), it follows that most of the HPLC detection systems for these molecules involve either refractive index (RI) determination (Kishi and Kawana 2001; Kobayashi *et al.* 2001), evaporative light scattering detectors (Wasik *et al.* 2007) or HPLC coupled to mass spectrometry (Koyama *et al.* 2005).

Other non-chromatographic methods for the determination of sweeteners in foods such as immunochemical assays and measurement in an enzyme-linked immunosorbent assay have also been reported (Wasik *et al.* 2007).

REFERENCES

Ager, D. J., Pantaleone, D. P., Henderson, S. A., Katritzky, A. R., Prakash, I. & Walters, D. E. (1998) Commercial, synthetic non-nutritive sweeteners. *Angewandte Chemie International Education in English* 37 (13–24), 1802–1817.

Anderson, J. & Young, L. (2008) Sugar and Sweeteners. CSU Ext. Fact Sheet number, 9.301. Colorado State University.

AOAC (2000a) Official Method 979.08 2000. Official Methods of Analysis of the Association of Official Analytical Chemists, 29.1, 14, 2.

Audrieth, L. F. & Sveda, M. (1944) Preparation and properties of some N-substituted sulfamic acids. *Journal of Organic Chemistry* 39, 89–101.

Barceloux, D. G. (2008) *Medical Toxicology of Natural Substances: Foods, Fungi, Medicinal Herbs, Toxic Plants, and Venomous Animals*. John Wiley & Sons, Hoboken, NJ, pp. 22–33.

Benford, D. J., DiNovi, M. & Schlatter, J. (2006) Safety Evaluation of Certain Food Additives: Steviol Glycosides; WHO Food Additives Series. World Health Organization Joint FAO/WHO Expert Committee on Food Additives (JECFA) 54, 140.

Benson, G. A. & Spillane, W. J. (1976) Structure-activity studies on sulfamate sweeteners. *Journal of Medical Chemistry* 19, 869–872.

Boyce, M. C. (1999) Simultaneous determination of antioxidants, preservatives and sweeteners permitted as additives in food by mixed micellar electrokinetic chromatography. *Journal of Chromatography A*, 847 (1–2), 369–375.

Brandle, J. (2004) FAQ: Stevia, Nature's Natural Low Calorie Sweetener. Agriculture and Agri-Food, Canada.

Bridel, M. & Lavielle, R. (1931) Sur le principe sucré des feuilles de Kaâ-hê-é (Stevia rebaundiana B). *Academie des Sciences Paris Comptes Rendus* 192, 1123–1125.

Brown, A. (2008) *Understanding Food Principles and Preparation*, 2nd edition. Thomson Wadsworth, Belmont, CA.

Carpenter, R. P., Lyon, D. H. & Hasdell, T. A. (1999) *Guidelines for Sensory Analysis in Food Product Development and Quality Control*, 2nd edition. Aspen Publishers, Maryland, USA.

Cohen, S., Arnold, L. & Emerson, J. (2006) Safety of Saccharin. *Agro Food Industry Hi-Tech* 19 (6), 24–28.

Dossantos, A. J. M. G. (1993) Potentiometric determination of saccharin in dietary products using mercurous nitrate as titrant. *Talanta* 40 (5), 737–740.

Erlandsen, H., Patch, M. G. & Gamez, A. (2003) Structural studies on phenylalanine hydroxylase and implications toward understanding and treating phenylketonuria. *Pediatrics* 112, 1557–1565.

Fatibello-Filho, O., Nobrega, J. A. & Guarita-Santos A. J. M. (1994) Flow-injection potentiometric determination of saccharin in dietary products with relocation of filtration unit. *Talanta* 41 (5), 731–734.

Geuns, J. M. C. (2003) Molecules of interest: Stevioside. *Phytochemistry* 64 (5), 913–921.

Guo-Fu, H. (1995) Fluorophore-assisted carbohydrate electrophoresis-technology and applications. *J. Chromatography A* 705, 89–103.

Hellekant, G. & Danilova, V. (1996) Species differences toward sweeteners. *Food Chemistry* 56, (3), 323–328.

Ikan, R. (ed.) (1991) 1-flavonoides, synthesis of naringin dihydrochalcone: a sweetening agent. In: *Natural Products: A Laboratory Guide*. Academic Press, London, pp. 17–18.

Jenkins, D. J., Wolever, T. M., Taylor, R. H., Barker, H., Fielden, H., Baldwin, J. M., Bowling, A. C., Newman, H. C., Jenkins, A. L. & Goff, D. V. (1981) Glycemic index of foods: a physiological basis for carbohydrate exchange. *American Journal of Clinical Nutrition* 34, 362–366.

Kauschus, U. & Their, H. P. (1985) The composition of soluble polysaccharides in fruit juices. *Lebensmittel-Unterssuchung Forschung* 181, 395–399.

Kishi, H. & Kawana, K. (2001) Determination of sucralose in foods by anion-exchange chromatography and reverse-phase chromatography. *Shokuni Eiseigaku Zasshi* 42 (2), 133–138.

Kobayashi, C., Nakazato, M., Yamajima, Y., Ohno, I., Kawano, M. & Yasuda, K. (2001) Determination of sucralose in foods by HPLC. *Shokuni Eiseigaku Zasshi* 42 (2), 139–143.

Koyama, M., Yoshida, K., Uchibori, N., Wada, I., Akiyama, K. & Sasaki, T. (2005) Analysis of nine kinds of sweeteners in foods by LC/MS. *Shokuhin Eiseigaku Zasshi*, 46, 72–78.

Kroger, M., Meister, K. & Kava, R. (2006) Low-calorie sweeteners and other sugar substitutes: A review of the safety issues. *Comprehensive Reviews in Food Science and Food Safety* 5 (2), 25–47.

Mattes, R. D. & Popkin, B. M. (2009) Non-nutritive sweetener consumption in humans: effects on appetite and food intake and their putative mechanisms. *American Journal of Clinical Nutrition* 89 (1), 1–14.

Mazurkiewicz, J., Rebilas, K. & Tomasik, P. (2006) Dextran: low-molecular saccharide sweetener interactions in aqueous solutions. *Food Hydrocolloids* 20, 21–23.

Negash, N., Moges, G. & Chandravanshi, B. S. (1997) liquid membrane electrode based on Brilliant Green-hydrogen phthalate ion pair. *Chimica Analityczna* 42 (4), 579–588.

Nofre, C. & Tint, J. (2000). Neotame: discovery, properties, utility. *Food Chem.* 69, 245–257.

Northam, E. A. (2004) Neuropsychological and psychosocial correlates of endocrine and metabolic disorders: a review. *Journal of Pediatric and Endocrinology Metabolism* 17, 5–15.

Nunes, A. P. M., Ferreira-Machado, S. C., Nunes, R. M., Dantas, F. J. S., De Mattos, J. C. P. & Caldeira-de-Araujo, A. (2007) Analysis of genotoxic potentiality of stevioside by comet assay. *Food and Chemical Toxicology* 45, 662–666.

Pesek, J. J. & Matyska, M. T. (1997) Determination of aspartame by high-performance capillary electrophoresis. *Journal of Chromatography A*, 781 (1–2), 423–428.

Pezzuto, J. M., Compadre, C. M., Swanson, S. M., Nanayakkara, D. & Kinghorn, A. D. (1985) Metabolically activated steviol, the aglycone of stevioside, is mutagenic. *Proceedings of National Academy of Science, USA* 82 (8), 2478–2482.

Porikos, K. P. & Koopmans, H. S. (1988) The effect of non-nutritive sweeteners on body weight in rats. *Appetite*, 11, Suppl. 1, 12–15.

Portmann, M.-O. & Kilcast, D. (1996) Psychophysical characterization of new sweeteners of commercial importance for the EC food industry. *Food Chemistry* 56 (3), 291–301.

Procinska, E., Bridges, B. A. & Hanson, J. R. (1991) Interpretation of results with the 8-azaguanine resistance system in *Salmonella typhimurium*: no evidence for direct acting mutagenesis by 15-oxosteviol, a possible metabolite of steviol. *Mutagenesis* 6 (2), 165–167.

Prodolliet, J. & Bruelhart, M. (1993) Determination of aspartame and its major decomposition products in foods. *Journal of the Association of Official Analytical Chemists (JAOAC) Int* 76 (2), 275–282.

Renwick, A. G. (2006) The intake of intense sweeteners – an update review. *Food Additives and Contaminants* 23 (4), 327–338.

Rogers, P. J., Carlyle, J-A., Hill, A. J. & Blundell, J. E. (1988) Uncoupling sweet taste and calories: Comparison of the effects of glucose and three intense sweeteners on hunger and food intake. *Physiology and Behavior* 43 (5), 547–552.

Sardesai V. M. & Waldshan, T. H. (1991) Natural and synthetic intense sweeteners. *Journal of Nutritional Biochemistry* 2 (5), 236–244.

Shallenberger, R. S. (1996) The AH,B glycophore and general taste chemistry. *Food Chemistry* 56 (3), 209–214.

Shallenberger, R. S. & Acree, T. E. (1967). A molecular theory of sweet taste. *Nature* 216, 480–482.

Sjoeberg, A. M. K. (1988) Liquid chromatographic determination of saccharin in beverages and desserts: complementary collaborative study. *Journal of the Association of Official Analytical Chemists (JAOAC) Int.*, 71, 1210–1212.

Sjoeberg, A. M. K. & Alanko, T. A. (1987) Liquid chromatographic determination of saccharin in beverages and sweets NMKL collaborative study. *Journal of the Association of Official Analytical Chemists*, 70, 58–60.

Szczeniac, A. S. (1973) Instrumental methods of texture measurements. In: *Texture Measurements of Foods*, Kramer, A. D & Szczeniac A. S. (eds), D. Reidel Publishers, Dordrecht, Holland.

Thompson, C. O., Trenerry, V. C. & Kemmery, B. (1995) Determination of cyclamate in low joule foods by capillary zone electrophoresis with indirect ultraviolet detection. *Journal of Chromatography A*, 704 (1) 203–210.

Tomasik, P. (2003) *Chemical and Functional Properties of Food Saccharides*. CRC Press, Boca Raton, Florida.

Viannnasoares, C. D. & Martins, J. L. S. (1995) Derivative ultraviolet spectrophotometric determination of saccharin in artificial sweeteners. *Analyst* 120 (1), 193–195.

Walker, J. C., Zaugg, S. E. & Walker, E. B. (1997) Analysis of beverages by capillary electrophoresis. *Journal of Chromatography A*, 781 (1–2), 481–485.

Wasik, A., McCourt, J. & Buchgraber, M. (2007) Simultaneous determination of nine intense sweeteners in foodstuffs by high performance liquid chromatography and evaporative light scattering detection—Development and single-laboratory validation. *Journal of Chromatography A* 1157, 187–196.

Weihrauch, M. R. & Diehl, V. (2004) Artificial sweeteners—do they bear a carcinogenic risk? *Annals of Oncology* 15, 1460–1465.

Willets, P., Anderson, S., Brereton, P. & Wood, R. (1996) Determination of intense sweeteners in foodstuffs: collaborative trial. *Journal of the Association of Public Analysts*, 32, 53–88.

Zhang, W., Piculel, L., Nilsson, S. & Knutsen, S. H. (1994) Cation specificity and cation binding to low sulfated carrageenans. *Carbohydrate Polymers* 23 (2), 105–110.

FURTHER READING

AOAC (2000b) Official Method 980 18 2000. *Official Methods of Analysis of the Association of Official Analytical Chemists*, 47.6. 14, 49.

Daniel, J. W., Renwick, A. G., Roberts, A. & Sims, J. (2000) The metabolic fate of sucralose in rats. *Food and Chemical Toxicology* 38 (S2), S115–S121.

Henkel, J. (1999) Sugar substitutes: Americans opt for sweetness and lite. *FDA Consumer Magazine* 33(6), 12–16.

Hutteau, F. & Mathlouthi, M. (1998) Physicochemical properties of sweeteners in artificial saliva and determination of a hydrophobicity scale for some sweeteners. *Food Chemistry* 63 (2), 199–206.

Kretchmer, N. & Hollenbeck, C. B. (eds) (1991) *Sugars and Sweeteners*. CRC Press, Boca Raton, Florida.

Lin, S. Y. & Cheng, Y. D. (2000) Simultaneous formation and detection of the reaction product of solid-state aspartame sweetener by FT-IR/DSC microscopic system. *Food Additives and Contaminants* 17 (10), 821–827.

Magnuson, B. A., Burdock, G. A. & Doull, J. (2007) Aspartame: a safety evaluation based on current use levels, regulations, and toxicological and epidemiological studies. *Critical Reviews in Toxicology* 37 (8), 629–727.

Marrs, W. M., Sworn, G. & Hart, R. J. (1989) Improved method for electrophoresis of polysaccharide. Research Report No. 635, Leatherhead Food RA, Leatherhead, UK.

Mazur, R. H. (1984) Discovery of aspartame. In: *Aspartame: Physiology and Biochemistry*, Stegink, L. D. & Filer Jr. L. J. (eds), Marcel Dekker, New York, pp. 3–9.

Olney, J. W., Farber, N. B., Spitznagel, E. & Robins, L. N. (1996) Increasing brain tumor rates: is there a link to aspartame? *Journal of Neuropathology and Experimental Neurology* 55 (11), 1115–123.

Rastogi, S., Zakrzewski, M. & Suryanarayanan, R. (2001) Investigation of solid-state reactions using variable temperature X-ray powder diffractometry. I. Aspartame hemihydrate. *Pharmaceutical Research* 18 (3), 267–273.

Rowe, R. C. (2009) Aspartame. *Handbook of Pharmaceutical Excipients*. Pharmaceutical Press, New Yorkshire, UK, pp. 11–12.

Smith, D. V. & Margolskee, R. F. (2001) Making sense of taste. *Scientific American* 284 (3), 32–39.

Soffritti, M., Belpoggi, F., Esposti, D. D., Lambertini, L., Tibaldi, E. & Rigano, A. (2006) First experimental demonstration of the multipotential carcinogenic effects of aspartame administered in the feed to Sprague-Dawley rats. *Environmental Health Perspectives* 114 (3), 379–385.

Stegink, L. D. (1987) The aspartame story: a model for the clinical testing of a food additive. *American Journal of Clinical Nutrition* 46 (1), 204–215.

Swithers, S. E. & Davidson, T. L. (2008) A role for sweet taste: calorie predictive relations in energy regulation by rats. *Behavioral Neuroscience* 122 (1), 161–173.

Trocho, C., Pardo, R. & Rafecas, I. (1998) Formaldehyde derived from dietary aspartame binds to tissue components in vivo. *Life Sciences* 63 (5), 337–349.

5 Fragrances, Flavouring Agents and Enhancers

Abstract: The food industry would not be complete without the so-called flavours or flavour enhancers and fragrances, as they play a vital role in imparting taste and smell to food items. Like all other additives, these exist in either artificial (in which case they are sourced from chemical processes such as distillation) or natural forms (sourced from either plant or animal raw materials). Flavourings may also exist as nature-identical compounds if they are a product of synthesis or isolation using chemical procedures. Flavour enhancers, which either enhance or sharpen the already-existing flavour, also exist and include monosodium glutamate (MSG), monopotassium glutamate (MPG), calcium diglutamate (CDG) and monoammonium glutamate (MAG).

Keywords: flavour; flavour enhancers; food flavourings; free amino acids; hydrolysate proteins

5.1 INTRODUCTION TO FLAVOURS AND FLAVOURING AGENTS

In simple terms, flavouring agents are extracts or concentrates. They are metabolites found in living media (tissues and cells) and their formation is complex, governed by genetic factors and well as being influenced by environmental factors. As the name suggests, a flavouring agent or system refers a substance that gives or modifies the characteristic and natural taste of a certain food or drink, causing it to become more appealing, sweet or sour as determined by the sensory mechanism of either a tongue (taste buds), smell or touch (Scheme 5.1) (Auvray and Spence 2008).

In Chapter 4, which dealt with the chemistry of sweeteners, the word 'taste' was used frequently. To differentiate this from the word 'flavour', which is used very often in this chapter, there is a need to define the two terms such that the difference is made clear. Taste comprises the rudimentary sensations of sweet, sour, bitter and salty, whereas flavour is the combined sensation of taste and the olfactory perception of food.

Flavour enhancers, on the other hand, are substances that are commonly added to food products for the purpose of imparting more taste (savoury) as well as adding nutritive value and appeal, especially to food items that are either tasteless or have an unsatisfactory taste (Lõliger 2000; Rangan and Barceloux 2009; Jinap and Hajeb 2010). Food additives classified as either fragrances, flavouring agents or flavour enhancers have been used extensively over the years. In the subsequent sections of this chapter, fragrances, flavour enhancers and agents will all be referred to as flavouring agents.

Chemistry of Food Additives and Preservatives, First Edition. Titus A. M. Msagati.
© 2013 John Wiley & Sons, Ltd. Published 2013 by John Wiley & Sons, Ltd.

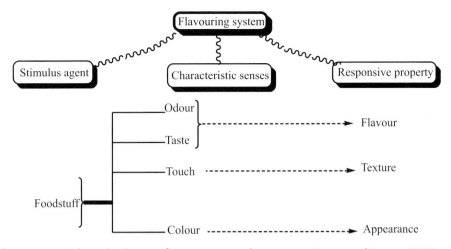

Scheme 5.1   Relationship between flavour, texture and appearance (Auvray and Spence 2008).

Further Thinking

Flavourings are normally incorporated in foodstuffs in small amounts to impart a particular desired taste to foods.

Flavourants as food additives play an important role in rendering a new taste to foodstuffs or in enhancing the already-existing flavour, if it is not sharp enough. They may be used as substitutes in cases of flavour loss during the processing, or even replace the missing components from the overall flavour of food items. They can also be used to mask undesirable flavours in certain food products.

Further Thinking

Flavour enhancers are added to foods to either sharpen or enforce the flavour, without introducing their own characteristic flavour.

5.2 CLASSIFICATION OF FOOD FLAVOURINGS

Generally foodstuffs rich in free amino acids such as glutamic acids as well as those composed of hydrolysate proteins are the foods which have, for a long time, found extensive application as flavouring agents across all civilisations (Rangan and Barceloux 2009; Jinap and Hajeb 2010). With regard to flavour enhancers, L-glutamic acid salt of sodium

(L-glutamate monosodium salt) has been reported in numerous studies and has been used extensively for this purpose. (Jinap and Hajeb 2010).

Generally speaking, there are many different methods of classifying flavouring agents. For convenience, in this chapter we will consider two main classification criteria according to the source of their formation: natural or artificial. This formation can be natural through biogenetic pathways using known precursors.

5.2.1 Mode of formation

Alternatively, flavouring agents may be classified according to their mode of formation: either biological, chemical or physical processes which act on either natural or artificial starting materials. Flavourings can also be subdivided according to starting materials, and includes flavouring molecules formed as a result of the following processes:

- Natural plant metabolism which remains intact as part of the plant chemistry (even when the tree is harvested). Some of the members of this group are volatile and some are non-volatile, for example essential oils type of flavouring compounds.
- Enzymatic reactions, for example alliaceous vegetables and mustard.
- Microbial activities such as fermentation, for example wine, tea or dairy products.
- Heating or cooking processes, such as those giving appealing flavours in cooked, barbecued or smoked meat or roasted coffee.
- Oxidative reactions, for example the flavour of stale light-coloured beer which results from the chemical oxidative reactions due to the formation of an aldehydic compound known as trans-2-nonenal, which is responsible for an unpalatable flavour.

5.2.2 Source of formation

Anther classification regimes groups flavouring agents into three main classes: (1) natural flavouring systems; (2) synthetic or chemically isolated molecules (which chemically possess identical properties to the natural flavouring agents); and (3) artificial flavouring agents (which possess no identical features to natural flavouring agents). Artificial flavouring agents produced through industrial processes are actually a direct breakdown of products or complex interaction products, depending on the raw materials used in the particular processing and the conditions imposed (Ohloff 1972).

5.2.3 Use of flavouring agents

Yet another classification regime is based on the use of flavouring agents (Hall and Marwin 1981), including: condiments (e.g. mustard and vinegar); spices (e.g. ginger, black pepper); essential oils (e.g. celery and cinnamon); and aromatic molecules (e.g. vanillin (Figure 5.1), anethol, menthol, citral, isobutyl-1- methoxy pyrazine). Within the classification which is based on usage, artificial flavourants are grouped into three categories. The first is the solid form of flavourants (these are mostly encapsulated flavourants), and they are advantageous in that they are highly volatile (e.g. dimethyl sulphide, methyl mercaptan) and can therefore only be encapsulated to provide flavours in their solid form. The second category comprises semisolid pastes, mainly oleoresins and fruit concentrates. These are attractive in that they form easy dispersions and provide uniform flavouring. The third category is made up of

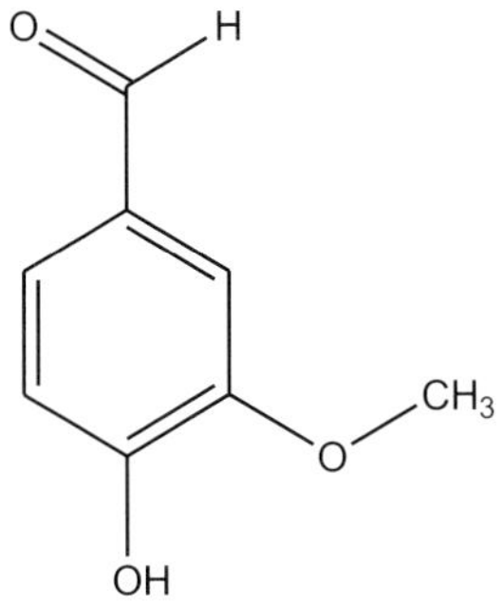

Fig. 5.1 Chemical structure of vanillin.

liquid forms which exist in an emulsified state with a compatible solvent base. These are useful in beverages and other liquid foodstuffs.

5.2.4 Functional group responsible for flavour

There are a number of functional groups in molecules that provide a characteristic aroma/flavour/taste in various food items. These include benzoaldehyde (almond), 2-methoxy-3-isobutyl-pyrazine (green pepper), pyrazin and thiol functionalities, 4-hydroxy-3-methoxy-benzoaldehyde (vanilla), 2-trans-6-cis-nonadienol (cucumber) and 2-pentylfuran (a reversal flavour of soybean). There exists a diverse range (although many are esters; Figure 5.2, Table 5.1) of known flavouring agents, such as isoamyl acetate, ethyl propionate, methyl antranilate, methyl salicylate and allyl hexanoate (Winterhalter and Rouseff 2002). Other molecules used as flavouring agents are aldehydes such as cinnamic aldehyde and benzaldehyde (Figure 5.2).

5.3 CHEMISTRY OF FOOD FLAVOURINGS

The precursors of flavour compounds are formed through chemical processes, some involving both enzymatic (e.g. those referred to as browning reactions) and non-enzymatic processes . Other types of reactions such as Strecker degradation, pyrazine formation, oxazole formation and thiazole formation are also known to generate flavour precursors.

There are a number of reactions that involve the browning aspect of foodstuffs but also provide characteristic flavours. These include: Maillard browning, whereby reducing sugars react with amine to give brown pigments and flavours; caramelisation browning, which converts sugars at high temperatures to brown pigments and special characteristic flavours; and enzymatic browning reactions, which convert phenolic compounds in the presence of polyphenoloxidase to brown pigments and flavours.

5.3.1 Maillard browning (non-enzymatic)

Maillard processes (Figure 5.3) are complex non-enzymatic reactions of amines or amino acids, usually from proteins and carbonyl compounds from sugars at elevated temperatures and high pH, to form a number of products which have an impact on the flavour and

Table 5.1 Carotenoid oxidative products which produce characteristic aroma.

Parent compounds	Chemical structure	Plant source
Dehydrolycopene	6-methyl-3,5-heptadien-2-one	Tomato
α-carotene	α-Ionone	Vanilla, black tea
Neoxanthin	β-Ionone	Black tea
Lycopene	Pseudo-ionone	Tomato
β-carotene	β-cyclocitral	Tomato
Neoxanthin	β-Damascenone	Coffee
Lycopene	6-methyl-5-hepten-2-one	Tomato
Neoxanthin	1,2-Dihydro-1,1,6-trimethylnaphthalene	Strawberry

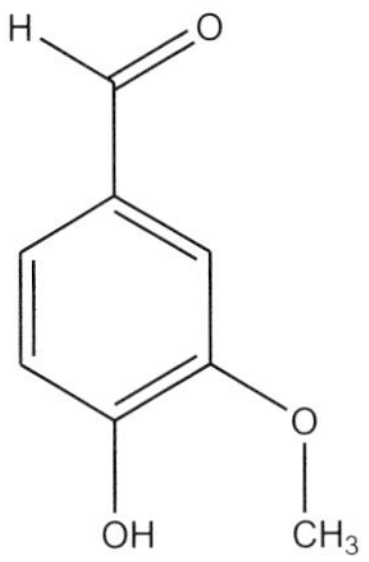

4-hydroxy-3-methoxy-benzoaldehyde
(found in vanilla)

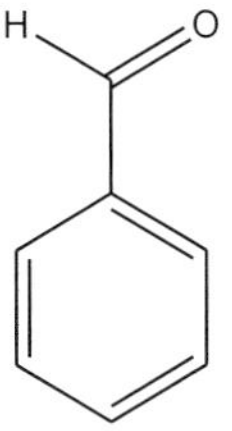

benzoaldehyde
(found in almond)

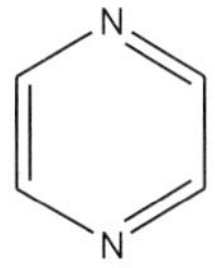

pyrazine

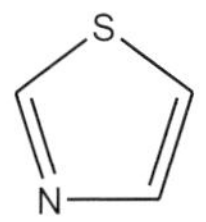

Thiazol

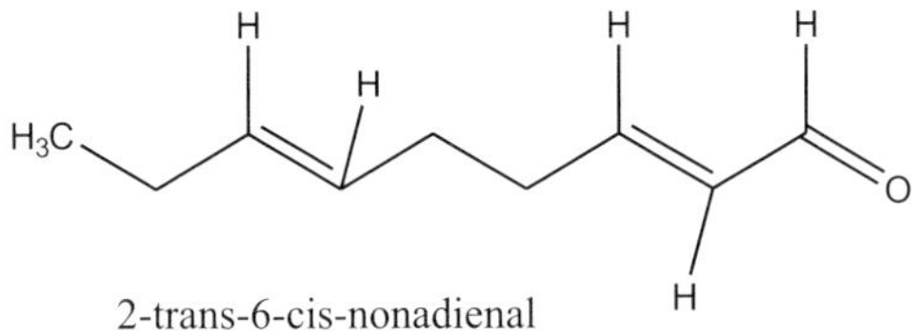

2-trans-6-cis-nonadienal
(found in cucumbers)

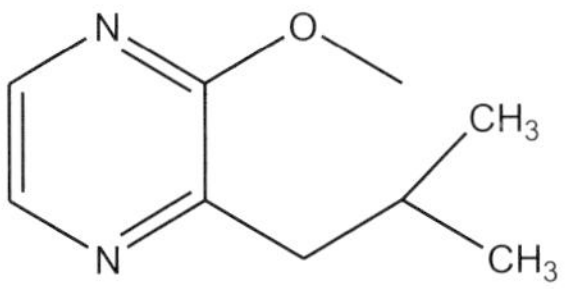

2-methoxy-3-isobutyl-pyrazine
(found in green pepper

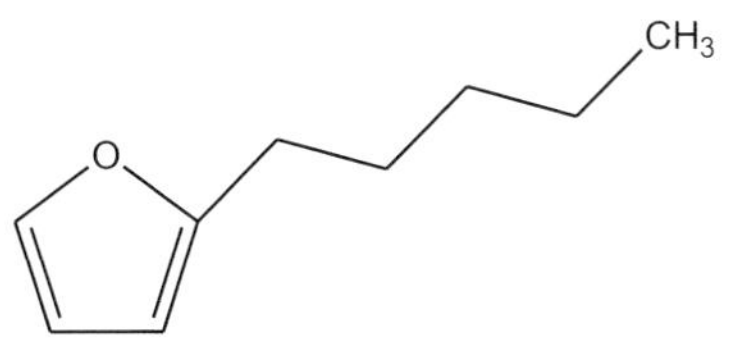

2-pentylfuran/2-pentenyl furan
(found in soybean oil)

Fig. 5.2 Chemical structures and functional groups responsible for flavour production in foodstuffs.

appearances of the foodstuffs (Grandhee and Monnier 1991). Maillard reactions occur in three main steps as follows.

5.3.1.1 *Step 1: Initiation step*

The initiation step results in the formation of N-substituted glycoside. The initiation reaction involves a condensation reaction between a reducing sugar and a primary amino acid to

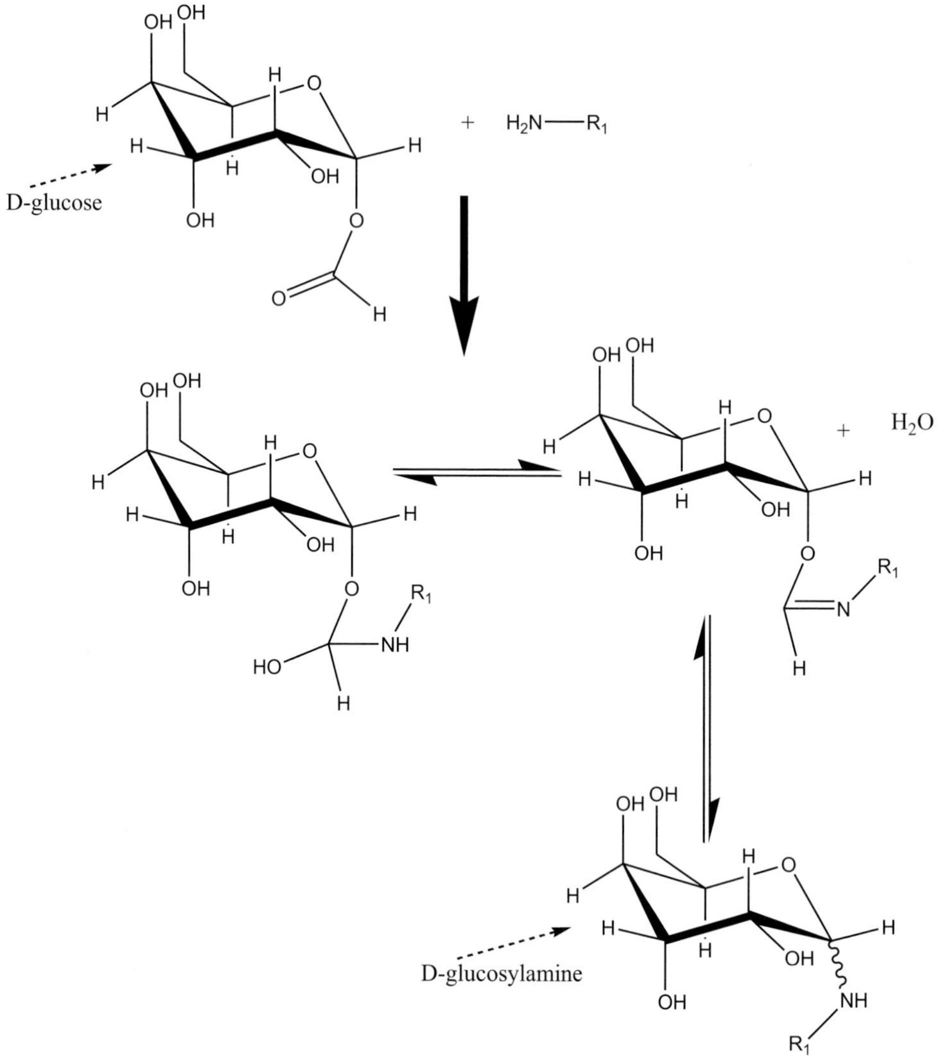

Fig. 5.3 The Maillard initiation step (Grandhee and Monnier 1991).

produce an imine which undergoes cyclisation/enolisation, resulting in the formation of N-substituted glycoside (a sugar molecule attached to NR_1 group; Scheme 5.2; Grandhee and Monnier 1991).

5.3.1.2 Step 2: Ketosamine compound formation

The initial formation of N-substituted glycoside is subsequently followed by the formation of a ketosamine compound (due to the instability of the N-glycosimine formed in the first step) by cyclisation/enolisation or by the isomerisation of the generated immonium ion. This alkali-catalysed isomerisation reaction is referred to as an Amadori rearrangement (Scheme 5.3; Grandhee and Monnier 1991).

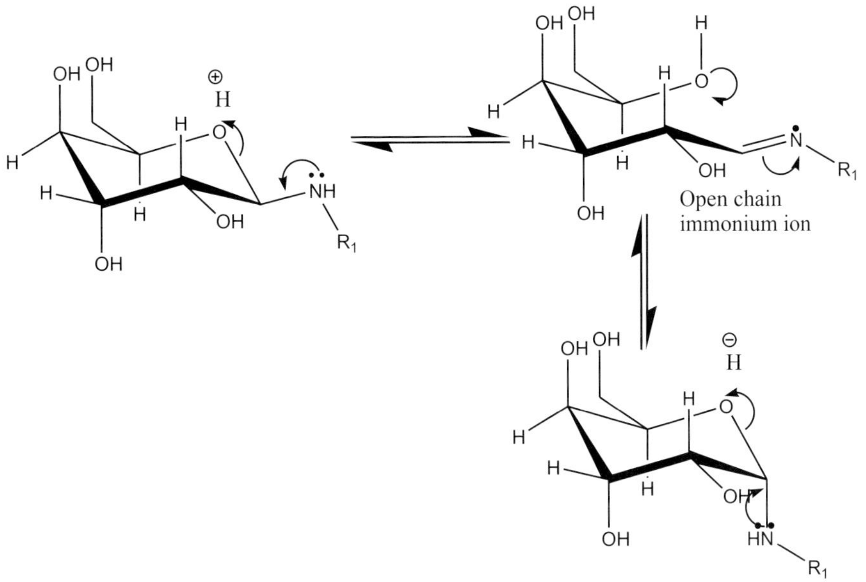

Scheme 5.2   Glycosilation mechanism (Grandhee and Monnier 1991).

The glycosylamine and Amadori products are unstable intermediates which form during the Maillard process and their concentrations are highly dependent on the reaction conditions such as pH, temperature and time. Normally at pH range 4–7, these products undergo degradation to form deoxyosone compounds such as 1-deoxydicarbonyl and 3-deoxydicarbonyl compounds (Grandhee and Monnier 1991). The deoxyosone products formed are known to be very reactive α-dicarbonyl compounds, and react further to form other secondary products. The 1-deoxydicarbonyl compounds form secondary products such as furanose (a compound with characteristic aroma), pentoses and hexoses. The 3-deoxydicarbonyl product is known to form secondary products such as pyrroles, pyridines and formylpyrrole (Scheme 5.4). The pentose products formed may further react with imines to generate orange dye products which influence the characteristic colour of the food item.

5.3.1.3   Step 3: Dehydration process

The third step involves a dehydration process (Scheme 5.5) whereby ketosamine products dehydrate to form reductones and dehydro reductone caramel products as well as short

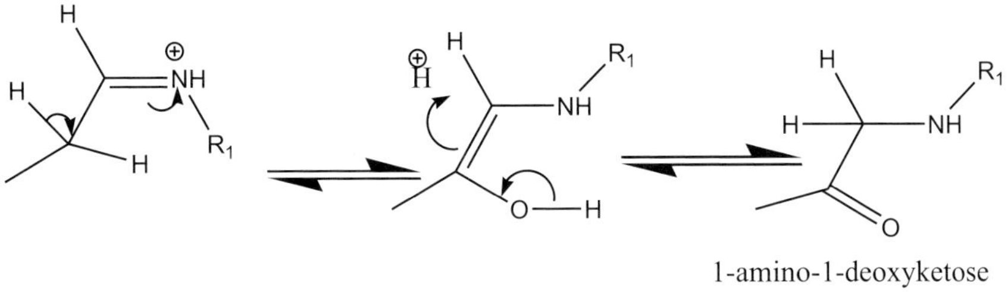

1-amino-1-deoxyketose

Scheme 5.3   The Amadori rearrangement step/cyclisation (Grandhee and Monnier 1991).

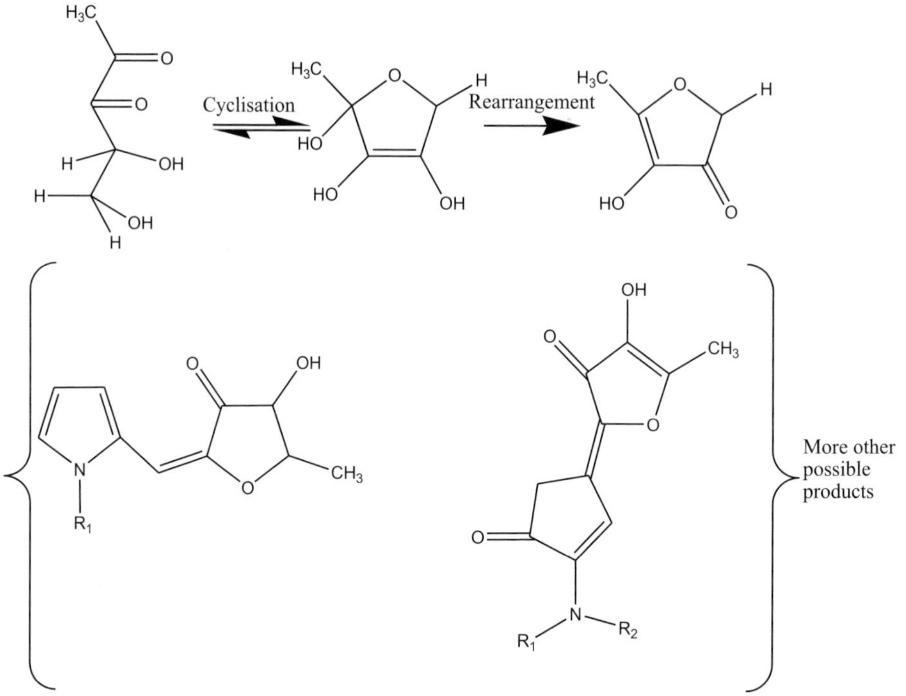

Scheme 5.4   Colour generation reactions in foods (Grandhee and Monnier 1991).

chain hydrolytic fission products. Example of these products include diacetyl, acetol or pyruvaldehyde, which then undergo the Strecker degradation (see Section 5.3.4; Strecker 1850; Miller 1953; Farmer and Patterson 1991; Rada-Mendoza *et al.* 2006).

5.3.1.4   Controlling and monitoring Maillard reactions

For a Maillard reaction to proceed at the expected mutarotation rate, the sugar reactants must be reducing sugars and in acyclic forms (e.g. sucrose). The order of reactivity for various

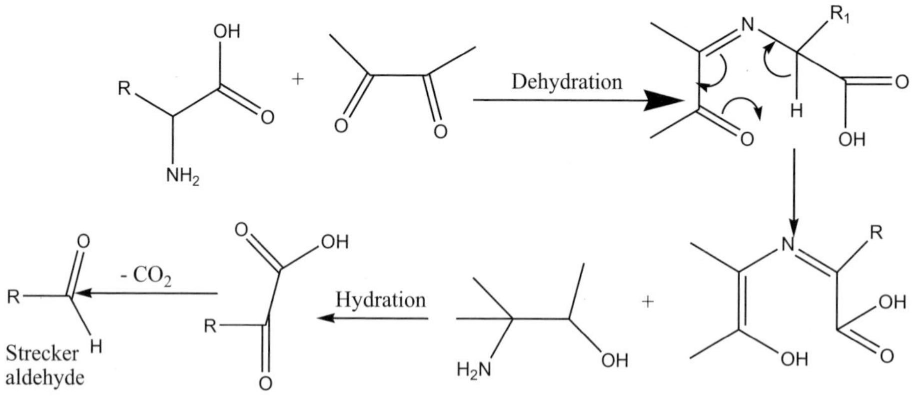

Scheme 5.5   Dehydration step (Strecker 1850; Miller 1953; Farmer and Patterson 1991; Rada-Mendoza *et al.* 2006).

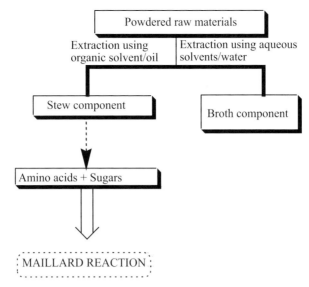

Scheme 5.6   Isolation of flavour precursors of foods of animal origin (Grandhee and Monnier 1991).

forms of sugars is such that pentose exceeds that of hexose, which also exceeds that of disaccharides. Comparison within pentoses may reveal that fructose sugars have increased reaction rates due to a greater extent of open chain form which is more than that of aldoses. When hexoses are compared, the reactivity decrease follows the order: D-galactose > D-mannose > D-glucose. Temperature and pH are the two factors which have the greatest affect on the extent (percentage) open chain form and the rate of mutarotation increase. The other reactants in Maillard reactions (amino compounds) normally act as nucleophiles. Being basic with hydroxyl moiety, they react strongly with reducing sugars which results in the production of aroma/flavour compounds.

The Maillard reaction can be monitored by UV-Vis spectrophotometry such that, during the initial stage where only products which lack chromophores (e.g. Schiff base, Amadori/Henys) are formed, there will be no UV absorption. The intermediate stage is however characterised by strong UV chromophores such as dicarbonyl compounds, dimethyl formamide etc., and demonstrates strong UV absorption. During the final stage, when an intense brown colour is formed, an absorption band at the visible wavelength (~420 nm) can be observed.

Scheme 5.6 shows the non-enzymatic Maillard process for the isolation of flavour precursors in beef products. When too much heat is used to process meat, meat proteins lump together due to the Maillard reaction in the same way as for bakery products. The explanation for this phenomenon is that strong heat together with the water present in meat itself will catalyse the conversion of collagen compounds present in meat into gelatin compounds, thus causing browning lumps.

The Maillard reaction can explain the phenomena that lead to the formation of colours and flavours in foods (Harrison and Dake 2005). In Scheme 5.7, solvent extraction (to partition the amino acids in the organic/oil phase and sugars in the aqueous/water phase) may be used to extract flavouring compounds. The two groups of compounds extracted (amino acids and sugars) will undergo Maillard reactions whereby the amino acid and a reducing sugar will react at elevated temperatures and high pH (where amino acids ionise) to produce a

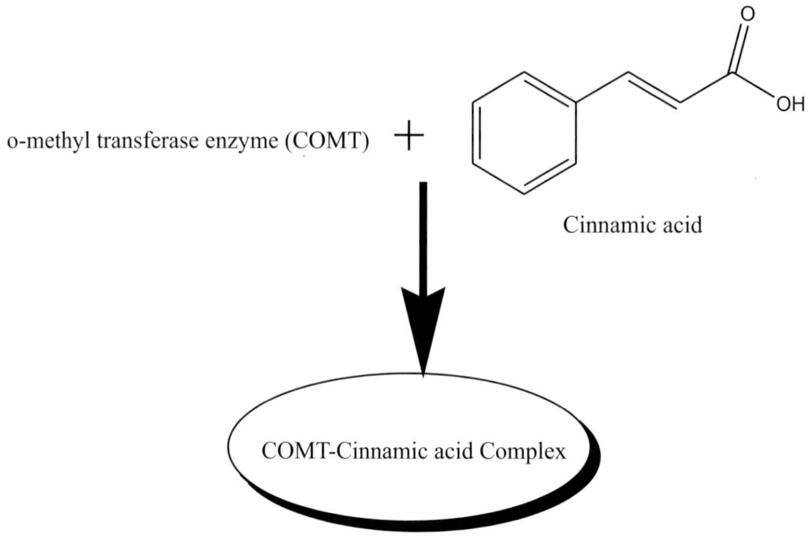

Scheme 5.7   Prevention of browning reactions using COMT enzyme (Grandhee and Monnier 1991).

range of flavours. pH in Maillard reactions is known to influence the ratio of the product formed, the rate of colour formation and the 2,3-enolisation pathways (Figure 5.4). In this process, the carbonyl moiety of the sugar reacts with the amino functionality of the amino acid, resulting in a number of molecules of characteristic aroma and flavour. The resulting flavour is determined by the type of amino acid used in the Maillard reaction.

5.3.2   Enzymatic browning phenomena

Browning in foods is an enzymatic controlled phenomena which takes place in foodstuffs such as fruits and vegetables when their tissues are cut or peeled and become exposed to air. Browning occurs due to the conversion of phenolic compounds in these foodstuffs to brown melanoidin compounds using Cu cofactors at pH range 5–7 (Grandhee and Monnier 1991).

In Scheme 5.8, the formation of quinone in the scheme is dependent on both polyphenoloxidase (PPO) enzyme and oxygen. This browning reaction is kick-started (catalysed) by the enzyme and all subsequent reactions proceed spontaneously, independent of PPO or oxygen. The phenolic substrates which play an important role in browning reactions include flavonoids and a variety of cinnamic acid derivatives (Figure 5.5a–g).

Browning reactions can be controlled in a number of ways such as de-oxygenation by immersion in water or brine and by vacuum treatment. Other approaches include the application of heat (blanching); pH adjustment; use of chelators such as EDTA (ethylenediaminetetraacetic acid); use of enzyme inhibitors or enzyme treatment (e.g. oxygenases, o-methyl transferases or proteases; see example in Scheme 5.7) with the aim of modifying the substrates; use of complexing agents such as cyclodetrins; and use of reducing agents such as sulphites, cysteine, hexylresorcinol, glutathione and ascorbic acid analogues (Figures 5.6a, b and 5.7).

Figure 5.6b shows how hexylresorcinol (HR) and glutathione functions to prevent browning reactions in foodstuffs: DNA damage, caused by hydrogen peroxides in white blood cells (mainly lymphocyte), is inhibited. This particular DNA strand has been reported to initiate

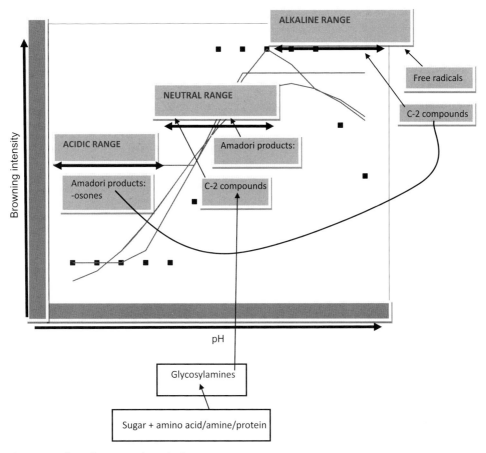

Fig. 5.4   Effect of pH on melanoidin formation.

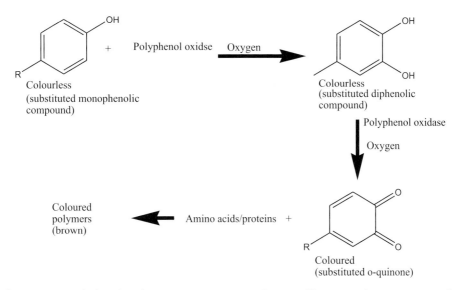

Scheme 5.8   Polyphenol oxidase (PPO) (tyrosinase) catalysation of browning phenomenon (Grandhee and Monnier 1991).

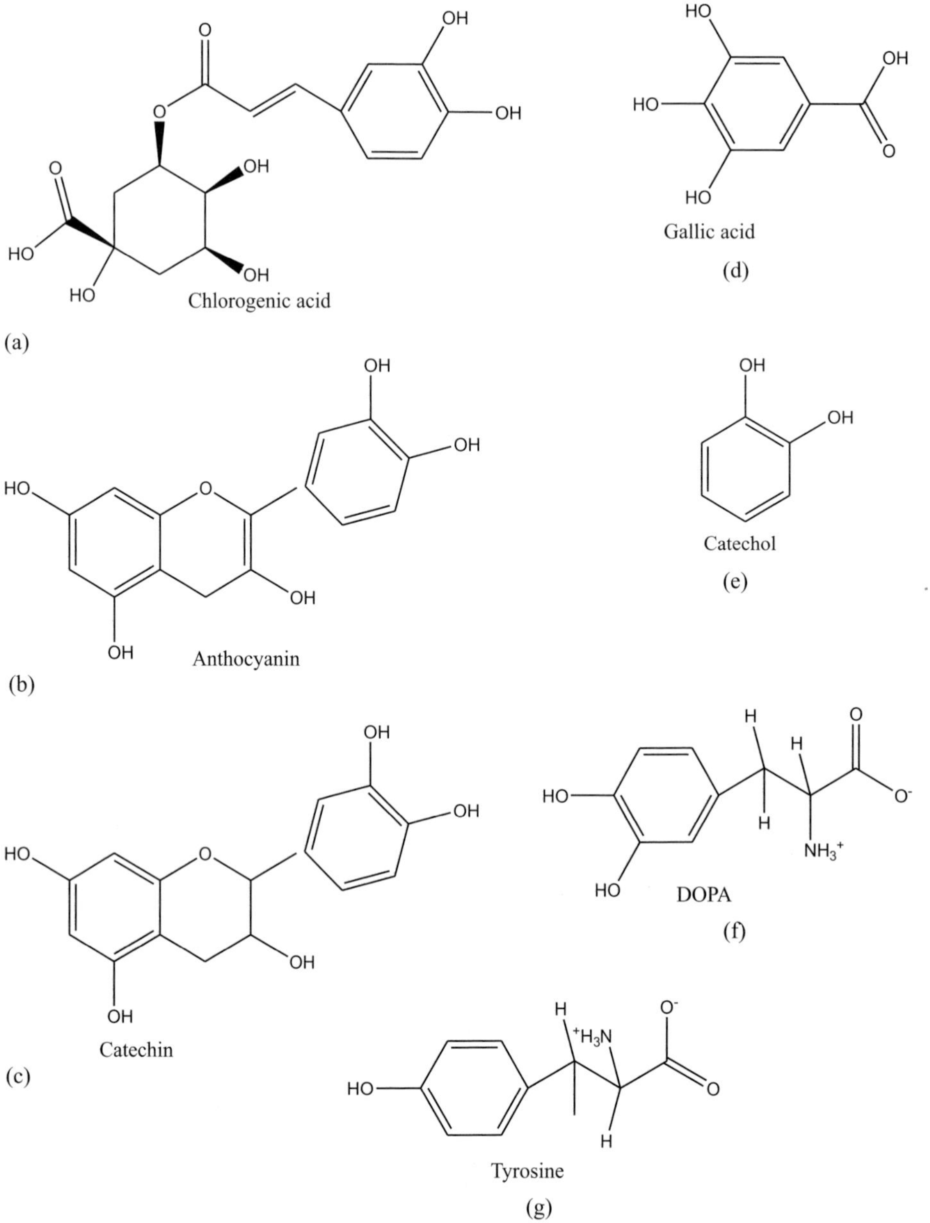

Chlorogenic acid

(a)

Gallic acid

(d)

Anthocyanin

(b)

Catechol

(e)

Catechin

(c)

DOPA

(f)

Tyrosine

(g)

Fig. 5.5 (a–g) Examples of phenolic substrates used in browning reactions (Grandhee and Monnier 1991).

and code for the synthesis of melanin (Halliwell and Gutteridge 1989). Hydrogen peroxide a molecule with high ability to penetrate cell cytoplasmic membrane thus inflicting serious damage to the gene is a nuisance in biological systems (Duthie *et al.* 1997; Halliwell and Gutteridge 1986). Other molecules such as glutathione inhibit melanin synthesis (browning reactions) in the reaction of tyrosinase and L-DOPA (L-3,4-dihydroxyphenylalanine) by interrupting its function (Yen *et al.* 2003; Matsuki *et al.* 2008).

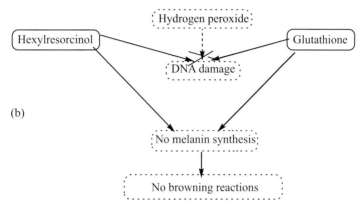

(a)

(b)

Fig. 5.6   (a) Structure of hexylresorcinol (HR) and (b) example of prevention of browning using HR and glutathione (Grandhee and Monnier 1991).

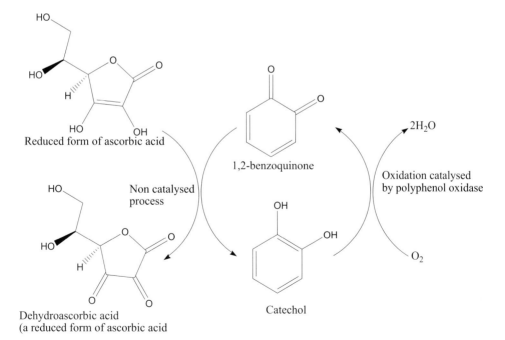

Fig. 5.7   The role of ascorbic acid in the prevention of colour formation in foods (Grandhee and Monnier 1991).

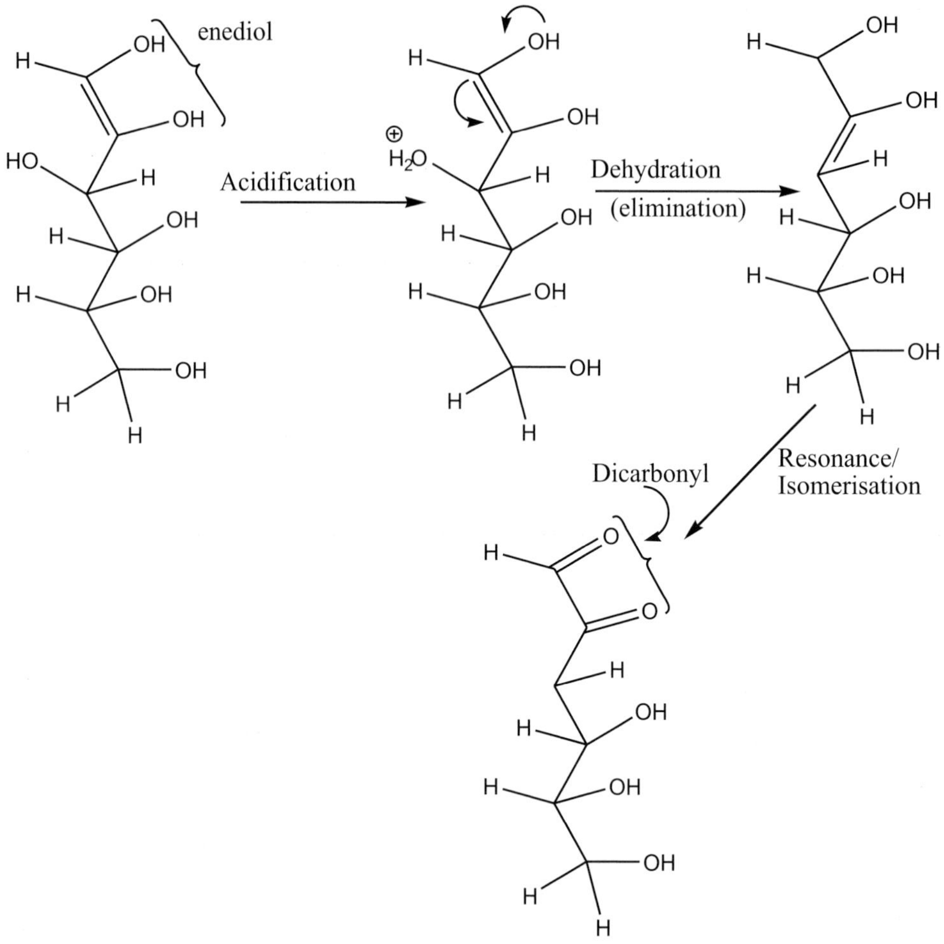

Scheme 5.9 Dehydration mechanism in caramelisation reactions (Kroh *et al.* 1996; Rada-Mendoza *et al.* 2006).

5.3.3 Caramelisation reactions

Caramelisation is a pyrolytic process in which sugars undergo the formation of pigments and flavour compounds. An elevated temperature accelerates sugar reactions such as isomerisation, water elimination and oxidation (Kroh *et al.* 1996; Rada-Mendoza *et al.* 2006). Normally caramelisation takes place at temperatures above the boiling point of water ($\sim$150 °C). Water is therefore eliminated (Scheme 5.9; dehydration of mechanism) leaving the reaction composition with high sugar content and no amine; no Maillard reaction can therefore take place (Kroh *et al.* 1996; Rada-Mendoza *et al.* 2006). This environment favours the formation of enediols and dicarbonyls, which are responsible for the aroma and pigments of caramel.

Caramelisation reactions generally result in a pleasant aroma/flavour and colour in many food items such as coffee or that produced by maltol (Figure 5.8). In some cases however, caramelisation produces an undesirable colour and an unpleasant smell and flavour, such as that of burnt sugar (Kroh *et al.* 1996; Rada-Mendoza *et al.* 2006).

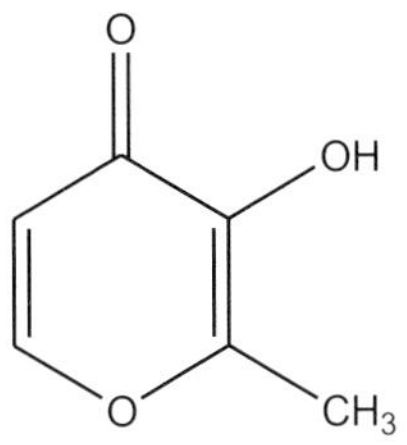

Fig. 5.8 Chemical structure of maltol (Rada-Mendoza *et al.* 2006).

5.3.4 Strecker reactions

The Strecker chemical reaction is a transamination reaction which involves α-dicarbonyl compounds such as the deoxyosones formed in the Maillard reactions and amines or α-amino acids to yield aminoketones, aldehydes and CO_2 through an imine intermediate (Scheme 5.10; (Strecker 1850; Miller 1953).

The Strecker aldehydes and aminoketones formed are normally associated with strong characteristic aromas. For example, ethanal, methylpropanal and 2-phenylethanal (all Strecker aldehydes) have fruity/sweet, malty and flowery/honey-resembling aromas, respectively. Similarly, a condensation of two aminoketone molecules gives pyrazine condensation products with a strong characteristic aroma (Scheme 5.11; Strecker 1850; Miller 1953).

5.3.5 Smoke flavourings

Smoke flavourings are normally regarded as additives of natural origin obtained from wood through thermal decomposition processes which take place in controlled environments (controlled temperature and air). In this process, apart from the smoke flavouring compounds, water and tar are also formed as a two-phase product (Miler 1962; Tilgner 1962; Daun 1972).

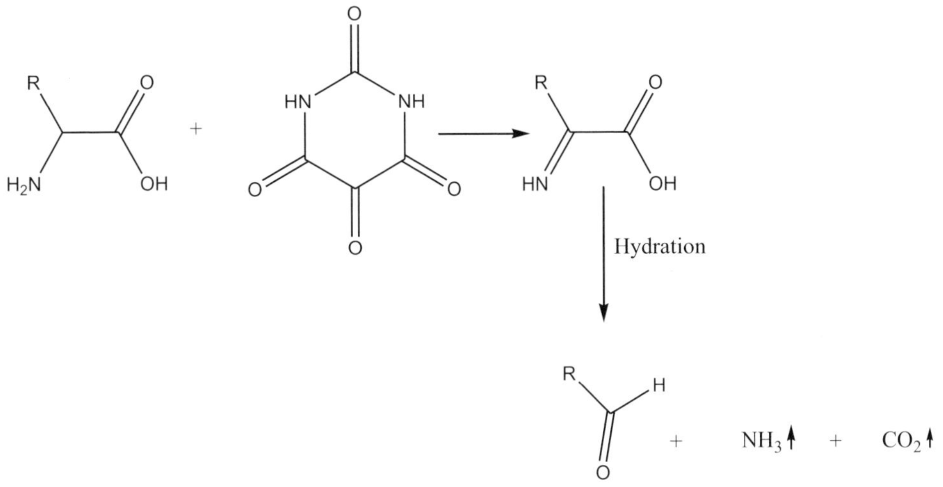

Scheme 5.10 The Strecker reactions (Strecker 1850; Miller 1953).

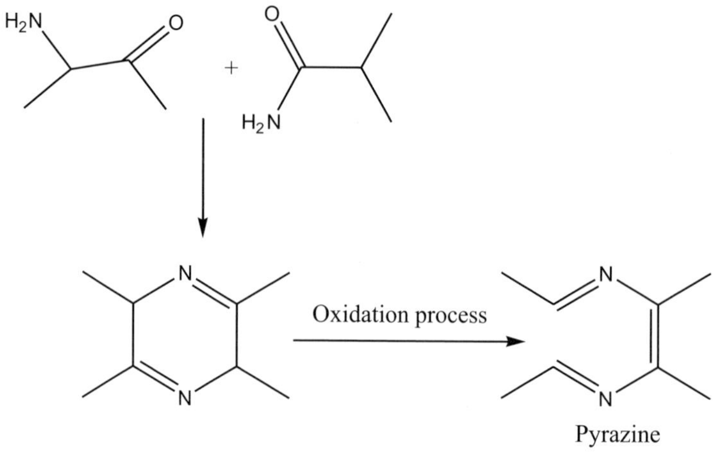

Scheme 5.11 The formation of pyrazine (Strecker 1850; Miller 1953).

Smoke flavourings are attractive as food additives because they are regarded as safe (there are no reports of toxicities) and economical (Borys 1979, 1998, 2001; Gomaa *et al.* 1993; Yabiku *et al.* 1993; Pszczoła 1995). Different types of wood have different chemical compositions in terms of their cellulosic, hemicellulosic and lignin contents and the ratio of these polymeric molecules to each other. This range of properties also creates differences in the pyrolytic conditions for various wood materials; the composition of compounds in the smoke favouring is also highly variable (Hruza *et al.* 1974; Rusz and Miler 1977; Chen and Maga 1993; Guillen and Ibargoitia 1996a, b). Smoke flavourings are composed of a mixture of volatile (important for smoke-curing aroma and flavour) and non-volatile compounds with various chemistries. The latter includes phenol, syringol and guaiacol and their derivatives as well as carbonyls, catechol and naphthalene derivatives (Kim *et al.* 1974; Hamm 1976; Guillen and Ibargoitia 1996a, b; Borys 1998).

5.3.6 Summary of flavour precursor mechanisms

Natural flavour compounds, such as the volatile flavouring compounds that develop in natural sources such as ripe fruits or flowers (plants), are due to the plant metabolic processes and are enzymatically controlled.

However, some of the flavour compounds present in fresh food items such as raw meat require a physical process such as heating before producing compounds which give an acceptable characteristic aroma. For example, 3, 5-dimethyl-1, 2, 4-trithiolane (Figure 5.9), a compound responsible for aroma in boiled animal meat and meat products, is produced after reaction steps where heat is involved (Scheme 5.12; Farmer and Patterson 1991).

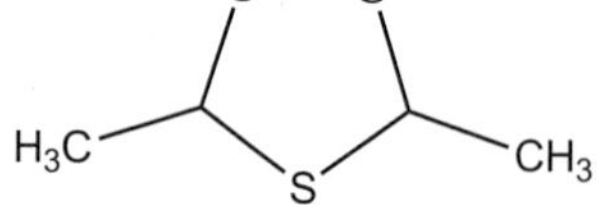

Fig. 5.9 Chemical structure of 3, 5-dimethyl-1, 2, 4-trithiolane.

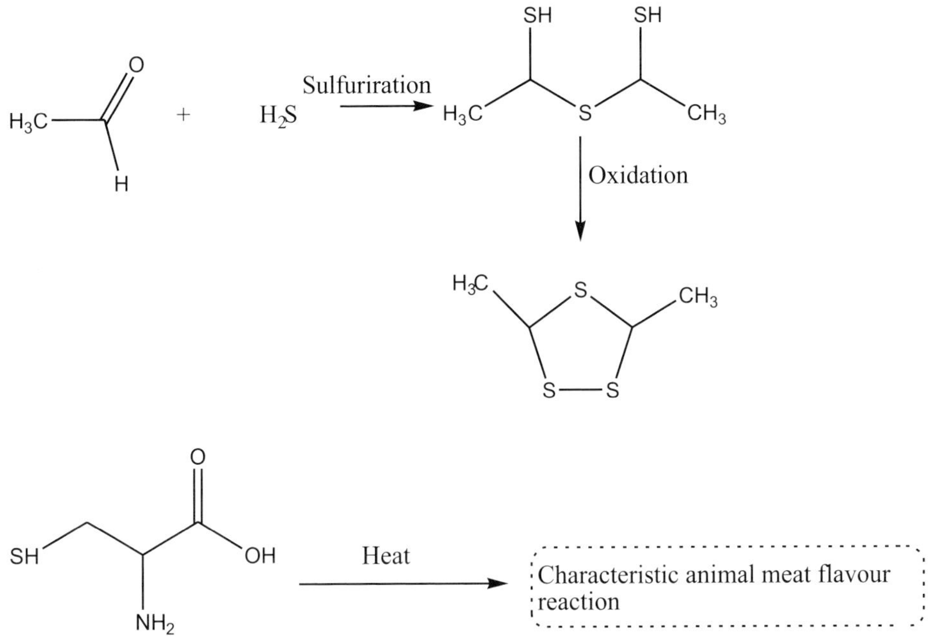

Scheme 5.12   Reactions generating animal meat flavour compounds in boiled products (Farmer and Patterson 1991).

5.4 QUALITY CONTROL OF FLAVOUR COMPOUNDS

Methods for quality control of flavourings in foods are based on comparative approaches with regard to sensory, analytical and microbiological measurements with a set of known standards and specifications. Individual flavourings are generally identified using analytically accepted measurements of density, refractive index, optical rotation and melting point. These measurements are attractive in that they are quick, simple, direct and economical to perform.

5.4.1 How safe are flavour compounds, enhancers and precursors?

Flavourings are intended to provide a special taste or aroma in foodstuffs. If excessive amount of these substances are added to foods, they will spoil the taste and give an undesirable strong smell. Flavourings are used in minute quantities, which also helps to avoid health hazards. A stringent monitoring regime for flavourings is therefore not only unnecessary but also uneconomical in terms of time and resources. In the food industry, flavourings consisting of an active moiety are contained in solvents which are used only as carriers. Stringent toxicological examination of flavourings is not an issue, since these agents are self-limiting.

However, due to the influx of so many flavourings, there is a need to examine the safety of newly introduced flavourants and flavour enhancers. With regards to safety, it is interesting to note that the naturality or originality of a molecule considered as a flavouring agent does not automatically qualify it as safe to consume. On the contrary, artificial flavours are regarded to be much safer for consumption than natural flavours. This is because the production of artificial flavours adheres to high standards of purity and their ingredients are normally consistent as required by regulations (Smitha 2005). Naturally flavouring systems are in most cases subject to natural presence of unwanted molecules which may be toxic but the artificially produced flavours on the other hand such impurities are removed during the production process and they also go through stages of testing and quality assurance to certify them as safe for human consumption (Smitha 2005).

Flavourings such as monosodium glutamate have been reported to cause some diseases in human beings, especially those who are asthmatic (Tarasoff and Kelly 1993; Geha *et al*. 2000). Moreover, several flavouring substances have been listed by European Union (EU) countries as toxicologically unacceptable and have already been banned. These flavourings include 4-allyl-1, 2-dimethoxybenzene, 1-allyl-4-methoxybenzene, N-(4-hydroxy-3-methoxybenzyl)-8-methylnon-6-enamide, propyl-4-hydroxybenzoate, pentane- 2, 4-dione, acetamide and 2-butylbuta-1, 3-diene (Barlow and Schlatter 2010).

5.5   ANALYTICAL METHODS FOR THE ANALYSIS OF FOOD FLAVOURINGS

Most methods for the analysis of aroma compounds in foodstuffs involve gas chromatography or mass spectrometry. This is due to the fact that the majority of aroma compounds are very volatile and the methods for their isolation are simple (Maarse and Visscher 1996).

5.5.1   Gas chromatography and olfactory method

There has been a sharp increase in the application of combined methods involving gas chromatography with olfactory (GC/O) for the analysis of aroma compounds. The two main GC/O methods that have been developed for the highly useful separation of the volatile compounds present in the aroma/flavouring molecules are: the aroma extract dilution analysis (AEDA; Ullrich and Grosch 1987) and the combined hedonic and response measurement (CHARM; Acree *et al*. 1984).

AEDA is normally performed by two 'sniffers'. A stepwise equimolar dilution of the sample extract is obtained by vacuum distillation, and sample extracts are sniffed at increasing rates of dilution until the compounds are no longer perceived (Blank *et al*. 1992; Cerny and Grosch 1992; Reiners and Grosch 1998; Stephan and Steinhart 1999). AEDA results are presented in plots known as the flavour dilution chromatograms (FD), which are plots of the flavour dilution factor against refractive index value (RI).

CHARM involves four–five times the number of sniffers of that of AEDA (Van Ruth *et al*. 1996; Pollien *et al*. 1997; Prost *et al*. 1998; Stephan and Steinhart 1999). The results of CHARM are normally presented as plots of CHARM value versus RI value, and are known as CHARM-aromagrams or diagrams.

Other techniques associated with GC/O methods but which do not involve dilution exist and are mainly for the measurement of aroma intensity. They include the Osme method which has been reported in a number of publications (McDaniel *et al*. 1990; Miranda-Lopez *et al*. 1992; Guichard *et al*. 1995) and the finger span method (Etievant *et al*. 1999).

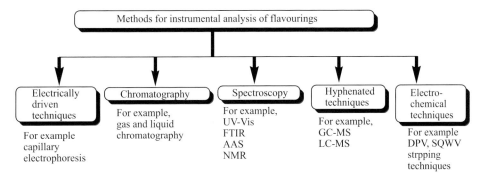

Fig. 5.10 Methods used in the analysis of flavourings (Guth and Grosch 1990; Semmelroch and Grosch 1995; Semmelroch *et al.* 1995; Rychlik and Grosch 1996).

5.5.2 Odour activity values

Other strategies for the characterisation of aroma are based on the determination of the odourant concentrations in the food as well as the determination of the odourant thresholds. A measurement of the magnitude of odour activity values (OAVs), introduced by Rothe and Thomas (1963), has been instrumental in the analysis of aroma in foods as it takes into consideration both concentration and threshold (Guth and Grosch 1990; Semmelroch and Grosch 1995; Semmelroch *et al.* 1995; Rychlik and Grosch 1996). These measurements are based on the assumption that the perceived intensity is proportional to the aroma.

5.5.3 Internal standards

To quantify compounds that have been recognised using GC/O methods, to be able to calculate OAVs of aroma compounds in foods and to perform the determination of the yields of volatiles in headspace samplers, GC/MS methods are highly powerful. They are recommended since they are capable of detecting very low concentrations (to the order of ng/L or ng/kg). The internal standards method of quantification is the most widely used. The choice of an internal standard should include sample compounds close to their physico-chemical properties such as functional group, retention time and boiling point.

In general, there are a number of techniques that have been devised and are used for the analysis of flavourings. Figure 5.10 summarises the instrumental methods (Guth and Grosch 1990; Semmelroch and Grosch 1995; Semmelroch *et al.* 1995; Rychlik and Grosch 1996).

REFERENCES

Acree, T. E., Barnard, J. & Cunningham, D. G. (1984) A procedure for the sensory analysis of gas chromatographic effluents. *Food Chemistry* 14, 273–286.

Auvray, M. & Spence, C. (2008) The multisensory perception of flavor. *Consciousness and Cognition* 17, 1016–1031.

Barlow, S. & Schlatter, J. (2010) Risk assessment of carcinogens in food. *Toxicology and Applied Pharmacology* 243, 180–190.

Blank, I., Sen, A. & Grosch, W. (1992) Potent odorants of the roasted powder and brew of Arabica coffee. *Zeitschrift fuer Lebensmittel-Umtersuchung und-forschung* 195, 239–245.

Borys, A. (1979) Skład chemiczny Bieszczackiego Rafinatu Dymu Węądzarniczego. *Rocz. Inst. Przem.*, XVI, 95–99.

Borys, A. (1998) A study of the chemical composition of liquid smokes from ash wood. *Rocz. Instyt. Przem.*, XXXV, 151–161.

Borys, A. (2001) The studies on the chemical composition and properties of the selected liquid smokes used in Polish meat industry. *Rocz. Inst. Przem.*, XXXVIII, 113–124.

Cerny, C. & Grosch, W. (1992) Evaluation of potent odorants in roasted beef by aroma extract dilution analysis. *Zeitschrift fuer Lebensmittel-Untersuchung und-forschung* 194, 322–325.

Chen, Z. & Maga, J. A. (1993) Wood smoke composition. In: *Food Flavours, Ingredients and Compositions*, Charalambous, G. (ed.), Elsevier, pp. 1001–1007.

Daun, H. (1972) Sensory properties of phenolic compounds isolated from curing smoke as influenced by its generation parameters. *Lebensmittel-Wissenschaft und Technologie* 5, 102–105.

Duthie, S. J., Collins, A. R., Duthie, G. G. & Dobson, V. L. (1997) Quercetin and myricetin against hydrogen peroxideinduced DNA damage (strand breaks and oxidized pyrimidines) in human lymphocytes. *Mutation Research* 393, 223–231.

Etievant, P., Callement, G., Langlois, D., Issanchou, S. & Coquibus, N. (1999) Odor intensity evaluation in gas chromatography-olfactometry by finger span method. *Journal of Agriculture and Food Chemistry* 47, 1673–1680.

Farmer, L. J. & Patterson, L. S. (1991) Compounds contributing to meat flavor. *Food Chemistry* 40 (2), 201–205.

Geha, R. S., Beiser, A. & Ren, C. (2000) Review of alleged reaction to monosodium glutamate and outcome of a multicenter double-blind placebo-controlled study. *Journal of Nutrition* 130 (suppl 45), 1058S–1062S.

Gomaa, E. A., Gray, J. I., Rabies, S., Lopez-Bate, C. & Borne, A. M. (1993) Polycyclic aromatic hydrocarbons in smoked food products and commercial liquid smoke flavoring. *Food Additives and Contaminants* 11, 669–684.

Grandhee, S. K. & Monnier, V. M. (1991) Mechanism of formation of the Maillard protein cross-link pentosidine. Glucose, fructose and ascorbate as pentosidine precursors. *Journal of Biological Chemistry* 266 (18), 11649–11653.

Guichard, H., Guichard, E., Langlois, D., Issanchou, S. & Abbott, N. (1995) GC sniffing analysis: olfactive intensity measurement by two methods. *Zeitschrift fur Lebensmittel-Untersuchung und-forschung* 201, 344–350.

Guillen, M. D. & Ibargoitia, M. L. (1996a) Volatile components of aqueous liquid smokes from Vitis vinifera L. shoots and Fagus sylvatica L. wood. *Journal of the Science of Food and Agriculture* 72, 104–110.

Guillen, M. D. & Ibargoitia, M. L. (1996b) Relationships between the maximum temperatures reached in the smoke generation processes from Vitis vinifera L. shoot sawdust and composition of the aqueous smoke flavoring preparations obtained. *Journal of Agriculture and Food Chemistry* 44, 1302–1307.

Guth, H. & Grosch, W. (1990) Deterioration of soya-bean oil: quantification of primary flavor compounds using a stable isotope dilution assay. *Lebensmittel-Wissenschaft und Technologie* 23, 513–522.

Hall, R. L. & Marwin, E. J. (1981) The role of flavour in food processing. *Food Technology* 22, 47–52.

Halliwell, B. & Gutteridge, L. M. C. (1986) Oxygen free radicals and iron in relation to biology and medicine: some problems and concepts. *Archive Biochemistry Biophysica* 246, 501–514.

Hamm, R. (1976). *Analysis of smoke and smoked foods.* In IUFOSTIUPAC symposium advances in smoking of foods, Warsaw, 8–10 September, pp. 1655–1665.

Harrison, T. J. & Dake, G. R. (2005) An expeditious, high-yielding construction of the food aroma compounds 6-acetyl-1,2,3,4-tetrahydropyridine and 2-acetyl-1-pyrroline. *Journal of Organic Chemistry* 70 (26), 10872–10874.

Hruza, D. E., Praag, M. & Heinsohn, H. (1974) Isolation and identification of the components of the tar of hicory wood smoke. *Journal of Agriculture & Food Chemistry* 22, 123–126.

Jinap, S. & Hajeb, P. (2010) Glutamate. Its applications in food and contribution to health. *Appetite* 55, 1–10.

Kim, K., Kurata, T. & Fujimaki, M. (1974) Identification of flavor constituents in carbonyl, non-carbonyl neutral and basic fractions of aqueous smoke condensates. *Agricultural Biology and Chemistry* 38, 53–63.

Kroh, L. W., Jalyschko, W., Haseler, J. (1996) Non-volatile reaction products by heat-induced degradation of alpha-glucans .1. Analysis of oligomeric maltodextrins and anhydrosugars. *Starch-Starke* 48 (11–12), 426–433.

Löliger, J. (2000) Function and importance of glutamate for savory foods. *Journal of Nutrition* 130, 915S–920S.

Maarse, H. & Visscher, C. A. (1996) *Volatile compounds in food. Qualitative and quantitative data*, 7th edition. TNO Biotechnology and Chemistry Institute, Zeist.

Matsuki, M., Watanabe, T., Ogasawara, A., Mikami, T. & Matsumoto, T. (2008) Inhibitory mechanism of melanin synthesis by glutathione. *Yakugaku Zasshi* 128 (8), 1203–1207.

McDaniel, M. R., Miranda-Lopez, R., Watson, B. T., Michaels, N. J. & Libbey, L. M. (1990) Pinot noir aroma: a sensory/gas chromatographic approach. In *Flavors and Off-flavors*, Charalambous, G. (ed.), Elsevier BV, Amsterdam.

Miler, K. (1962) *Possibilities of curing smoke generation free from 3, 4-benzopyrene and 1, 2, 5, 6-dibenzanthracene*. PhD thesis, Politechnika Gdanska, Gdanask, Poland.

Miller, S. L. (1953) A Production of amino acids under possible primitive Earth conditions. *Science* 117, 528–529.

Miranda-Lopez, R., Libbey, L. M., Watson, B. T. & McDaniel, M. R. (1992) Odor analysis of pinot noir wines from grapes of different maturities by a gas chromatography-olfactometry technique (OSME). *Journal of Food Science* 57, 985–993.

Ohloff, G. (1972). Classification and genesis of food flavours. *Flavour Industry* 3, 501–508.

Pollien, P., Ott, A., Montigon, F., Baumgartner, M., MunozBox, R. & Chaintreau, A. (1997) Hyphenated headspace gas chromatography sniffing technique: Screening of impact odorants and quantitative aroma-gram comparisons. *Journal of Agricultural and Food Chemistry* 45 (7), 2630–2637.

Prost, C., Serot, T. & Demaimay, M. (1998) Identification of the most potent odorants in wild and farmed cooked turbot (Scophtalamus maximus L). *Journal of Agricultural and Food Chemistry* 46 (8), 3214–3219.

Pszczoła, D. E. (1995) Tour highlights production and uses of smoke based flavors. *Food Technology* 49, 70–74.

Rada-Mendoza, M., Villamiel, M., Molina, E. & Olano, A. (2006) Effects of heat treatment and high pressure on the subsequent lactosylation of beta-lactoglobulin. *Food Chemistry* 99 (4), 651–655.

Rangan, C. & Barceloux, D. G. (2009) Food additives and sensitivities. *Disease-a-month* 55 (5), 292–311.

Reiners, J. & Grosch, W. (1998) Odorants of virgin olive oils with different flavor profiles. *Journal of Agriculture and Food Chemistry* 46, 2754–2763.

Rothe, M. & Thomas, B. (1963) Aromastofe des Brotes. *Zeitschrift fuer Leebensmittel-Untersuchung und forschung* 119, 302–309.

Rusz, J. & Miler, K. B. M. (1977) Physical and chemical processes involved in the production and application of smoke. *Pure & Appied Chemistry* 49, 1639–1654.

Rychlik, M. & Grosch, W. (1996) Identification and quantification of potent odorants formed by toasting of wheat bread. *Lebensmittel Wissenschaft and Technologie* 29, 515–525.

Semmelroch, P. & Grosch, W. (1995) Analysis of roasted coffee powders and brews by gas chromatography-olfactometry of head-space samples. *Lebensmittel Wissenschaft and Technologie* 28, 310–313.

Semmelroch, P., Laskawy, G., Blank, I. & Grosch, W. (1995) Determination of potent odorants in roasted coffee by stable isotope dilution assays. *Flavor Fragrance Journal* 10, 1–7.

Smitha, R. L., Cohenb, S. M., Doullc, J., Ferond, V. J., Goodmane, J. I., Marnettf, L. J., Portogheseg, P. S., Waddellh, W. J., Wagneri, B. M., Hallj, R. L., Higleyk, N. A., Lucas-Gavinl, C. & Adamsm, T. B. (2005) A procedure for the safety evaluation of natural flavor complexes used as ingredients in food: essential oils. *Food and Chemical Toxicology* 43 (3), 345–363.

Stephan, A. & Steinhart, H. (1999 Identification of character impact odorants of different soybean lecithins. *Journal of Agriculture and Food Chemistry* 47, 2854–2859.

Strecker, A. (1850) Uber die künstliche Bildung der Milchsäure und einen neuen, dem Glycocoll homologen Körper. *Annals of Chemistry and Pharmacology* 75 (1), 27–45.

Tarasoff, L. & Kelly, M. F. (1993) Monosodium L-glutamate: a double-blind study and review. *Food Chemistry and Toxicology* 31, 1019–1035.

Tilgner, D. J. (1962) Advances in the engineering of the smoke curing process. II Intern. Session 15–19, 11 1960, Gdansk.

Ullrich, F. & Grosch, W. (1987) Identification of the most intense volatile flavor compounds formed during autoxidation of linoleic acid. *Zeitschrift fur Lebensmittel-Uutersuchung und-forschung* 184, 277–282.

Van Ruth, S. M., Roozen, J. P. & Cozijnsen, J. L. (1996) Gas chromatography/sniffing port analysis evaluated for aroma release from rehydrated French beans (phaseolus vulgaris). *Food Chemistry* 56, 343–346.

Winterhalter, P. & Rouseff, R. (eds) (2002) Carotenoid-derived aroma compounds: An introduction. Vol. 802. American Chemical Society, Symposium series. DOI: 10.1021/bk-2002-0802.

Yabiku, H. Y., Martins, M. S. & Takahashi, M. Y. (1993) Levels of benzo [a] pyrene and other polycyclic aromatic-hydrocarbons in liquid smoke flavor and some smoked foods. *Food Additives and Contaminants* 10 (4), 399–405.

Yen, G. C., Duh, P. D. & Lin, C. W. (2003) Effects of resveratrol and 4-hexylresorcinol on hydrogen peroxide-induced oxidative DNA damage in human lymphocytes. *Free Radical Research* 37 (5), 509–514.

FURTHER READING

Yu, H., Martins, S. M. & Takahashi, M. Y. (1993) Levels of benzo(a)pyrene and other polycyclic aromatic hydrocarbons in liquid smoke flavor and some smoked foods. *Food Additives and Contaminants* 10, 503–521.

6 Food Acids and Acidity Regulators

Abstract: In food processing industries (or even in homes) food acids are included in foods to accomplish a number of tasks including sharpening the flavours as well as perform the functions of preservatives and antioxidants. Food acids used most frequently for these purposes include acetic acid (vinegar), citric acid, tartaric acid, malic acid, fumaric acid, and lactic acid. Other organic acids added to foods (tartaric acid, malic acid and lactic acid as well as some plant juices) are also used to give a sour taste to foods. Another food acid, fumaric acid (found naturally in some mushrooms), is mainly used in beverages and baking powders where it contributes to the sour taste of foods products. Acidity regulators are substances such as organic or mineral acids, bases, neutralising agents, or buffering agents and are included in food products to modify and control the pH of foods.

Keywords: food acids; food acidity regulators; sharp taste; food pH

6.1 WHAT ARE FOOD ACIDS AND ACID REGULATORS?

Further Thinking

Food acids and acid regulators are additives that impart sharp tastes to foods such as those found in acidic food items (e.g. oranges with citric acid), apples, tomatoes, yogurt, etc. Food acids also perform other functions, such as helping the setting process of gels and as preservatives.

Food acids and acid regulators perform dual functions in that they act as preservatives (antimicrobial) and they also sharpen the taste or flavour of foodstuffs by imparting tartness. Acid regulators are normally added to foodstuffs to adjust food pH (to regulate the acidity and alkalinity of foods). Fruits and vegetables are among the food products where acids and acid regulators are massively used. The addition of food acids in foodstuffs is normally performed simultaneously with the adjustment of the total soluble solids for the purpose of matching the Brix/acid value with that of the food item being processed.

Chemistry of Food Additives and Preservatives, First Edition. Titus A. M. Msagati.
© 2013 John Wiley & Sons, Ltd. Published 2013 by John Wiley & Sons, Ltd.

Further Thinking

Acids, bases and buffers are always added to foods in order to control the acidity or alkalinity of foodstuffs to ensure the safety and stability of flavour as well as acting as food preservative because they discourage the growth of some microbes. Moreover, bicarbonate additives such as bicarbonate of soda as well as phosphates are used in foods to provide a buffering environment.

There are a number of other roles that food acids and regulators play apart from those mentioned above. These various roles include:

- flavouring to providing the desired intensity of taste which will exactly match, enhance or modify the natural or original flavour of that particular foodstuff;
- lowering the acidity/alkalinity of food products to minimise/prevent/retard the growth of microorganisms or inhibit the germination of spores;
- buffering effect by maintaining the acidity/alkalinity of foods;
- providing the proper environment for metal ion chelation, an important phenomenon in the minimisation of lipid oxidation; and
- alteration of the structure of foods such as gels made from gums and other proteins.

To perform such a wide variety of functions, food acids contain a variety of functionalities and diversity in their chemistry. Some are more suited for the role of flavour sharpeners and others as acidifiers, metal chelators, antimicrobial agents and solubilisers.

The most commonly used food acids are acetic acid, citric acid, malic acid, phosphoric acid, fumaric acid and tartaric acid (Palmer and List 1973; Seuss and Martin 1993). All of these food acids are generally recognised as safe (GRAS), with the exception of fumaric acid.

6.2   TYPES OF FOOD ACIDS

Table 6.1 lists the many food acids which have applications in the food industry (Palmer and List 1973; Seuss and Martin 1993).

6.2.1   Citric acid

The citric acid used in foods is normally prepared by fermentation processes of carbohydrates by fungal microbes known as *Aspergillus niger*. The type of carbohydrate usually used for the fermentation process is molasses. Other sources of citric acid include citrus fruits such as lemons. The functions that citric acid performs in foods are many and diverse such as: providing sharp tastes to foods; acting as an antioxidant (especially in fatty foods); and serving as a preservative agent for foods such as meat. Citric acid can be used to prevent oxidation reactions which result in the browning phenomena in foods such as salads.

Table 6.1   Physical and chemical properties of food acids

Food acid	CFR and physical state	pKa	Taste	Application products	References
Citric acid	GRAS-crystalline; powder	3.14; 4.17; 6.39	A burst of tartness	Carbonated and non-carbonated beverages, wines, jams and jellies, canned products	Palmer and List 1973; Tavares *et al.* 2005
Fumaric acid	Food additive-white granules or crystalline; powder	3.03; 4.04	Tart	Frozen concentrates, cider and apple drinks	Palmer and List 1973
Malic acid	GRAS-crystalline, powder	3.4; 5.11	Smooth tartness	Fruit-flavoured sodas	Kenney 1991; Kunkee 1991
Phosphoric acid	GRAS-liquid	2.12; 7.21; 12.67	Acrid	Buffering agent in jams and jellies	Castro *et al.* 2000
Glucano-delta-lactone (GDL)	GRAS-white crystalline; powder	3.7	Neutral taste, but acid taste upon hydrolysis	Salad dressing	Lucey *et al.* 1998
Tartaric acid	GRAS-crystalline, powder	4.34; 2.98	Extremely tart	Cranberry and grape-flavoured fruits	Kenney 1991
Acetic acid	GRAS-clear colourless liquid	4.75	Tart and sour	Pickled fruits	Seuss and Martin 1993

6.2.2   Lactic acid

Lactic acid is also produced via fermentation processes, although synthetic routes are also used. It also has many uses in foodstuffs; for example, it forms one of the raw materials in the production of certain emulsifiers which are used in dairy products, sweets and in the baking industry.

6.2.3   Acetic acid

Acetic acid such as that found in vinegar has been used in foods and many other applications for longer than can be remembered. It finds application within the food industry mainly as a flavouring agent (in the form of sodium diacetate), in confectionaries and as a preserving agent due to its superb antimicrobial properties. Acetic acid is normally produced by fermentation processes as well as synthetically.

6.2.4   Malic acid

Malic acid, as found in a number of fruits such as bananas, apples and tomatoes, is found in many of the low-energy drinks. It is generally produced industrially using maleic anhydride as a starting material.

6.2.5 Fumaric acid

Fumaric acid is known for its strong taste when present in foods. It has also low solubility in aqueous media, and is therefore not a preference in many food applications. It is produced by synthesis using malic acid as a starting material.

6.2.6 Tartaric acid

Tartaric acid in the form of its salts, such as potassium hydrogen tartrate, finds most of its applications in the production of emulsifiers used in the baking industry and in confectionaries. It is produced synthetically using maleic anhydride, and can also be obtained naturally by extraction from wine products.

6.2.7 Phosphoric acid

Phosphoric acid in the form of its salts is used in foods to provide a buffering environment, for example in baking powders. It is also used as emulsifying agent and as a flavourant in some cola drinks.

6.3 USES OF FOOD ACIDS

6.3.1 As acidity regulators

The pH of food determines its taste; it is known that acidic foods (e.g. citrus fruits and sour milk) taste sour while alkaline foods (e.g. baking soda) taste bitter (Seuss and Martin 1993). Acidity regulators have therefore been employed in food processing industries to act as buffering agents. They adjust and maintain the acidity or alkalinity of a specific type of food product at a desired level for the purpose of either giving a specific taste or for preventing microbial attack (Sofos and Busta 1981; Blocher and Busta 1985).

The most commonly used acidity regulators are the weak organic acids including citric acid (in fruits, vegetables and soft drinks; Palmer and List 1973); acetic acid (in margarine, butter and curry powder; Tavares *et al.* 2005); fumaric acid (in bread, wine and jams; Kenney 1991); lactic acid (in milk and cheese; Datta *et al.* 1995); tartaric acid (in juices and bakery products; Palmer and List 1973); malic acid (in jams and tinned fruits; Kenney 1991); and calcium acetate, which also plays the important roles of antioxidant booster, gel property enhancer and thickening agent (Tavares *et al.* 2005).

6.3.2 As preservatives

The organic acids used as food acidity regulators are referred to as weak since they do not dissociate fully in solution; instead, they exist in an equilibrium between the undissociated and dissociated forms of the respective acids (Figure 6.1a–d; Booth and Kroll 1989; Brul and Coote 1999). In acidic media however, the predominant state of these acidic molecules is the undissociated state. This is the most effective state for the inhibitory activity of the acids, as it attains the neutrality form (at low pH) needed for it to diffuse across and thereby disrupt the hydrophobic microbial cell membrane. The acid gets inside the cytoplasm of the alkaline call, causing the molecule to dissociate into charged states and thus preventing

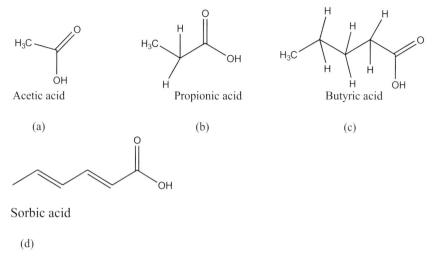

Fig. 6.1 Structures of weak organic acids and acid regulators (Palmer and List 1973; Seuss and Martin 1993): (a) acetic acid; (b) propionic acid; (c) butyric acid and (d) sorbic acid.

back-crossing to the outside the cell (the charged species cannot cross the cell membranes; Freese *et al.* 1973; Bracey *et al.* 1998; Stratford and Anslow 1998). The driving force for the movement of neutral species of the weak acids across the hydrophobic plasma membrane into the microbial cell is the pH gradient (lower outside but higher inside), and this movement will continue until equilibrium is established (Booth and Kroll 1989).

The disruption of the microbial cell membrane by the weak organic acids not only inhibits the growth of microbes, but also interferes with processes involving important cellular metabolic reactions, leading to cell death (Krebs *et al.* 1983; Salmond *et al.* 1984; Eklund 1985; Cole and Keenan 1987; Holyoak *et al.* 1996; Bracey *et al.* 1998).

REFERENCES

Blocher, J. C. & Busta, F. F. (1985) Multiple modes of inhibition of spore germination and outgrowth by reduced pH and sorbate. *Journal of Applied Bacteriology* 59, 467–478.

Booth, I. R. & Kroll, R. G. (1989) The preservation of food by low pH. In: *Mechanism of Action of Food Preservation Procedures*, Gould G. W. (ed.), Elsevier, London, pp. 119–160.

Bracey, D., Holyoak, C. D. & Coote, P. J. (1998) Comparison of the inhibitory effect of sorbic acid and amphotericin B on *Saccharomyces cerevisiae*: is growth inhibition dependent on reduced intracellular pH? *Journal of Applied Microbiology* 85 (6), 1056–1066.

Brul, S. & Coote, P. (1999) Preservative agents in foods: mode of action and microbial resistance mechanisms. *International Journal of Food Microbiology* 50 (1–2), 1–17.

Castro, J. B., Bonelli, P. R., Cerrella, E. G. & Cukierman, A. L. (2000) Phosphoric acid activation of agricultural residues and bagasse from sugar cane: influence of the experimental conditions on adsorption characteristics of activated carbons. *Industrial & Engineering Chemistry Research* 39 (11), 4166–4172.

Cole, M. B. & Keenan, M. H. J. (1987) Effects of weak acids and external pH on the intracellular pH of *Zygosaccharomyces boilii*, and its implications in weak-acid resistance. *Yeast* 3, 23–32.

Datta, R., Tsai, S.-P., Bonsignore, P., Moon, S.-H. & Frank, J. R. (1995) Technological and economic potential of poly (lactic acid) and lactic acid derivatives. *FEMS Microbiology Review* 16 (2–3), 221–231.

Eklund, T. (1985) The effect of sorbic acid and esters of parahydroxybenzoic acid on the proton motive force in *Escherichia coli* membrane vesicles. *Journal of General Microbiology* 131, 73–76.

Freese, E., Sheu, C. W. & Galliers, E. (1973) Function of lipophilic acids as antimicrobial food preservatives. *Nature* 241, 321–325.

Holyoak, C. D., Stratford, M., McMullin, Z., Cole, M. B., Crimmins, K., Brown, A. J. P. & Coote, P. (1996) Activity of the plasma membrane H^+ – ATPase and optimal glycolytic flux are required for rapid adaptation and growth in the presence of the weak acid preservative sorbic acid. *Applied Environmental Microbiology* 62, 3158–3164.

Kenney, B. F. (1991) Determination of organic acids in food samples by capillary electrophoresis. *Journal of Chromatography A*, 546, 423–430.

Krebs, H. A., Wiggins, D., Sole, S. & Bedoya, F. (1983) Studies on the mechanism of the antifungal action on benzoate. *Biochemical Journal* 214, 657–663.

Kunkee, R. E. (1991) Some roles of malic acid in the malolactic fermentation in wine making. *FEMS Microbiology Letters* 88 (1), 55–72.

Lucey, J. A., Tamehana, M., Singh, H. & Munro, P. A. (1998) A comparison of the formation, rheological properties and microstructure of acid skim milk gels made with a bacterial culture or glucono-δ-lactone. *Food Research International* 31 (2), 147–155.

Palmer, J. K. & List, D. M. (1973) Determination of organic acids in foods by liquid chromatography. *Journal of Agriculture and Food Chemistry* 21 (5), 903–906.

Salmond, C. V., Kroll, R. G. & Booth, I. R. (1984) The effect of food preservatives on pH homeostasis in *Escherichia coli*. *Journal of General Microbiology* 130, 2845–2850.

Seuss, I. & Martin, M. (1993) The influence of marinating with food acids on the composition and sensory properties of beef. *Fleischwirtschaft* 73 (3), 292–295.

Sofos, J. N. & Busta, F. F. (1981) Antimicrobial activity of sorbate. *Journal of Food Protection* 44, 614–622.

Stratford, M. & Anslow, P. A. (1998) Evidence that sorbic acid does not inhibit yeast as a classic weak acid preservative. *Letters in Applied Microbiology* 27, 203–206.

Tavares, R. G., Schmidt, A. P., Abud, J., Tasca, C. I. & Souza, D. O. (2005) In vivo quinolinic acid increases synaptosomal glutamate release in rats: reversal by guanosine. *Neurochemical Research* 30, 439–444.

FURTHER READING

Tavares-Araújo, C. S., Lira de Carvalho, J., Ribeiro-Mota, D., de Araújo C. L. & Coelho, N. M. M. (2005) Determination of sulphite and acetic acid in foods by gas permeation flow injection analysis. *Food Chemistry* 92 (4), 765–770.

7 Food Colour and Colour Retention Agents

Abstract: Not all food additives are meant to be exclusively nutritious; in the case of food colouring and colour retention agents, food additives are introduced in foods to give them a more attractive look and, in some instances, to replace lost or fading natural colours during preparation, transportation or storage. Some of these agents are natural (e.g. carotenoids, anthocyanins and betalains) and others are prepared chemically or artificially (e.g. azo-compounds, amaranth, brilliant blue, indigo carmine, new red, ponceau 4R, sunset yellow, tartrazine and allura).

Keywords: food colours; natural food colourants; synthetic food colourants

7.1 WHY ADD COLOURANTS TO FOODS?

The use of food colourants to make food more attractive and appetising has been in practice for centuries (Ghorpade *et al.* 1995). Colour is an important attribute as well as a selection criterion when it comes to food choices; it enhances the appeal towards foods, thus influencing preference, pleasantness and acceptability of food products (Clydesdale 1993; Hallagan *et al.* 1995; Delgado-Vargas *et al.* 2000). It has actually been said that colour is the most outstanding parameter by which the quality of food and flavour are judged (Altinoz and Toptan 2003).

In general, food colours are made up of mixtures containing a number of major ingredients including the main colouring compound, inorganic salt and volatiles. The main purpose of adding colours to foods is to restore colour lost during processing, storage or transportation; to make the food look more appealing to consumers; to stabilise and properly reinforce the already existing natural colours; and preserve the identical appearance across the processed batch.

7.2 CLASSIFICATION OF FOOD COLOURANTS

There are two criteria by which food colourants are classified: one classification is based on whether they are natural or artificial and the other criteron is their chemistry, which is based on the functional groups responsible for the colouring effect. Based on the former criterion, food colourants are classified as one of three main groups: natural food colourants, nature-identical colourants and artificial/synthetic colourants.

Chemistry of Food Additives and Preservatives, First Edition. Titus A. M. Msagati.
© 2013 John Wiley & Sons, Ltd. Published 2013 by John Wiley & Sons, Ltd.

Based on their chemistry, the major foods colours can be grouped within the following classes: flavonoids (main sources fruits and vegetables); indigoid (main source beetroot); and carotenoids (main sources carrots, tomatoes and oranges).

7.2.1 Natural food colourants

This class includes compounds originating from plant sources, mainly anthocyanins which are obtained from red fruits. Natural food colouring agents are composed of a number of major and minor pigment classes. One of the known properties of the natural colourants is their instability towards pH, heat or light (Barz 1980; Peisker 1990; Brown *et al.* 1991). This property is in sharp contrast to synthetic colourants, which are stabilised by the presence of other molecules. To counter this limitation, technologies involved in plant cell and tissue culture, microbial fermentation and gene manipulation have been applied to that of mass production of stable pigments. These approaches have not yet been approved in terms of toxicological considerations of the products.

Some of the natural sources of food colourants include the anthocyanins with characteristic colour ranging from red to blue found in mature fruit (e.g. strawberry, blueberries, cherries, grapes), vegetables (e.g. onions, cabbages), seeds (e.g. purple sunflower) and flowers. Another source is betanin which has a characteristic red colour as found in red beets (beetroot). Caramel pigment is obtained through catalytic heating of carbohydrates. Some pigments such as carminic acid or carmine are obtained from insects (female cochineal insects). Other sources include the carotenoids, β-carotene (e.g. bixin, norbixin or annatto extract); lycopene; lutein (xanthophyll and canthaxanthin). Other sources include the widely known natural green pigments of chlorophyll and chlorophyllin (a water-soluble pigment). Curcumin is another source, extracted from a root tuber of a plant known as curcuma. Curcumin is also the major pigment of turmeric.

7.2.2 Nature-identical food colourants

Members of this class are actually compounds synthesised to the chemical identity of the natural colourants. Examples include β-carotene, canthaxanthin and riboflavin.

Generally, the majority (if not all) of the natural and nature-identical colours are hydrophobic, that is, mostly insoluble in water. This is a property which means that adding them directly to foods is impractical. One way of introducing them into foods is to convert them into their sodium or potassium salt forms, making them hydrophilic and hence soluble in water. Another approach that is normally used is to dissolve them in a hydrophobic medium such as oil, and then introduce them into water-soluble platforms which can be introduced in foods.

7.2.3 Synthetic/artificial food colourants

As their name suggests, artificial food colourants are a product of chemical processes in which molecules which are capable of imparting colours to foods are produced or synthesised. Examples of synthetic food colourants include tartrazine and carmoisine.

The majority of synthetic colourants are hydrophilic and thus water soluble, a property which means they can be introduced in foods without the need for pre-processing. The main

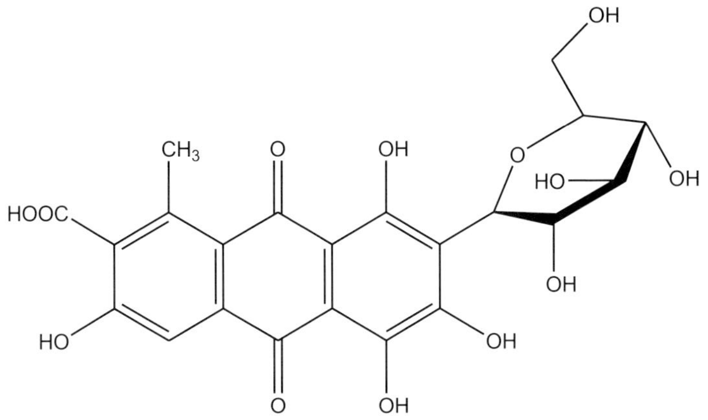

Fig. 7.1 Chemical structure of carminic acid, responsible for the red colour in cochineal extracts (Maguregui *et al.* 2007).

classes of synthetic food colours are azo dyes (e.g. amaranth); quinoline (e.g. quinoline yellow); xanthene (e.g. erythrosine); triarylmethanes and indigoid (e.g. indigo carmine).

7.2.4 Classification based on the nature of chromophores

Food colourants may also be classified according to the functional groups responsible for their characteristic colours (Kvavadze *et al.* 2009). Almost all the functional chromophores are characterised by intensive unsaturation in their chemical structures, which play a significant role in imparting characteristic colours to foodstuffs. The chromophores responsible for characteristic colours in food products are listed in Table 7.1 (Ames and Hofmann 2001).

7.3 OVERVIEW OF COLOURANTS

7.3.1 β-carotene

Some of the natural colourants such as β-carotene, which is obtained naturally from a number of fruits and vegetables, can also by synthesised or produced artificially (Hallagan *et al.* 1995). This pigment is an isomer of carotene which has a characteristic colour of orange–yellow. The compound is also used in the food industry as a food supplement, as it also shows characteristic antioxidant activity.

7.3.2 Cochineal extract and carmine

Carmine pigment (Figure 7.1) is obtained naturally as a concentrated solution after removing alcohol from an aqueous alcoholic extraction of the dried insect *coccus cacti*. This insect is known to be rich in carminic acid which is the principle colouring pigment. The cochineal extract of carmine is therefore responsible for the characteristic orange–red colour in foodstuffs (Maguregui *et al.* 2007).

Table 7.1 Chromophores responsible for colours in foods

Chromophore	Class	Chemical structure	Characteristic colour
Anthocyanins (Natural)	Anthocyanions	X' refers to an acidic functionality	Red, blue or purple
Riboflavin	Iso-alloxazine		Yellow

X' refers to an acidic functionality

	R	R_1
Pelargonidin	H	H
Cyanidin	OH	H
Delphinidin	OH	OH
Peonidin	OCH_3	H
Petunidin	OCH_3	OH
Malvidin	OCH_3	OCH_3

Saffron   Carotenoid   Red

Crocetin

Crocin

(Continued)

Table 7.1 (Continued)

Chromophore	Class	Chemical structure	Characteristic colour
Turmeric	None	Curcumin	Orange-yellow
Titanium dioxide	Inorganic colouring matter	TiO_2	White opaque
Dehydrated beets (beet powder)	Betalains	Betanis;	Red

Annatto	Carotenoid	Orange-yellow

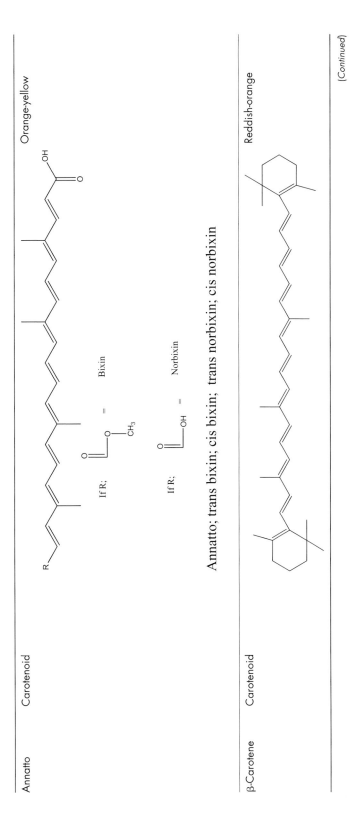

Annatto; trans bixin; cis bixin; trans norbixin; cis norbixin

β-Carotene	Carotenoid	Reddish-orange

(Continued)

137

Table 7.1 (Continued)

Chromophore	Class	Chemical structure	Characteristic colour
β- apo-8'-carotenal	Carotenoid		Reddish-orange
Canthaxanthin	Carotenoid		Reddish-orange
Carmine; cochineal extract	None	Carmine	Orange-red

Caramel	Complex structures existing mainly in three polymeric classes (caramelans, caramelens and caramelins):	Reddish-brown to brown-black
	(i) Caramelans with molecular formula, $C_{24}H_{36}O_{18}$	
	(ii) Caramelens with molecular formula, $C_{36}H_{50}O_{25}$	
	(iii) Caramelins with molecular formula, $C_{125}H_{188}O_{80}$	

| Paprika and paprika oleoresin (natural) | Carotenoid | Deep red |

Capsaicin

Capsanthin

Capsorubin

(Continued)

Table 7.1 (Continued)

Chromophore	Class	Chemical structure	Characteristic colour
Azo dyes	Diimide	Methylene blue	Red
	Diazenes	Based on -N=N- azo structure	Predominantly red

	R_1	R_2	R_3	R_4	R_5
Sudan I	OH	H	H	H	H
Sudan II	OH	H	CH₃	H	CH₃
Sudan III	OH	Ph-N=N	Ph-N=N	Ph-N=N	Ph-N=N
Sudan IV	OH	CH₃	oCH₃Ph-N=N-	H	H
Sudan red 7B	EtNH	H	Ph-N=N-	H	H
Sudan red B	OH	H	mCH₃-Ph-N=N-	CH₃	H
Sudan red G	OH	CH₃-O	H	H	H
Para red	OH	NO₂	NO₂	NO₂	NO₂

NB: o = otho substitution; m= meta substitution

140

Sudan orange G

Butter yellow

Orange

Yellow

7.3.3 Turmeric

This pigment is a product from a root tuber of a plant known as *Curcuma longa*. Solvent extraction of this rhizome yields turmeric oleoresin, containing the colouring molecule known as curcumin which is responsible for the characteristic colour orange–yellow. Turmeric is also beneficial to health as it has been proven to be a therapeutic agent as well as an antioxidant (Jensen 1982; Ravindranath and Chandrasekhara 1982).

7.3.4 Canthaxanthin

This colouring agent can be prepared synthetically or extracted from natural sources (mainly from algae of *Daphnia species*, brine shrimp and *Cantharellus cinnabasinus*). This pigment is responsible orange–red characteristic colours in foodstuffs. Canthaxanthin is however known to have a negative effect on the human retina, as deposition of this compound in the retina causes dysfunction (Hallagan *et al.* 1995) which may persist for some time. The reaction is known to be reversible within a few months.

7.3.5 Anthocyanins

These compounds occur in abundance in the plant kingdom (e.g. in sunflower, ginger and cranberries). They are found mainly as glycosides, existing in various combinations and providing various characteristic colours ranging from red, blue or purple (Hallagan *et al.* 1995). Examples of pigments that belong to the anthocyanins include cyaniding, delphinidin, pentunidin and malvidin. Anthocyanins are known to be readily absorbed and metabolised (Hallagan *et al.* 1995).

7.3.6 Caramel

Caramel is normally obtained from a controlled heat treatment of some of the carbohydrates, and is responsible for characteristic reddish-brown to brown-black colouration.

7.3.7 Titanium dioxide (TiO2)

TiO_2 occurs naturally in three different crystalline forms and to avoid such a mixture, when used as food additive TiO_2 is normally prepared artificially through synthesis. The synthetic compound, which is a white opaque powder, is used as a colouring agent in cheese and confectionary products.

7.3.8 Paprika and paprika oleoresin

This compound is obtained naturally from the dried pods of mild capsicum (*Capsicum annuum*). The compound is responsible for a characteristic deep red colouration, due to its main ingredients of capsanthin and capsorubin. These compounds are used as colouring agents in meat products, vegetable oils and a number of canned products. The compounds also find use as spices (Hallagan *et al.* 1995).

7.4 CHEMISTRY OF FOOD COLOURANTS

7.4.1 Stabilisation

There are normally no rules for pigment stabilisation in foodstuffs due to the different chemistries that exist within the food colouring agents and the fact that they are used as food additives under a wide range of physical and chemical conditions. However, for pigments with a sandwiched metallic element such as chlorophyll (which has magnesium at the centre), stabilisation of the colour may be attempted by substituting the metal at the centre with other metals, for example substituting magnesium with maybe zinc or copper (Taylor 1984). In certain instances however, it has been observed that some metals bring about the degradation of pigments including anthocyanins. The degradation is evidenced when metal is substituted, which causes the anthocyanins to change from red to blue. In the case of chlorophyll, its bright-green colour disappears when a metal is substituted. A study by Jackman *et al.* (1996) has shown that anthocyanins interact strongly with malonic, malic or citric acid in the cell vacuole and their stability can be explained by the presence of pectin in their structures (Jackman *et al.* 1996).

7.4.2 Mechanisms

The mechanism by which colouring and colour retention agents work involves the binding to oxygen atoms. This causes oxidation in foods, resulting in loss or fading of the natural colour thus preventing the destructive process before it commences.

7.4.3 Measurement of colour and physico-chemical parameters

Purified pigments are characterised by parameters such as colour strength, Brix measurements, sugar, pH, viscosity bulk density and storage stability.

Colour strength is defined as the absorbance at 535 nm of the 1% (v/v) extracts solution and may be measured using a spectrophotometer. Colour measurement is performed using a colourimeter with the spectral data measured from the ultraviolet to the visible region of the electromagnetic radiation spectrum (i.e. 380–780 nm) at specified increments, e.g. every 10 nm.

The total soluble solids and water content (or Brix measurement) of extracted liquids is normally determined using a refractometer, while the water content may be measured using a specialised titrator. The monosaccharide sugars D-glucose/D-fructose measurements are normally performed using specialised kits. pH and viscosity measurements are made using ordinary analytical pH meters and a viscosimeter, respectively.

7.5 EXTRACTION FROM NATURAL SOURCES

7.5.1 Anthocyanin pigments

The extraction of anthocyanin pigments from natural sources of plant origin involves the use of acidified methanol followed by vacuum treatment, which minimises pigment degradation through oxidation (Prata and Oliveira 2007). The process may be repeated a number of times and the extracts are pooled together to maximise the yield. The extracts may then be analysed for the monomeric contents of anthocyanins (Giusti and Wrolstad 2001) and for the spectral measurements to determine the spectral characteristic of the pigments. The pigment content is normally calculated as cyanidin-3-rutinoside, using an extinction coefficient of 28.800 L/(cm mg) and molecular weight of 445.2 g (Pazmin-Duran *et al.* 2001).

In addition to these measurements, alkaline and acidic hydrolysis for the extracts is also normally performed. The alkaline hydrolysis of anthocyanins (saponification) is normally performed on the extracts to identify the presence of organic acids acylated to the anthocyanin (Hong and Wrolstad 1990). The saponification solution used is normally a 10% solution of NaOH performed at room temperature and in the absence of light. The neutralisation of the hydrolysate follows and is accomplished by using a dilute acid such as 2M HCl. Further purification may be achieved using chromatographic techniques.

In the same way, acid hydrolysis (especially of glycosilated molecules attached to the anthocyanins) is normally carried out in order to obtain anthocyanidin components in the extracts, important for a comparison with known anthocyanin extracts (Hong and Wrolstad 1990).

7.5.2 Azo-dye pigments

The extraction of azo-dyes may be achieved by using either ethyl acetate, acetonitrile-ethyl acetate (1:1, v/v) or ethyl acetate-cyclohexane (1:1). The extracts are then cleaned up by gel permeation chromatography (GPC); Suna and Wang 2007).

7.6 QUALITY ASSURANCE OF FOOD COLOURANTS

7.6.1 Quality measurements

Due to the many colouring agents and colour retention molecules that enter the market, both natural and synthetic, it is necessary to monitor the quality of such products. The most common quality assurance that is normally performed on food colourants is the peroxide value (PV) measurement. In some instances, the PV measurement is preceded by the determination of hydroperoxide isomers. This is generally performed using high-performance liquid chromatography (HPLC) with a diphenyl-1-pyrenyphosphine (DPPP) post-column fluorescence detection technique (Ohshima *et al.* 1996, 1997).

The hydroperoxide isomers are then monitored by both an ultraviolet (UV) detector at a wavelength of 234 nm as well as by a fluorescence detector. For the latter, the fluorescence intensity of the DPPP oxide is followed at the emission wavelength of 380 nm and excitation wavelength of 352 nm. These hydroperoxide isomers may be further confirmed by mass spectrometry techniques (Ohshima *et al.* 1996, 1997). The determination of the peroxide value may also be accomplished photometrically using the ferric thiocyanate method (Hapman and Mackay 1949; Pan *et al.* 2005).

7.6.2 Safety of food colourants

Dyes used in the food industries are normally sulphonate derivatives of aromatic amines. Sulphonation makes the colour additive more polar, causing it to be poorly absorbable in the alimentary canal system. For this reason, any negative effects on health due to food colouring agents are highly unlikely, unless used in a high dose (Chung and Cerniglia 1992).

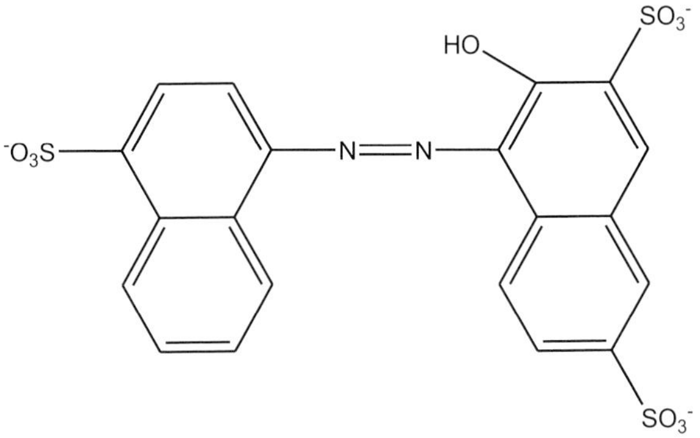

Fig. 7.2   Chemical structure of amaranth (Collins *et al.* 1975; Clode *et al.* 1987).

Lockey (1959) however indicated that there was an association between the occurrence of asthma, hypersensitivity and urticaria with the synthetic aniline dye tartrazine, even though the findings were not conclusively verified.

Amaranth colourants (Figure 7.2) were previously associated with embryo toxicity, renal pelvis hyperplasia and calcification, leading to a ban on their use in foods (Collins *et al.* 1975; Clode *et al.* 1987). Another dye that has been banned is the lipid-soluble formulations of erythrosine because it was linked with the promotion of thyroid tumours in male rats fed this dye. The water-soluble components of erythrosine are however allowed for use in foods, because they are poorly absorbed in the digestive system (Borzelleca *et al.* 1987; Lakdawalla and Netrawali 1988). Carminic acid (red), annatto (orange) and saffron (yellow) have also been associated with disease conditions such as anaphylaxis, urticaria and angioedema (DiCello *et al.* 1999; Moneret-Vautrin *et al.* 2002).

7.7   ANALYTICAL METHODS

Several methods for the analysis of colourants in food items are well known. Since these compounds have strong UV-absorbing chromophores, the reversed-phase liquid chromatography technique is the method of choice in many cases (Nagase *et al.* 1989; Zhang *et al.* 2005, 2006, 2007).

Other methods such as micellar electrokinetic capillary chromatography coupled with UV detection (Mejia *et al.* 2007) and HPLC coupled to atmospheric pressure ionisation mass spectrometry (LC–API-MS/MS) for the simultaneous determination of a number of colourants in foods have been reported (Calbiani *et al.* 2004). Gel permeation chromatography (GPC) with mass spectrometric detection is another hyphenated technique that has been reported for the successful analysis of colourants in food (Suna and Wang 2007).

Electrochemical methods such as the adsorptive voltammetry are known to be useful in the determination of colourants, since these compounds contain electro-active functionalities (Ni *et al.* 1996).

REFERENCES

Altinoz, S. & Toptan, S. (2003) Simultaneous determination of Indigotin and Ponceau-4R in food samples by using Vierordt's method, ratio spectra first order derivative and derivative UV spectrophotometry. *Journal of Food Composition and Analysis* 16 (4), 517–530.

Ames, J. M. & Hofmann, T. (eds) (2001) Chemistry and physiology of selected food colorants. *American Chemical Society Symposium Series*. Volume 775. DOI: 10.1021/bk-2001-0775.

Barz, W. M. (1980) Degradation of flavonoids and isoflavonoids. In *Pigments in Plants*, 2nd edition, Czygan, F.-C. (ed.), Gustav Fischer, Stuttgart, Germany and New York, USA, pp. 21 1–223.

Borzelleca, J. F., Capen, C. C. & Hallagan, J. B. (1987) Lifetime toxicity/carcinogenicity study of FD and C Red No. 3 (erythrosine) in mice. *Food and Chemical Toxicology* 25, 735–737.

Brown, S. B., Houghton, J. D. & Hendry, C. A. F. (1991) Chlorophyll breakdown. In: *Chlorophylls*, Scheer, H. (ed.), CRC Press, Boca Raton, pp. 465–489.

Calbiani, F., Careri, M., Elviri, L., Mangia, A., Pistara, L. & Zagnoni, I. (2004) Development and in-house validation of a liquid chromatography-electrospray-tandem mass spectrometry method for the simultaneous determination of Sudan I, Sudan II, Sudan III and Sudan IV in hot chilli products. *Journal of Chromatography A*, 1042 (1–2), 123–130.

Chung, K. T. & Cerniglia, C. E. (1992) Mutagenicity of azo dyes: structure-activity relationships. *Mutation Research* 277, 201–207.

Clode, S. A., Hooson, J., Grant, D. & Butler, W. H. (1987) Long-term toxicity study of amaranth in rats using animals exposed in utero. *Food and Chemical Toxicology* 25, 937–946.

Clydesdale, F. M. (1993) Color as a factor in food choice. *Critical Reviews in Food Science and Nutrition* 33 (1), 83–101.

Collins, T. F., Keeler, H. V. & Black, T. N. (1975) Long-term effects of dietary amaranth in rats. I. Effects on reproduction. *Toxicology* 3, 115–128.

Delgado-Vargas, F., Jimenez, A. R. & Paredes-Lopez, O. (2000) Natural pigments: Carotenoids, anthocyanins, and betalains: Characteristics, biosynthesis, processing, and stability. *Critical Reviews in Food Science and Nutrition* 40, 173–289.

DiCello, M. C., Myc, A. & Baker J. R. Jr. (1999) Anaphylaxis after ingestion of carmine colored foods: two case reports and a review of the literature. *Allergy Asthma Proceedings* 20, 377–382.

Ghorpade, V. M., Deshpande, S. S. & Salunkhe, D. K. (1995) Food colours. In: *Food Additive Toxicology*, Maga, J. A. & Tu, A. T. (eds), Marcel Dekker, New York, pp. 179–233.

Giusti, M. M. & Wrolstad, R. E. (2001) Characterization and measurement of anthocyanins by UV–visible spectroscopy. In: *Current Protocols in Food Analytical Chemistry*, Wrolstad, R. E. & Schwartz S. J. (eds), John Wiley & Sons, Inc., New York, pp. F1.2.1–F.1.2.13.

Hallagan, J. B., Allen, D. C. & Borzelleca, J. F. (1995) The safety and regulatory status of food, drug and cosmetics color additives exempt from certification. *Food and Chemical Toxicology* 33 (6), 515–528.

Hapman, R. A. & Mackay, K. (1949) The estimation of peroxides in fats and oils by the ferric thiocyanate methods. *Journal of American Oil Chemists' Society* 26, 360–363.

Hong, V. & Wrolstad, R. E. (1990) Use of HPLC separation/photodiode array detection for the characterization of anthocyanins. *Journal of Agriculture and Food Chemistry* 38, 708–715.

Jackman, R. L. (1996) Anthocyanins and betalains. In: *Natural Food Colorants*, 2nd edition, Hendry, C. A. F. & Houghton, I.D. (eds), Blackie, London.

Jensen, N. J. (1982) Subchronic oral toxicity of turmeric oleoresin in pigs. *Mutation Research* 105, 393–396.

Kvavadze, E., Bar-Yosef, O., Belfer-Cohen, A., Boaretto, E., Jakeli, N., Matskevich, Z. & Meshveliani T. (2009) 30,000-year-old wild flax fibers. *Science* 325 (5946), 1359–1359.

Lakdawalla, A. A. & Netrawali, M. S. (1988) Mutagenicity, comutagenicity, and antimutagenicity of erythrosine (FD and C red 3), a food dye, in the Ames/Salmonella assay. *Mutation Research* 204, 131–139.

Lockey, S. D. (1959) Allergic reactions due to FDandC yellow no. 5 tartrazine, an aniline dye used as a coloring and identifying agent in various studies. *Annals of Allergy* 17, 719–25.

Maguregui, M. I., Alonso, R. M., Barandiaran, M., Jimenez, R. M., Garcia, N. (2007) Micellar electrokinetic chromatography method for the determination of several natural red dyestuff and lake pigments used in art work. *Journal of Chromatography A*, 1154 (1–2), 429–436.

Mejia, E., Ding, Y. S., Mora, M. F. & Garcia, C. D. (2007) Determination of banned sudan dyes in chili powder by capillary electrophoresis. *Food Chemistry* 102 (4), 1027–1033.

Moneret-Vautrin, D. A., Morisset, M. & Lemerdy, P. (2002) Food allergy and IgE sensitization caused by spices: CICBAA data. *Allergie et Immunologie (Paris)*, 34, 135–140.

Nagase, M., Osaki, Y. & Matesuda, T. (1989) Determination of methyl yellow, sudan-i and sudan-ii in water by high-performance liquid-chromatography. *Journal of Chromatography A* 465 (92), 434–437.

Ni, Y. N., Bai, J. L. & Jin, L. (1996) Simultaneous adsorptive voltammetric analysis of mixed colorants by multivariate calibration approach. *Analytica Chimica Acta* 329 (1–2), 65–72.

Ohshima, T., Hopia, A., German, B. & Frankel, E. N. (1996) Determination of hydroperoxides and structures by high-performance liquid chromatography with post-column detection with diphenyl-1-pyrenylphosphine. *Lipids* 31, 1091–1096.

Ohshima, T., Ushio, H. & Koizumi, C. (1997) Analysis of polyunsaturated fatty acid isomeric hydroperoxides by high-performance, liquid chromatography with post-column fluorescence detection. In *Flavor and Lipid Chemistry of Seafoods*, Shahidi, F. & Cadwallader, K. R. (eds), American Chemical Society, ACS Symposium Series, Washington, DC, Vol. 674, pp. 198–217.

Pan, X., Ushio, H. & Ohshima, T. (2005) Effects of molecular configurations of food colorants on their efficacies as photosensitizers in lipid oxidation. *Food Chemistry* 92, 37–44.

Pazmin-Duran, E. A., Giusti, M. M., Wrolstad, R. E. & Gloria, M. B. A. (2001) Anthocyanins from banana bracts (Musa X paradisiaca) as potential food colorants. *Food Chemistry* 73, 327–332.

Peisker, C., Thomas, H., Keller, F. & Matile, P. (1990) Radiolabelling of chlorophyll for studies of catabolism. *Plant Physiology* 136, 544–549.

Prata, E. R. B. A. & Oliveira, L. S. (2007) Fresh coffee husks as potential sources of anthocyanins. LWT-Food Science and Technology 40 (9), 1555–1560.

Ravindranath, V. & Chandrasekhara, N. (1982) Prophylactic and therapeutic actions of supplemental β-carotene in mice inoculated with C3HBA adenocarcinoma cells; lack of therapeutic action of supplemental ascorbic acid. *Journal of the National Cancer Institute* 69, 73–79.

Suna, H. & Wang, F. L. A. (2007) Determination of banned 10 azo-dyes in hot chili products by gel permeation chromatography–liquid chromatography–electrospray ionization-tandem mass spectrometry. *Journal of Chromatography A*, 1164, 120–128.

Taylor, A. (1984) Natural colours in food. In *Developments in Food Colors*, Walford, I. (ed.), Elsevier, London, pp. 159–207.

Zhang, Y. P., Zhang, Y. J., Gong, W. J., Gopalan, A. I. & Lee, K. P. (2005) Rapid separation of Sudan dyes by reverse-phase high performance liquid chromatography through statistically designed experiments. *Journal of Chromatography A*, 1098, 183–187.

Zhang, Y. T., Zhang, Z. J. & Sun, Y. H. (2006) Development and optimization of an analytical method for the determination of Sudan dyes in hot chilli pepper by high-performance liquid chromatography with on-line electrogenerated BrO–luminol chemiluminescence detection. *Journal of Chromatography A*, 1129 (1), 34–40.

Zhang, Y., Wu, H. L., Xia, A. L., Han, Q. J., Cui, H. & Yu, R. Q. (2007) Interference-free determination of Sudan dyes in chilli foods using second-order calibration algorithms coupled with HPLC-DAD. *Talanta* 72 (3), 926–931.

8 Flour Treatment/Improving Agents

Abstract: Flour which is normally used for the preparation of various foods such as bakery products also needs to have its properties improved in many ways. Properties such as colour and fineness of the grains need to be improved to enable the final product to have the desired texture, softness and colour and hence meet the requirements of the consumer. There are a number of flour treatment agents (also called flour improving agents), categorised as either flour bleaching agents, flour maturing agents or flour processing agents.

Keywords: bleaching agents; buffering agents; firming agents; flour treatment agents; maturing agents; sequestrants

8.1 WHAT ARE FLOUR TREATMENT/IMPROVING AGENTS?

Flour treatment agents are additives that are used in the food industry which play a variety of roles and have different functionalities. Flour improving agents can be classed as one of four categories: (1) flour maturing agents; (2) flour bleaching agents; (3) flour processing agents; and (4) general flour treatment agents such as those which play a role as buffering agents, dough conditioners, firming agents, sequestrants, leavening agents, yeast food, etc.

8.2 FLOUR MATURING AGENTS

These additives are normally included to assist with glutein development. Glutein is a protein composite which is characteristic of foods originating from wheat and other plant species related to wheat. Glutein imparts a characteristic elasticity to flour doughs, necessary in the rising mechanism of doughs.

Examples of flour maturing agents are azodicarbonamide; simple peroxides such as t-butyl hydroxyperoxide, methyl ethyl ketone peroxides and acetone peroxides; and flour lipids (oxidised methyl linoleate). Other maturing agents include formamidine disulphide and potassium bromate ($KBrO_3$).

Chemistry of Food Additives and Preservatives, First Edition. Titus A. M. Msagati.
© 2013 John Wiley & Sons, Ltd. Published 2013 by John Wiley & Sons, Ltd.

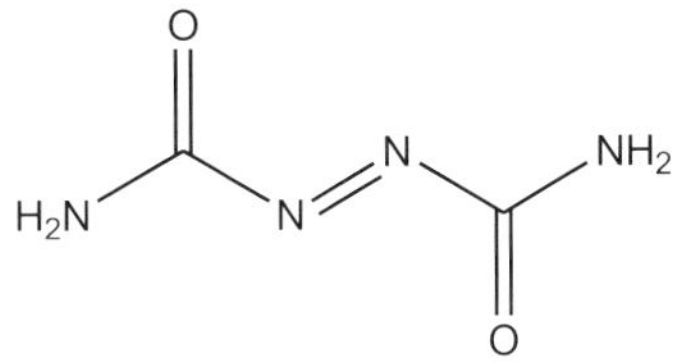

Fig. 8.1 Chemical structure of azodicarbonamide (Oser *et al.* 1965).

8.2.1 Chemistry of azodicarbonamide

This compound is also known as azodicarbonamide, azodicarboxylic acid diamide or azob-isformamide (Figure 8.1). It is mainly used in the food industry as a flour improvement agent which functions as a flour maturing agent (Oser *et al.* 1965).

Azodicarbonamide is used as a maturing agent in food products that are prepared from flour such as dough. When this maturing agent is mixed with dough which is known to contain iodate, a fast sulphhydryl (–SH) oxidising agent, it oxidises the sulphhydryl groups according to the possible reactions depicted in Figure 8.2 (Oser *et al.* 1965; Becalski *et al.* 2004).

The formation of hydrazodicarbonamide (biurea), which is a metabolite of azodicar-bonamide, may strongly suggest that the oxidation of the sulphhydryl groups in the flour by azodicarbonamide is more from Equation (8.2) rather than by Equation (8.1). The reaction described by Equation (8.2) is an oxidation-reduction equation while Equation (8.1) is an addition type of a reaction. This reaction improves the rheological properties of the flour doughs greatly (Becalski *et al.* 2004).

The other maturing agents, the oxidised methyl linoleate and the simple peroxides function in a similar mechanism. When oxygen is in abundance in the flour, the oxidised lipids and

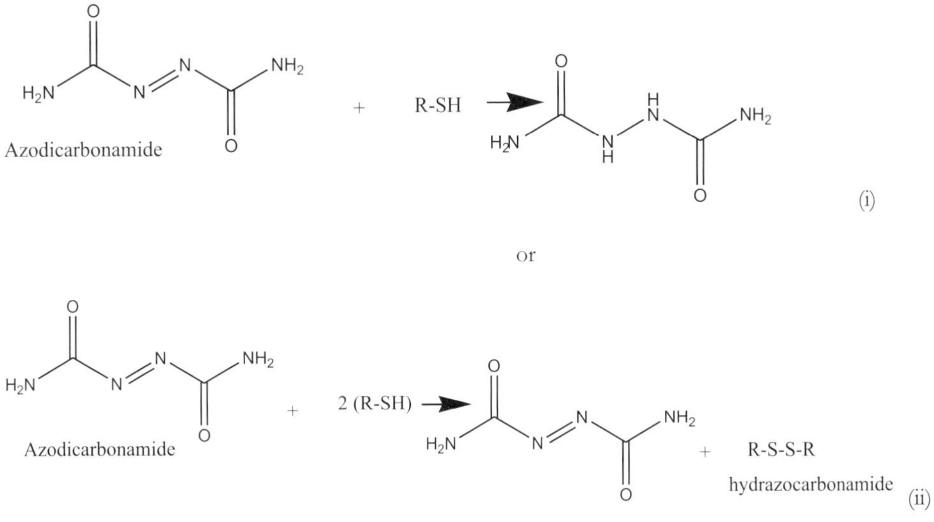

Fig. 8.2 Reaction mechanisms of azidocarbonamide (Oser *et al.* 1965; Becalski *et al.* 2004).

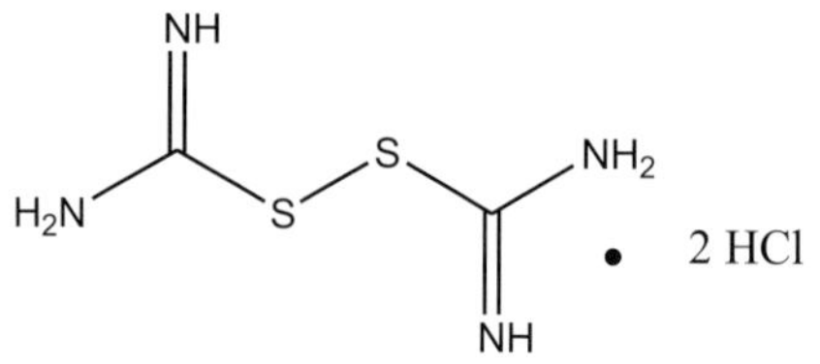

Fig. 8.3   Chemical structure of formamidine disulphide hydrochloride (Sullivan and Dahle 1966).

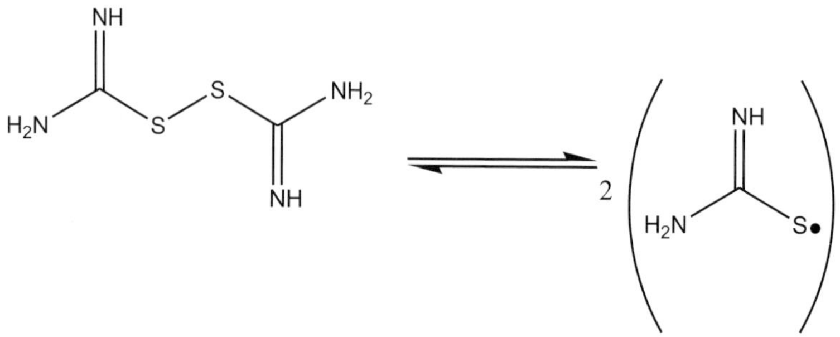

Fig. 8.4   Formation of a radical by cleavage of formamidine disulphide hydrochloride (Sullivan and Dahle 1966).

the oxidised peroxides will react with both sulphhydryl groups with the result of improving the flour properties (Becalski *et al.* 2004).

8.2.2   Formamidine disulphide hydrochloride

The chemical name for this compound is 1, 1'-dithioformamidine dihydrochloride and its chemical structure is depicted in Figure 8.3 (Sullivan and Dahle 1966).

Formamidine disulphide hydrochloride owes its excellent flour maturing properties to its ability to block the sulphhydryl group (-SH) in the flour doughs, which occurs according to the following reactionsdepicted in Figure 8.4.

Formamidine disulphide hydrochloride can undergo cleavage/splitting, which will create a radical (Figure 8.4; Sullivan and Dahle 1966). After splitting, the compound reacts with the R-SH group as shown in Figure 8.5 (Sullivan and Dahle 1966). In the presence of excess R-SH, there can be an exchange between the disulphide and the disulphhydryl as illustrated in Figure 8.6 (Sullivan and Dahle 1966).

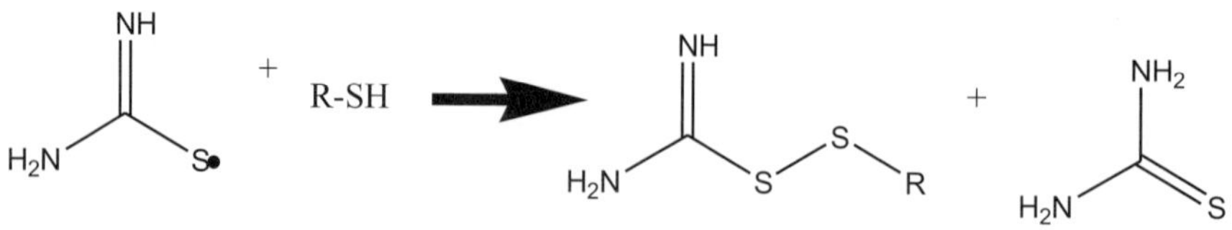

Fig. 8.5   R-SH radical reaction (Sullivan and Dahle 1966).

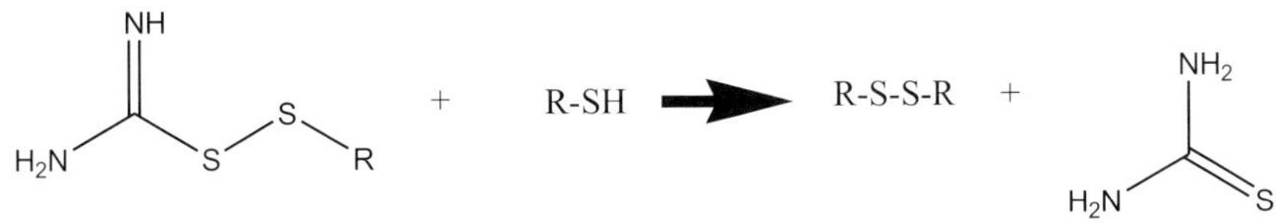

Fig. 8.6 Disulphide –disulphhydryl exchange reactions (Sullivan and Dahle 1966).

8.2.3 Potassium bromate

Potassium bromate plays an important role in food industries as a maturing agent in flour as well as a dough conditioner (Chipman *et al.* 1998).

8.3 FLOUR BLEACHING AGENTS

These are food additives which aim to whiten the flour or oxidise the flour grains to assist with the development of glutein. Flour bleaching agents can be classified as either natural bleaching agents (mainly enzymes such as lipoxygenases; Scheme 8.1) or synthetic bleaching agents including organic peroxides (e.g. benzoyl peroxide), calcium peroxide, chlorine and azodicarbonamide (Fukayama *et al.* 1986; Gelinas *et al.* 1998; Saiz *et al.* 2001; Becalski *et al.* 2004; Borrelli *et al.* 2006).

The use of natural bleaching agents as opposed to synthetic/chemical agents is attractive since it reduces health risks as well as the food appeal to consumers (Lamsal and Faubion 2009). However, due to the slow action of natural/enzymatic bleaching agents, the use of chemical agents has become popular as they quicken the flour-bleaching process (Pyler 1988). Among the chemical bleaching agents for flours that are currently in use is benzoyl peroxide which produces, among other products, benzoic acid.

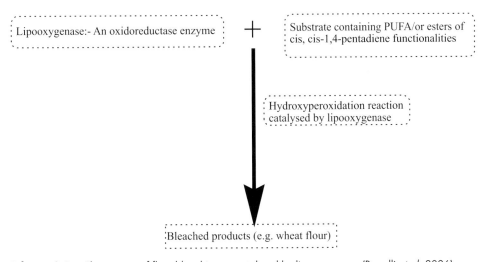

Scheme 8.1   The process of flour bleaching as catalysed by lipoxygenases (Borrelli *et al.* 2006).

8.3.1 Flour lipoxygenases

The enzyme-active soy flour (lipoxygenase) is one of the natural agents normally added to flours to improve the properties of dough. Lipoxygenase performs the important function of catalysing the oxidation of polyunsaturated fatty acids into hydroperoxides and other intermediate free radicals, which in turn oxidise the lipophilic colouring materials/pigments and thiol groups and hence whiten/bleach the flour (Drapron and Godon 1987; Gelinas *et al.* 1998; Borrelli *et al.* 2006).

However, lipoxygenases have a negative effect on bread flavour due to the fact that the process of oxidation of polyunsaturated fatty acids, as catalysed by lipoxygenases, leads to other unwanted oxidative rancid products (Gelinas *et al.* 1998). Other natural enzymatic agents have been researched for use instead of lipoxygenases, and these include oxidoreductases and lipases. A combination of peroxidase, lipase and linoleic acid has shown some promising results as a bleaching agent (Gelinas *et al.* 1998). Another general drawback of the majority of natural or enzymatic oxidation processes for flour is that they are time consuming, undesirable in a commercially driven industry (Lamsal *et al.* 2009).

8.3.2 L-threo-ascorbic acid

Another natural bleaching agent that is used for flour food products is L-threo-ascorbic acid (Vitamin C). This bleaching agent is normally used in the food industry to improve the rheological properties of dough flour. It functions by oxidising L-ascorbic acid to dehydroascorbic acid, which becomes the oxidising agent for the sulphhydryl groups in the flour (Elkassabany and Hoseney 1980; Grosch and Wieser 1999).

8.3.3 Benzoyl acid

This compound is a free-radical initiator known to catalyse pigment oxidation (e.g. carotenoid) by a free-radical mechanism. In the bleaching process, the oxidation of colouring molecules in the flour takes place due to the disruption of the conjugated double-bond system of the pigments (e.g. carotenoid) to a less conjugated colourless system (Saiz *et al.* 2001).

This compound is also known as benzoyl superoxide, dibenzoyl peroxide or dibenzoyldioxidane (Figure 8.7), and is synthesised through the reaction of benzoyl chloride, sodium hydroxide and hydrogen peroxide (Saiz *et al.* 2001).

There have been reports that benzoyl peroxide residues and those of its oxidation product (which is mainly benzoic acid) cause health problems. Benzoyl peroxide itself has been implicated as a carcinogen as it was reported to promote tumour development (Swauger

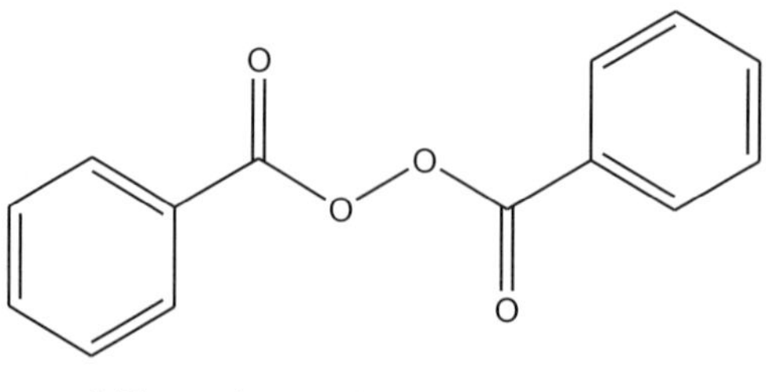

Fig. 8.7 Chemical structure of dibenzoyl peroxide (Saiz *et al.* 2001).

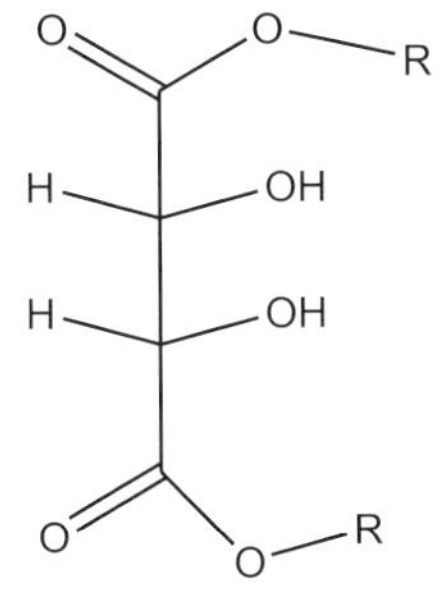

Fig. 8.8   Chemical structure of stearyl tartrate (Borrelli *et al.* 2006).

et al. 1991; Hazlewood and Davies 1996). Kraus *et al.* (1995) reported that the benzoyl peroxide activity caused the breakage of the genetic material (DNA strands) when cells were exposed to this compound, as it generated free-radical species (the benzoyloxyl radical species).

8.3.4   Stearyl tartrate

This compound is also known as stearyl palmityl tartrate (Figure 8.8). It is obtained through the esterification reaction of tartaric acid with stearyl alcohol (FAO/WHO 2001).

8.3.5   Magnesium-DL-lactate

Magnesium-DL-lactate (Figure 8.9) is mostly available as a salt (DL-lactic acid magnesium salt/magnesium di-DL-lactate). It is also known as magnesium DL (-)-2-hydroxypropionate and, as well as being a bleaching agent, it can also function as a dough conditioner and as a buffering agent in flour dough (FAO/WHO 2001).

8.3.6   Citric acid esters of mono- and di-glycerides

These compounds are also known as citric acid esters of mono- and diglycerides, or citro-glycerides. They can be produced through chemical reactions by the esterification reactions of glycerol with citric acid and fatty acids. They can also be produced through the reaction of a mixture of mono- and diglycerides of edible fatty acid with citric acid. It therefore consists of a mixture of ester products of citric acid and fatty acids with glycerol (FAO/WHO 2001).

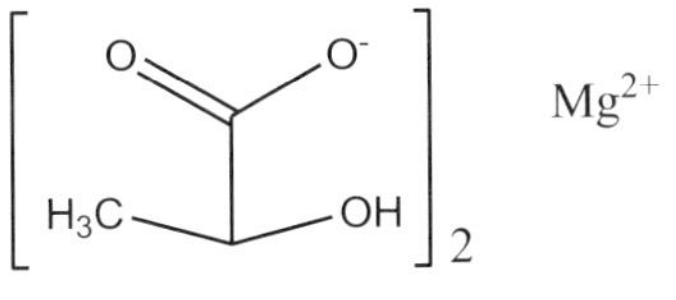

Fig. 8.9   Chemical structure of magnesium-DL-lactate (Borrelli *et al.* 2006).

8.4   FLOUR PROCESSING AGENTS

These include compounds which perform the function of reducing the processing time. An example of such compounds is L-cysteine, which acts to soften the dough (FAO/WHO 2001).

REFERENCES

Becalski, A., Lau, B. P., Lewis, D. & Seaman, S. W. (2004). Semicarbazide formation in azodicarbonamide-treated flour: a model study. *Journal of Agriculture and Food Chemistry* 52 (18), 5730–5734.

Borrelli, G. M., Ficco, D. B. M., Di Fonzo, N. & Fares, C. (2006) Effects of lipoxygenase and of chemical oxidising agent potassium iodate on rheological properties of durum dough. *International Journal of Food Science Technology* 41 (6), 639–645.

Chipman, J. K., Davies, J. E., Parsons, J. L., Neill, J. O. & Fawel, J. K. 1998. DNA oxidation by potassium bromate; a direct mechanism or linked to lipid peroxidation? *Toxicology* 126, 93–102.

Drapron, R. & Godon, B. (1987) Role of enzymes in baking. In: *Enzymes and their Role in Cereal Technology*, Kruger, J. E., Lineback, D. & Stauffer, C. E. (eds), American Association of Cereal Chemists Publishers, pp. 281–317.

Elkassabany, M. & Hoseney, R. C. (1980) Ascorbic acid as an oxidant in wheat flour dough. 11. Rheological effects. *Cereal Chemistry* 57, 88–91.

FAO/WHO (2001) Evaluation of certain food additives and contaminants. World Health Organization Technical Report Series, 901, i-viii, x, 1–107.

Fukayama, M. Y., Tan, H., Wheeler, W. B. & Wei, C. I. (1986) Reactions of aqueous chlorine and chlorine dioxide with model food compounds. *Environmental Health Perspectives* 69, 267–274.

Gelinas, P., Poitras, E., McKinnon, C. M. & Morin, A. (1998). Oxido-reductases and lipases as dough-bleaching agents. *Cereal Chemistry* 75, 810–814.

Grosch, W. & Wieser, H. (1999) Redox reactions in wheat dough as affected by ascorbic acid. *Journal of Cereal Science* 29, 1–16.

Hazlewood, C. & Davies, M. J. (1996) Benzoyl peroxide-induced damage to DNA and its components: direct evidence for the generation of base adducts, sugar radicals, and strand breaks. *Archives of Biochemistry and Biophysics* 332, 79–91.

Kraus, A. L., Munro, I. C., Orr, J. C., Binder, R. L., Leboeuf, R. A. & Williams, G. M. (1995) Benzoyl peroxide – an integrated human safety assessment for carcinogenicity. *Regulatory Toxicology and Pharmacology* 21, 87–107.

Lamsal, B. P. & Faubion, J. M. (2009) Effect of an enzyme preparation on wheat flour and dough color, mixing, and test baking. *LWT – Food Science and Technology* 42, 1461–1467.

Oser, B. L., Oser, M., Morgareidge, K. & Sternberg, S. S. (1965) Studies of the safety of azodicarbonamide as a flour-maturing agent. *Toxicology and Applied Pharmacology* 7, 445–472.

Pyler, E. J. (1988) Wheat and wheat flour. In: *Baking Science and Technology*, 3rd edition. Sosland, Kansas City, MO, pp. 300–377.

Saiz, A. I., Manrique, G. D. & Fritz, R. (2001) Determination of benzoyl peroxide and benzoic acid levels by HPLC during wheat flour bleaching process. *Journal of Agriculture and Food Chemistry* 49, 98–102.

Sullivan, B. & Dahle, L. K. (1966) Disulfide-sulfhydryl interchange studies of wheat flour. 1. The improving action of disulfide. *Cereal Chemistry* 43 (4), 373–383.

Swauger, J. E., Dolan, P. M., Zweier, J. L., Kuppusamy, P. & Kensler, T. W. (1991). Role of the benzoyloxyl radical in DNA damage mediated by benzoyl peroxide. *Chemical Research in Toxicology* 4, 223–228.

9 Anticaking Agents

Abstract: The caking of granulated food items or of amorphous and free-flowing powdered food materials, especially during storage, is always undesirable. Caking may affect food quality in many ways. For example, it may result in physical and morphological changes in a particular food item, which may cause deterioration or loss of quality. In the food industry, anticaking agents are normally introduced to counter this tendency. Granulated or powdered food materials such as milk powder, powdered sugar, tea and coffee can therefore remain intact without forming lumps or caking, meaning these products can be stored safely and easily as well as being transported without undergoing any physical or morphological transformation.

Keywords: amorphous powdered particles; anticaking; caking; crystalline powdered particles

9.1 THE CAKING PHENOMENA

The degree of caking is one of the parameters used to assess the quality of powdered food items (Aguilera *et al.* 1995). Other quality-defining parameters include hygroscopicity, dispersibility, the flowability of the food powders (Haugaard *et al.* 1978; Pisecky 1985) and the sticky point temperature (Lazar *et al.* 1956; Ranganna 1987).

The extent or degree of caking is normally given as a percentage of the respective food powder that remains on a specified sieve after the powdered food has been subjected to repeated drying processes followed by sieving (Pisecky 1985). Hygroscopicity on the other hand expresses the tendency of the powdered food products to absorb moisture when subjected to high-relative-humidity environments (Aguilera *et al.* 1995). The other parameter that defines the quality of powdered foods items, the dispersibility, which is a measure of the ability of powder to become wet without forming dry lumps in water (Pisecky 1985). As a quality parameter, flowability is actually a kinetic parameter which gives a measure of the time required for a specified measure of any powdered food product to percolate through specified slots with known sizes. Factors that influence the flowability of powders include particle size distribution and moisture content of the specified food powder (Pisecky 1985).

The 'sticky temperature' is the temperature at which particles of the powdered food product start to stick to each other due to heating (or an increase in temperature). This parameter is important as it reveals the effects of temperature and moisture content on the caking of the powders (Papadakis and Babu 1992).

Chemistry of Food Additives and Preservatives, First Edition. Titus A. M. Msagati.
© 2013 John Wiley & Sons, Ltd. Published 2013 by John Wiley & Sons, Ltd.

In the food industry, the caking phenomena is a nuisance as it causes low-moisture food items and those which are free-flowing, granulated or in powder form to undergo transformation into lumps. This change initially leads into an agglomerated solid and then to a sticky form, causing loss of food functionality as well as its quality (Aguilera *et al.* 1995).

Caking is not only relevant for ready or raw food items, but is also a problem during the food production process where caking and bridge formation are known to result in reduced production rates or can even halt the whole production process due to the increased moisture (Fitzpatrick *et al.* 2010). For this reason, knowledge of the tendency of the powered food item to absorb water is always needed to predict the kinetics of caking in the process; this may be a difficult task however, due to the fact that many food powder materials contain a variety of composition. There are also a number of uncertainties of the physical conditions, such as temperature and humidity.

There are other physical phenomena that may cause caking. These include recrystallisation, due to either the melting of fat or processes where there is surface wetting accompanied by moisture equilibration or cooling (Peleg 1983). In a true caking process, changes in low-moisture foodstuffs are related to a number of physical parameters such as temperature, moisture and the position within the powder. These take into consideration all the different caking stages, which include:

- bridging, due to surface deformation as well as the sticking of particles at contact points, without any decrease in system porosity (Schubert 1981; Ruzland 1991);
- agglomeration, irreversible consolidation of bridges while maintaining the high porosity of the particulate, thus creating particle clumps with structural integrity;
- compaction, linked to the loss of the integrity of the system due to thickening of interparticle bridges because of the flow, reduction in terms of interparticle spaces and deformation of particle clumps under pressure; and
- liquefaction or loss of interparticle bridges (Schubert 1981; Ruzland 1991; Figures 9.1 and 9.2).

Caking is measured by a unit called the 'caking index' (Aguilera *et al.* 1995) which refers to the state of the system at any time relative to an initial state. There are two morphological indicators of the state of the system: (1) the ratio of instant system porosity to initial system porosity, or $p(t)/p_0$; and (2) the ratio of interparticle bridge diameter to particle diameter, $D_{bridge}/D_{particle}$ (Aguilera *et al.* 1995; Foster *et al.* 2005; Christakis *et al.* 2006; Hartmann and Palzer 2011; see also Figures 9.1 and 9.2).

9.2   MECHANISMS OF CAKING

Several caking mechanisms of action have been proposed which include the van der Waals attractions between particles and high compaction pressures within the mixture. Another school of thought proposes that caking may be a result of the formation of a liquid bridge as the capillary forces present are strong enough to cake the powder. All these theories are based on two important facts about the mechanisms of caking: (1) they are based on the intermolecular structure of the powder particles (Palzer 2006); and (2) crystalline mechanisms differ from those of amorphous powder particles, especially when amorphous particles are subjected to higher humidity conditions.

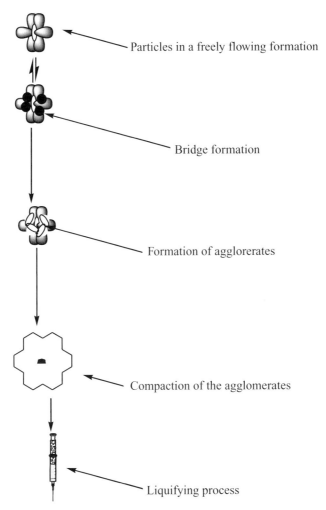

Particles in a freely flowing formation

Bridge formation

Formation of agglorerates

Compaction of the agglomerates

Liquifying process

Fig. 9.1 Important steps/phases during particle caking process (Aguilera *et al.* 1995; Foster *et al.* 2005; Christakis *et al.* 2006; Hartmann and Palzer 2011).

9.2.1 Crystalline powder caking mechanisms

Humidity plays an important role in the caking behaviour of crystalline materials; when the relative humidity is low, the tendency of crystalline materials to absorb water decreases and for above a critical relative humidity point, the tendency to absorb water increases. This tendency will have an effect on the liquid bridge formation as well as the partial dissolution of soluble constituents on the surface of the crystalline materials. The stability of the liquid bridges formed in which the low molecular substances are dissolved is entirely dependent on the capillary forces. In the case of an increased cohesive forces action brought about by a drying phenomenon, solidification of the remaining solids in the bridge may be observed. Moreover, the crystallisation of the solids dissolved in the liquid may be experienced in the form of solid crystal bridges occurring between the particles.

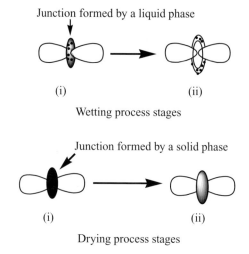

Wetting process stages

Drying process stages

Fig. 9.2   Solid bridge formation and caking (Aguilera *et al.* 1995; Foster *et al.* 2005; Christakis *et al.* 2006; Hartmann and Palzer 2011).

9.2.2 Amorphous materials caking mechanisms

Amorphous sugars and other powdered food products such as dried dairy products are composed of amorphous glassy ingredients, such as amorphous sugars. For example, skim milk and many spray-dried dairy powders are composed of lactose in its amorphous state, which contributes greatly to the formation of caking phenomena (Fitzpatrick *et al.* 2007). The influence of temperature and humidity in the caking behaviour of amorphous materials is always noticeable; the stickiness of the powdered particles commences when the temperature is above the glass transition temperature (known as the sticky point temperature, normally between 10 and 20°C above the glass transition onset temperature for low-molecular-weight carbohydrates) at a known value of humidity (Roos and Karel 1991; Aguilera *et al.* 1995; Kudra 2003; Adhikari *et al.* 2005; Paterson *et al.* 2005; Foster *et al.* 2006; Murti *et al.* 2006). The sticking together of powder particles at the sticky point is caused by the mobility of molecules within the amorphous material, which brings about the deformation of the surface. This enables molecular interactions to increase also the cohesion between them (Aguilera *et al.* 1995).

As already mentioned, the caking of powder materials which contains exclusively amorphous constituents is highly influenced by the moisture content as well as the temperature. The amorphous material behaves differently to that of crystalline substances under moist conditions. The bridging between individual particles tends to undergo a process in which molecules move into the gap between two neighbouring particles because of the existing differences in surface tension (Palzer 2004, 2005; Descamps and Palzer 2007).

In cases where the amorphous powder temperature is greater than its glass transition temperature (T_g), a discontinuous solid crystal phase will form within the amorphous particles due to the fact that the molecules of the amorphous substance will have attained enough kinetic energy to trigger the process of crystallisation (Jouppila and Roos 1994, 1997; Roos 1995, 2002, 2003).

Other factors such the molecular weight of the amorphous molecules play a crucial role in influencing the extent of the crystallisation process, such that the smaller the molecules the faster they move and pack themselves appropriately to form solids/crystals (Fitzpatrick 2007).

Further Thinking

Some industrially processed foodstuffs are composed of components mixed in the form of powders. Anticaking agents are needed to facilitate the homogeneous mixing of these ingredients during processing. Under normal conditions, anticaking agents do not have nutritive value; they are therefore used in small amounts. Examples of anticaking agents include magnesium carbonate (which is used to facilitate the flow characteristics of table salts), silicone dioxide, calcium silicate, dicalcium phosphate and sodium aluminosilicate as well as some natural anticaking agents such as kaolin, potato starch, talc and microcrystalline cellulose.

9.3   CLASSIFICATION OF ANTICAKING AGENTS

Anticaking agents can be divided into two main groups: synthetic and natural anticaking agents. Synthetic anticaking agents are predominantly manufactured from a number of raw materials such as silicon dioxide or solid saturated fatty acids (e.g. magnesium and calcium stearates). Examples of synthetic anticaking agents include calcium silicate ($CaSiO_3$), magnesium carbonate ($MgCO_3$), a baking soda (e.g. sodium bicarbonate or $NaHCO_3$), sodium ferrocyanide ($Na_4Fe(CN)_6$), potassium ferrocyanide ($K_4Fe(CN)_6$) and sodium aluminosilicate (Na_2AlSiO_3).

Natural anticaking agents include kaolin, talc and bentonite (all these being silicate materials). Talc is a powdered natural hydrated magnesium silicate containing varying proportions of materials as alpha-quartz, calcite, chlorite, dolomite, magnesite and phlogopite.

Anticaking agents are also used to perform other different, but just as important, roles. These include: mannitol which is also a texturising agent, a sweetening agent, an anti-sticking agent and a humectant; and alpha and microcrystalline cellulose, which are also used in the food industry as bulking agents, binders and stabilisers.

9.4   ANTICAKING AGENTS IN USE

9.4.1   The need for anticaking agents

As defined earlier, anticaking agents are compounds which prevent agglomeration of powdered food items, thus enabling a free-flowing state. Such agents include a variety of substances such as proteinous and polysaccharide polymers and inorganic materials such magnesium carbonate and silica. They are normally added to foodstuffs such as table salt, flours, coffee and sugar.

If industrially processed powdered foodstuffs are not supported by anticaking agents, the food particles will absorb moisture and stick together to form lumps, preventing the free movement of particles.

9.4.2 Mode of action of anticaking agents

The mechanism of anticaking agents involves either the adsorption of excess moisture or coating of the powder particles to give them a water-repellent character. Some anticaking agents are known to be water soluble, while others are soluble in organic solvents. Some are soluble in both aqueous as well as organic solvents, for example $CaSiO_3$ which is one of the most commonly used anticaking agents.

9.4.3 Foodstuffs which contain anticaking agents

A number of food products contain anticaking agents, such as milk and cream powders, baking powder, table salt, cocoa, coffee and hot chocolate (drinking chocolate), to mention a few examples.

REFERENCES

Adhikari, B., Howes, T., Lecomte, D. & Bhandari, B. R. (2005) A glass transition temperature approach for the prediction of the surface stickiness of a drying droplet during spray drying. *Powder Technology* 149, 168–179.

Aguilera, J. M., del Valle, J. M. & Karel, M. (1995) Caking phenomena in amorphous food powders. *Trends in Food Science and Technology* 6 (5), 149–155.

Christakis, N., Wang, J., Patel, M. K., Bradley, M. S. A., Leaper, M. C. & Cross, M. (2006) Aggregation and caking processes of granular materials continuum model and numerical simulation with application to sugar. *Advanced Powder Technology* 17 (5), 543–565.

Descamps, N. & Palzer, S. (2007) Modeling the sintering of water soluble amorphous particles, PARTEC 2007, Nuremburg, Germany.

Fitzpatrick, J. J., Hodnett, M., Twomey, M., Cerqueira, P. S. M., O'Flynn, J. & Roos, Y. H. (2007) Glass transition and the flowability and caking of powders containing amorphous lactose. *Powder Technology* 178, 119–128.

Fitzpatrick, J. J., Descamps, N., O'Meara, K., Jones, C., Walsh, D. & Spitere, M. (2010) Comparing the caking behaviours of skim milk powder, amorphous maltodextrin and crystalline common salt. *Powder Technology* 204, 131–137.

Foster, K. D., Bronlund, J. E. & Paterson, A. H. J. (2005) The contribution of milk fat towards the caking of dairy. *International Dairy Journal* 15, 85–91.

Foster, K. D., Bronlund, J. E. & Paterson, A. H. J. (2006) Glass transition related cohesion of amorphous sugar powders. *Journal of Food Engineering* 77, 997–1006.

Hartmann, M. & Palzer, S. (2011). Caking of amorphous powders – Material aspects, modeling and applications. *Powder Technology* 206, 112–121.

Haugaard, I. S., Krag, J., Pisecky, J. & Westergaard, V. (1978) *Analytical Methods for Dry Milk Powders.* Niro Atomizer, Denmark.

Jouppila, K. & Roos, Y. H. (1994) Glass transitions and crystallization in milk powders. *Journal of Dairy Science* 77, 2907–2915.

Jouppila, K. & Roos, Y. H. (1997). The physical state of armophous corn starch and its impact on crystallization. *Carbohydrate Polymers* 32, 95–104.

Kudra, T. (2003) Sticky region in drying - definition and identification. *Drying Technology.* 21 1457–1469.

Lazar, M. E., Brown, A. H., Smith, G. S., Wang, F. F. & Lindquist, F. E. (1956) Experimental production of tomato powder by spray drying. *Food Technology* 13, 129–134.

Murti, R., Paterson, A. H. J., Pearce, D. L. & Bronlund, J. E. (2006) Sticky point determination in dairy powders using the particle gun approach. Fifth World Congress on Particle Technology, Orlando, Florida, USA.

Palzer, S. (2004) Impact of glass transition of amorphous components on the agglomeration of food powders. 9th International Congress on Engineering and Food. Montpellier, France, pp. 225–230.

Palzer, S. (2005) The effect of glass transition on the desired and undesired agglomeration of amorphous food powders. *Chemical Engineering Science* 60, 3959–3968.

Palzer, S. (2006) Influence of supramolecular structure and storage conditions on the caking of powders. Fifth World Congress on Particle Technology, Orlando, Florida, USA.

Papadakis, S. E. & Babu, R. E. (1992) The sticky issues of drying. *Drying Technology* 10, 817–837.

Paterson, A. H. J, Brooks, G. F., Bronlund, J. E. & Foster, K. D. (2005) Development of stickiness in amorphous lactose at constant T–Tg levels. *International Dairy Journal* 15, 513–519.

Peleg, M. (1983) Physical characteristics of foof powders. In: *Physical Properties of Foods*, Peleg, M. & Bagley, E. (eds), AVI, Westport, pp. 293–323.

Pisecky, J. (1985) Standards, specifications and test methods for dry milk products. In: *Concentration and Drying of Foods*, Diarmuid, M. C. (ed.), Elsevier Science Publishing Co., New York, pp. 203–220.

Ranganna, S. (1987) *Handbook of Analysis and Quality Control for Fruit and Vegetable Products*. Tata McGraw-Hill Publications, New Delhi.

Roos, Y. H. (1995) *Phase Transitions in Foods*, Academic Press, San Diego, CA, USA.

Roos, Y. H. (2002) Importance of glass transition and water activity to spray drying and stability of dairy powders. *Lait* 82, 475–484.

Roos, Y. H. (2003) Thermal analysis, state transitions and food quality. *Journal of Thermal Analysis and Calorimetry* 71, 197–203.

Roos, Y. H. & Karel, M. (1991) Phase transitions of mixtures of amorphous polysaccharides and sugars. *Biotechnology Progress* 7, 49–53.

Ruzland, D. W. (1991) Fertilizer caking: Mechanisms, influential factors, and methods of prevention. *Nutrient Cycling in Agroecosystems* 30 (1), 99–114.

Schubert, H. (1981) Princniples of agglomeration. *International Chemical Engineering* 21, 363–377.

FURTHER READING

Jouppila, K., Kansikas, J. & Roos, Y. H. (1997) Glass transition, water plasticization and lactose crystallization in skim milk powders. *Journal of Dairy Science* 80, 3152–3160.

10 Humectants

Abstract: There is sometimes the requirement to keep products moist in the food industry, preventing any loss of quality. This is the case for food items such as cream in a cone, chocolate and cheese, in which additives preventing foodstuffs from drying (humectants) are incorporated. Humectants perform the function of moisturisers by attracting water molecules. In this chapter, various humectants (both natural and synthetic) that are used in the food industry will be discussed.

Keywords: food softness; moisturising; natural humectants

10.1 HUMECTANTS AND MOISTURE CONTROL

There are a number of food items which always need to be moist for the sake of safety, quality observance and to optimise shelf life for these particular food products. The controlled moisture content of these specified food products ensures quality and safety because they determine the microbial stability, physical properties, sensory properties and the rate of chemical changes that, if not controlled, are the cause of reduced shelf life (Labuza and Hyman 1998). The types of food items being referred to here include dry cereal with semi-moist raisins, ice cream in a cone, chocolate, hard candy with liquid centres and cheese, just to mention a few.

Factors such as water activity equilibrium and diffusion normally influence the moisture content of food products. This tendency needs to be controlled to ensure that the food moisture content is constant, such that the food quality parameters are maintained.

Among the strategies used to stabilise the moisture content of foodstuffs is the incorporation of food additives. These include non-ionic polyols such as sucrose, glycerin/glycerol and its triester (triacetin). These common humectant food additives are used for the purpose of controlling the viscosity and texture. They also add bulk, retain moisture, reduce water activity and perform the important function of improving food softness (Sloan and Labuza 1975; Fennema 1996; Lindsay 1996). One of the main advantages of these humectant food additives is that, since they are non-ionic, they are not expected to influence any variation of the pH of aqueous systems.

10.2 CLASSIFICATION OF HUMECTANTS

Humectants may be divided into two classes: natural and synthetic. Examples of natural humectants include the tartrates, glycerin and its triester triacetin, invert sugars

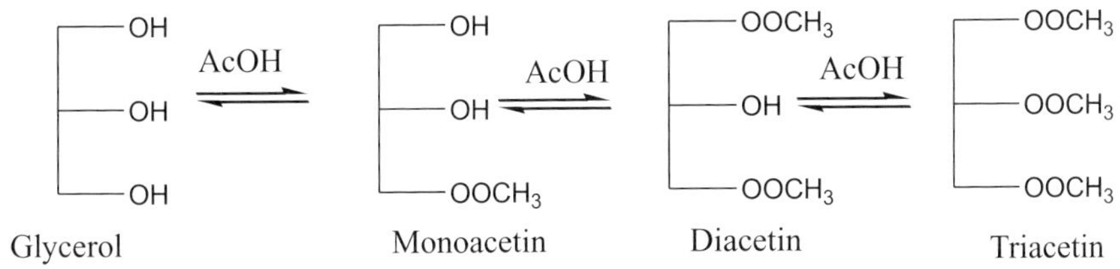

Fig. 10.1 Non-ionic polyols used as food humectants (Sritongtae *et al.* 2011).

(Figure 10.1) and honey. Synthetic humectants include monopropylene glycol, sorbitol and mannitol (Sritongtae *et al.* 2011).

10.2.1 Tartrate series

The tartrate series includes the mono- and disodium tartrates, the mono- and dipotassium tatrates, tartaric acid and sodium potassium tartrate. Tartaric acid, also called dihydroxybutanedioic acid, is a dicarboxylic acid and is naturally obtained as a by-product from wine industries. It can also be found in cocoa powders, fruit and tomatoes. The production of tartaric acid involves heating wastes from the wine-making process and neutralising them with lime water (calcium hydroxide), a step which will cause the precipitation of the insoluble calcium tartrate. This calcium tartrate will then react with sulphuric acid to produce tartaric acid. The other tartrates (sodium and potassium) may be produced in a similar way by a precipitation reaction (involving sodium and potassium hydroxides, respectively).

The presence of tartaric acid in food products does not pose any danger because of the fact that it is broken down efficiently by the intestinal microbes, meaning it cannot be absorbed by the gut. The remaining fraction is filtered in the kidney nephrons and excreted as part of the urine composition.

10.2.2 Glycerol or glycerin humectants

Glycerin or glycerol is widely used in the food industry where it plays a number of very important functions, including: as a humectant and softening agent in food products such as meats and cheeses; as a solvent and sweetener applicable in food preservation; as a solvent for flavours; and also as a food colouring agent. Glycerin is used in the manufacture of a number of other products including mono- and diglycerides (which are used in the food industry as emulsifiers) and polyglycerol esters (used as shortening and margarine). It is also used as a filler in low-fat food products (Holloway *et al.* 2010).

Glycerin can be obtained from either natural sources (plants and animals, as well as industrial sources such as the by-products of biodiesel and soap) or it can be synthesised. For glycerin to be suitable for human consumption, it must be ultra-pure (99.5%; Holloway *et al.* 2010). For glycerin to attain the high purity standard for use as a humectant food additive, it is recommended that the contents of ethylene glycol and diethylene glycol should not exceed the limits of 0.1% in glycerin, set by the American Food and Drug Administration (FDA) (USP 2008). Intensive purification procedures are needed to prepare this to the necessary standard.

A number of protocols have been devised for the purification of glycerol for use as a food additive (humectant) including distillation (Brockmann *et al.* 1987), solvent extraction and ion exchange (Aiken 2006). The major problem with the distillation approach is that this compound has a very high boiling point (290°C). Solvent extraction as a purification technique is not cheap in terms of the set-up, design as well as the process as a whole. A combination of purification techniques is usually employed to obtain the high purity (99.5%) of glycerol needed by the food industries to function as humectants (Scheme 10.1).

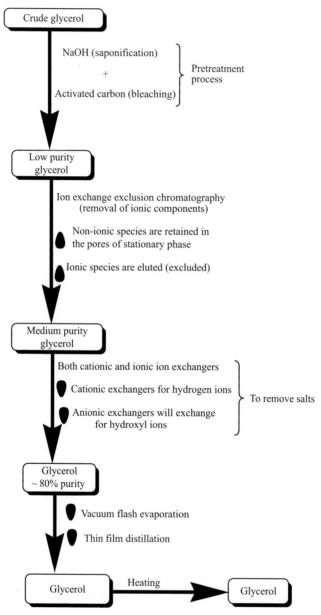

Scheme 10.1 Purification of glycerol extracted from natural sources (Brockmann *et al.* 1987; Aiken 2006).

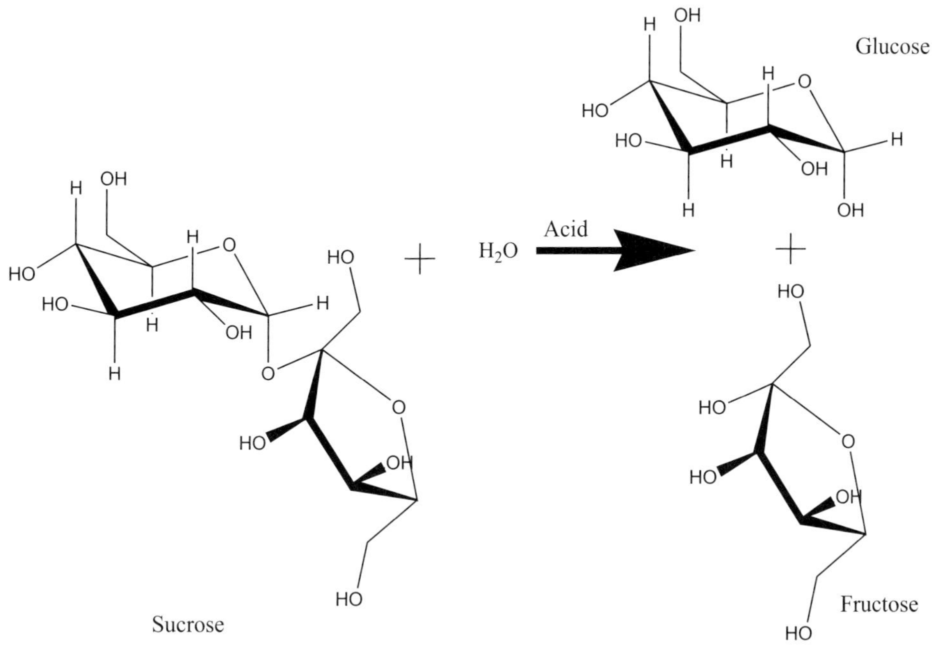

Scheme 10.2   Acid hydrolysis of sucrose to form invert sugars (Schiweck and Clarke 2007).

10.2.3 Invert sugars

Invert sugars (or sugar syrup) refers to a mixture of two monosaccharides: glucose and fructose. The two monosaccharides result from the hydrolysis of sucrose (a disaccharide), catalysed by either heating or by using sucrases or invertases enzymatic catalysts (Scheme 10.2). Both sucrases and invertases belong to the glycoside hydrolase category of enzymes (Schiweck and Clarke 2007). The reaction temperature can be maintained at 50–60°C.

10.2.4 Sorbitol/glucitol

Sorbitol or glucitol (Figure 10.2; El-Kabbani *et al.* 2004) is a sugar alcohol humectant used as a moisturising agent in creams. It is found naturally in a number of fruits including berries, grapes, plums and pears, although sorbitol can also be synthesised from the raw material D-glucose.

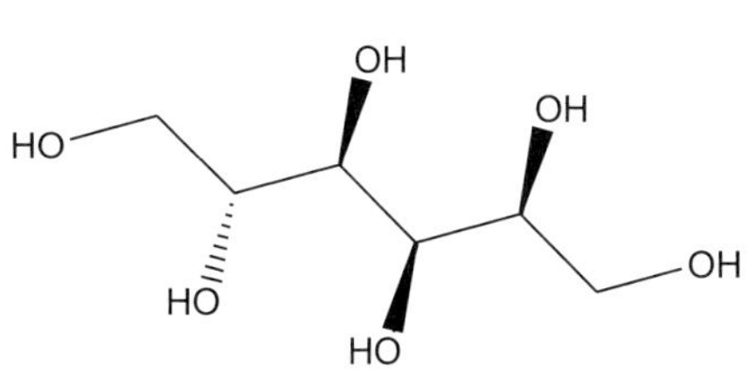

Fig. 10.2   The chemical structure of sorbitol (El Kabbani *et al.* 2004).

REFERENCES

Aiken, J. E. (2006) *Purification of glycerin*. US Patent No.7, 126, 032 B1.

Brockmann, R., Jeromin, L., Johannisbauer, W., Meyer, H., Michel, O. & Plachenka, J. (1987) *Glycerol distillation process*. US Patent No. 4, 655, 879.

El-Kabbani, O., Darmanin, C. & Chung, R. P. (2004). Sorbitol dehydrogenase: structure, function and ligand design. *Current Medicinal Chemistry* 11 (4), 465–476.

Fennema, O. (1996) Water and ice. In: *Food Chemistry*, Fennema, O. (ed.), Marcel Dekker, New York, pp. 17–94.

Holloway, G., Maheswaran, R., Leeks, A., Bradby, S. & Wahab, S. (2010) Screening method for ethylene glycol and diethylene glycol in glycerin-containing products. *Journal of Pharmaceutical and Biomedical Analysis* 51, 507–511.

Labuza, T. P. & Hyman, C. R. (1998) Moisture migration and control in multi-domain foods. *Trends in Food Science and Technology* 9, 47–55.

Lindsay, R. (1996) Food additives. In: *Food Chemistry*, Fennema, O. (ed.), Marcel Dekker, New York, pp. 767–823.

Schiweck, H. & Clarke, M. (2007) Günter Pollach Sugar. In: *Ullmann's Encyclopedia of Industrial Chemistry*, Wiley-VCH, Weinheim.

Sloan, A. E. & Labuza, T. P. (1975) Investigating alternative humectants for use in foods. *Food Product Development* 9 (9), 75–88.

Sritongtae, B., Mahawanich, T. & Duangmal, K. (2011) Drying of osmosed cantaloupe: effect of polyols on drying and water mobility. *Drying Technology* 29 (5), 527–535.

USP (2008) USP 31-NF 26, Glycerine, USP, Rockville, MD, USA, pp. 2286–2287.

11 Antifoaming Agents

Abstract: The foodstuffs we consume are composed of nutritious ingredients, contained in biomolecules such as proteins, vitamins and minerals, in addition to energy-providing biomolecules, carbohydrates and fats. In the processing of food products or beverages comprising proteins, there is a high tendency for foaming. Because of their amphiphilic structure which gives them the flexibility to unfold and adsorb at the interface, protein molecules act as surfactants and create strong intermolecular interactions. This generates a viscoelastic irreversibly adsorbed layer at the air–liquid surface, which stabilises the foam. The foam is a nuisance as it results in a severe loss of production capacity, deterioration in the quality of foods as well as a loss in volume efficiency. Antifoaming agents and techniques have therefore been introduced to counter this foaming process.

Keywords: amphiphilic structures; antifoaming agents; defoamers; foaming processes

11.1 SOURCES OF FOAM IN FOOD PROCESSING

Foaming phenomena arise due to the variation in surface tension caused by temperature gradients. These surface tension gradients instigate motion at the liquid surface and the lower bulky liquid, and the motion will travel from regions of lower surface tension to regions of higher surface tension (between the liquid surface and underlying bulk liquid).

Foam may be classified as one of two main types: (1) metastable foam produced in the presence of charged surfactant molecules, stabilised by double-layer forces, and (2) unstable foam produced in the presence of neutral or low-surface-active molecules. Metastable foam has a higher stability and lasts longer than unstable foam (Fruhner *et al.* 1999).

Foaming which arises during processing in the food industry is highly undesirable, such that food additives capable of reducing or stopping the formation of foam have been designed. These molecules are called antifoam agents, foam control agents or defoamers. They find application in the production process of a number of food items and beverages, especially carbonated drinks, and in the production of sugar from sugar beets. Fermentation processes (e.g. ethanol production using yeasts) are associated with foam production, and thus require defoamers to control foaming during the microbial fermentation activity until the end of the ethanol production process.

There are multiple sources of foam in the food production process. For example, in the production of sugar from sugar beets, foam may originate from non-sugar materials present in the raw materials. These non-sugar materials may include cellulose, lignin, protein, betaine, choline and saponin (Morgan and Moynihan 2000).

11.2 PROPERTIES OF ANTIFOAMING AGENTS

Defoamers or antifoaming agents are surface-active molecules which decrease the surface elasticity of liquids, thereby preventing the foam to attain a state of equilibrium between the surface elasticity and the antifoaming agent. This will destabilise the foam and interfere with the foam formation process.

Antifoaming agents include products made from oils, fatty acids, esters, polyglycols and siloxanes, alcohols, sulphites and sulphonates, although their antifoaming strength and properties vary (Currie 1953; Prins and Van Riet 1987). For any antifoaming agent to be considered suitable for use in the industrial production processes of food products, it must also be suitable for use with a living system without causing any harm or even having a potential harmful effect. It must also not react or interfere in any way with any of the analytical processing or measuring mechanisms such as probes or meters. All defoamers in bioprocesses must have the qualities to withstand the sterilisation conditions and retain their integrity.

11.3 MECHANISMS OF ANTIFOAMING AND FOAM DESTABILISATION

The mechanisms of defoamers are dependent on a number of factors such as nature of the antifoaming compound itself (its chemistry), the type of foam and the nature of the substances causing foaming (Kulkarni *et al.* 1977). Moreover, a surface-active defoamer molecule may destabilise foams using various mechanisms.

For example, antifoaming agents may act as hydrophobic bridges between the respective phases of film surfaces and thus discontinue the foam formation process (Van-Riet *et al.* 1984). When a hydrophobic defoamer is at the interface between the two respective liquid phases, it triggers the formation of a convex surface in the foam itself. This convex film will cause the liquid phases to flow away from the antifoamer (which now behaves as a hydrophobic bridge) because of the build up under the convex curved surface, which automatically causes the destabilisation and collapse of the foam (Figure 11.1). This mechanism can only work in cases where the surface tension of the foam exceeds that of the sum of the surface tension at the interface between the two liquid phases in contact (Van-Riet *et al.* 1984).

Another possible mechanism of foam destruction involves the displacement of the adsorbed protein molecules which cause the foam generation on the foam film surface, thus interrupting the stabilising action of the protein. This mechanism disturbs the equilibrium on the foam film, causing the destruction of the foam.

Another mode of defoamers spreads rapidly over the surface of the foam/film to squeeze the liquid, thus destroying the foam.

11.4 SYNTHETIC DEFOAMERS

Aqueous dispersions of antifoaming agents that display a high antifoaming and a long-term activity over a wide temperature range are most ideal for use as antifoaming agents in the food industry. The majority of such dispersions contain alcohols, which are solid at room

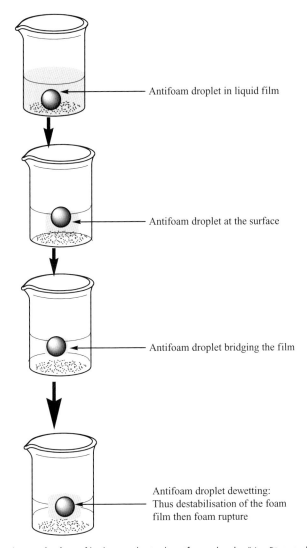

Antifoam droplet in liquid film

Antifoam droplet at the surface

Antifoam droplet bridging the film

Antifoam droplet dewetting:
Thus destabilisation of the foam
film then foam rupture

Fig. 11.1   The bridging of a foam film by a spherical antifoam droplet (Van-Riet *et al*. 1984).

temperature, and glycerol polyglycol ethers or polyglycerol polyglycol ethers which are liquid at room temperatures.

The most commonly used types of defoamers in bioprocesses include those which are silicone based, oil based, water based and silicone surfactants (e.g. polydimethylsiloxane (pDMS), polyoxyethylene (pEO) and polyoxypropylene (pPO) glycol based).

11.4.1   Silicone-based defoamers

These surface-active substances contain silicone compound as the active ingredient. These defoamers have excellent qualities in terms of knocking down surface foam as well as releasing the entrained air. The structural chemistry of silicones (Figure 11.2) is distinctive

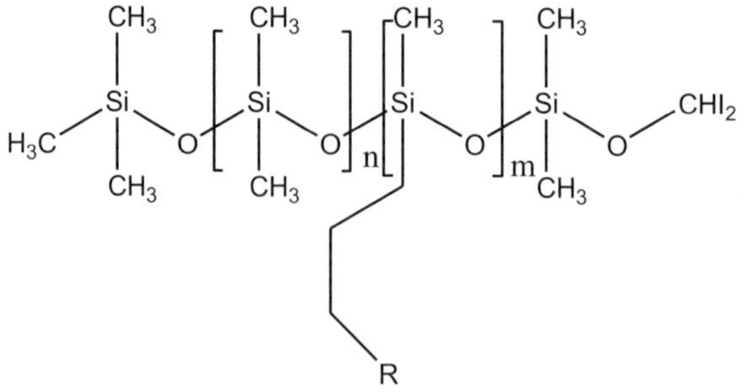

Fig. 11.2   General structure of silicone surfactant (copolymers of silicone and polyethers). R represents hydrophilic modification. For monomeric surfactant, $n=0$ and $m=1$ (Brandrup and Immergut 1989).

in that the Si–O–Si bond angles are larger than C–O–C bond angles; the Si–O bond lengths are also longer than C–O–C or C–C bonds. Moreover, there is a greater freedom of rotation around the Si–O bond than that which exists for C–C bond and also a freely rotating methyl groups which can orient towards interfaces (Brandrup and Immergut 1989).

Other attributes of silicones which qualify them as ideal defoamers include the fact that they are water repellent, heat stable and highly resistant to chemical attack. Silicones are also insoluble in both hydrophilic and hydrophobic solvents, but when dissolved in these solvents they only form a third phase which makes them very suitable as defoamers (O'Lenick 2000).

11.4.2   Oil-based defoamers

Oil-based surface-active antifoaming agents are known to contain an oil carrier. Types of oils suitable for this role are those which are insoluble in the phases where foam may be generated, for example, mineral oil or vegetable oil.

11.4.3   Silicone surfactant defoamers

Silicone surfactant (copolymers of silicone and polyethers) types of defoamers consist of a permethylated siloxane hydrophobic group (mainly polydimethylsiloxane or pDMS). This hydrophobic group is coupled to one or more polar groups such as polyoxyethylene (pEO) or polyoxypropylene (pPO), which contain polyethylene glycol and polypropylene glycole copolymers (Hill 2002).

11.4.4   Other chemical defoamers

Other chemical defoamers include the mixed hydroxyethers which are synthesised using epoxides of unsaturated fatty acid esters (as raw materials) and using polyglycol ethers as ring opening agents.

11.5   NATURAL DEFOAMERS

As well as synthetic antifoamers, there are also defoamers of natural origin such as the quillaia extracts. Quillaia extracts can be found naturally by performing aqueous extraction of the bark *Quillaja saponaria* molina plants (Leung 1980). Compounds obtained from aqueous extracts of quillaia include saponins, tannins, polyphenols and calcium oxalate, although polyphenols and tannins form the major components (Kensil *et al.* 1991). Quillaia extracts have also found use as foaming agents in soft drinks and other food items (WHO 1987; Gibney 1999).

REFERENCES

Brandrup, J. & Immergut, E. H. (1989) *Polymer Handbook*, 3rd edition. John Wiley & Sons, Inc., New York.

Currie, C. C. (1953) *Foams: Theory and Industrial Applications*. Reinhold, New York.

Fruhner, H., Wantke, D. & Lunkenheimer, K. (1999) Relationship between surface dilational properties and foam stability. *Colloids and Surfaces A: Physicochemical and Engineering Aspects* 162, 193–202.

Gibney, M. J. (1999) Dietary intake methods for estimating food additive intake. *Regulatory Toxicology and Pharmacology* 30, S31–S33.

Hill, R. M. (2002) Silicone surfactants new developments. *Current Opinion in Colloid and Interface Science* 7, 255–261.

Kensil, C. R., Patel, U., Lennick, M. & Marciani, D. (1991) Separation and characterization of saponins with adjuvant activity from Quillaja saponaria molina cortex. *Journal of Immunology* 146, 431–437.

Kulkarni, R. D., Goddard, E. D. & Kanner, B. (1977) Mechanism of antifoaming action. *Journal of Colloid and Interface Science* 59 (3), 468–476.

Leung, A. Y. (1980) *Encyclopedia of Common Natural Ingredients used in Foods, Drugs, and Cosmetics*. John Wiley & Sons, Inc., New York.

Morgan, B. P. & Moynihan, M. S. (2000) Steroids. In: *Kirk-Othmer Encyclopedia of Chemical Technology*, John Wiley & Sons, VCH Verlag.

O'Lenick, A. J. (2000) Silicone emulsions and surfactants. *Journal of Surfactants and Detergents* 3 (3), 387–393.

Prins, A. & Van-Riet, K. (1987) Proteins and surface effects in fermentation: Foam, antifoam and mass transfer. *Trends in Biotechnology* 5, 296–301.

Van-Riet, K., Prins, A. & Nieuwenhuijse, J. A. (1984) Some effects of foam control by dispersed natural oil on mass transfer in a bubble column. Third European Congress of Biotechnology, Munchen, Germany 3, 521–525.

WHO (1987) Guidelines for the study of dietary intakes of chemical contaminants. World Health Organization (WHO) Publication No. 87, Geneva, p. 54.

12 Minerals and Mineral Salts

Abstract: Not all food additives are necessarily organic in nature or are directly nutritive. Some, like the majority of inorganic mineral salts, are nutritional additives in the sense that their deficiency in our bodies leads to health problems such as anaemia (deficiency of iron) or goitre (iodine deficiency). Mineral salts also perform the function of preservatives (table salt or NaCl) and can also be used as a condiment and as a flavouring in many foodstuffs. Organic salts also function as food additives as well as flavourings, antioxidants, acidity regulators and emulsifying agents. This chapter focuses on the chemistry of both inorganic and organic mineral salt as food additives.

Keywords: inorganic mineral salts; organic mineral salts

12.1 THE IMPORTANCE OF MINERALS AND MINERAL SALTS

12.1.1 Minerals

Minerals are very important in our bodies for growth processes, development and maintenance of body structures. Other functions that minerals play include mobilisation and maintaining of the digestive juices and the fluids present within and outside somatic cells.

Unlike plants, human bodies do not make minerals (cations and anions) even though minerals are needed in our bodies for many metabolic functions. The source of minerals in our bodies is the plants we eat either directly or indirectly by eating meat and other meat products. Among the minerals which our bodies need are: calcium, chlorine (chloride), potassium, phosphorus, sulphur sodium, magnesium, manganese, selenium, iron, zinc, fluorine (fluoride), copper, iodine, cobalt, molybdenum, chromium and zinc.

Calcium is important for inducing bone strength, blood clotting (agglutination) and in the proper functioning of glands. Magnesium also performs some of the functions performed by calcium, such as its role in the formation of bones and the proper functioning of nerves.

Selenium serves as an antioxidant, is at the core of glutathione peroxidase (GPx-4 enzymes) and plays an important function in the muscles surrounding the heart. In GPx-4, selenium forms part of one of the amino acids known as selenocysteine, placed co-translationally along the bases sequence uracil-guanine-adenine (UDA) on the respective mRNAs.

Iron forms part of the haemoglobin, a red pigment that gives blood its red characteristic colour. This haemoglobin is essential in the distribution of oxygen in the body. Lack of iron

Chemistry of Food Additives and Preservatives, First Edition. Titus A. M. Msagati.
© 2013 John Wiley & Sons, Ltd. Published 2013 by John Wiley & Sons, Ltd.

in the body is characterised by a deficient disease disorder known as anaemia. Iron is also important in the processes involved in the production of energy.

Copper is also an important part of the haemoglobin, and it is also essential in the production of collagen (thus important in muscles). It also plays an important part in the functioning of the heart as well as in the production of energy.

Phosphorus is important in the generation of energy (adenosine triphosphate or ATP, adenosine diphosphate ADP) and in phosphorylation mechanisms, and is present in bones and in the DNA (DNA sugars are phosphates).

Cobalt forms part of B vitamins, for example, cobalamin (B12).

Iodine is essential for the proper functioning of the thyroid glands. A lack of iodine in the body causes a deficient disease known as goitre.

Chromium is used in the formulation treatment for diabetes, and is essential in processes involving the metabolism of carbohydrates, especially when working in conjunction with a hormone insulin.

Zinc plays an important function in a number of reproduction processes.

12.1.2   Mineral salts

A number of elements are needed to supplement our diets. These are an important source of the highly needed nutrients in our bodies, and also prevent the occurrence of some food-deficiency-related diseases. The elements referred to in this category include both the macro as well as the micro elements. The salts of most of the macro elements, which include sodium, potassium and calcium, play other roles such as antioxidants, emulsifiers and stabilisers. The mineral salts of the micro elements, also known as the essential elements, include iodine, iron, fluoride and zinc, which are actually used as food supplements.

Both inorganic and organic mineral salts are very important as food additives. Inorganic mineral salts are derived from mineral acids, while organic salts are derived from organic acids.

12.2   INORGANIC MINERAL SALTS

A number of inorganic mineral salts such as sodium chloride, as well as the phosphate salts of sodium, potassium, calcium, ammonium and magnesium, are used in food items for various purposes or as supplements. The majority of these are salts of mineral acids such as hydrochloric acid (chlorides) and phosphoric acid (phosphates).

12.2.1   Sodium chloride (table salt)

Table salt or sodium chloride (NaCl) is a common additive to many foodstuffs where it plays many roles such as flavouring agent, but also as a preservative agent against microbial attack of foodstuffs under storage. However, a number of scientific research studies have indicated the need to reduce salt (NaCI) intake by humans (Expert Panel 1980; Sebranek *et al*. 1983; Shank *et al*. 1983). In these studies it has been observed that excessive intake of NaCl has negative consequences health-wise and a reduction of NaCl in diets has been recommended (Thayer *et al*. 1987).

12.2.2 Phosphate salts

These are mainly salts of phosphoric acid and they play a number of important roles as food additives/supplements.

12.2.2.1 Mono-, di- and trisodium phosphate salts

Monosodium phosphate salts are used as food additives in food items such as cheese, meat and meat products, where they serve as antioxidants, stabilisers and as buffering media. An excessive intake of these salts is known to disturb the calcium/phosphorus equilibrium in the body. Disodium phosphate salts are used as emulsifiers in food items such as cheese, in addition to the roles also played by the monosodium phosphate salts. These salts are also included in dairy products, especially powdered milk where they prevent gelation. Trisodium phosphates are added to food products such as meat where they serve as antioxidants, stabilisers and buffering agents.

12.2.2.2 Mono-, di- and tripotassium phosphate salts

Phosphate salts of phosphoric acid perform similar functions to that of the phosphate salts of sodium.

12.2.2.3 Phosphate salts of calcium

Like sodium and potassium, inorganic phosphate salts of calcium exist as mono-, di- and tricalcium phosphate species. The anhydrous and monohydrate forms of calcium phosphate salts serve as leavening acids or as an ingredient of baking powder for bakery products. Dicalcium phosphate salts which also exist in the anhydrous as well as the dehydrate forms are used in the food industry as antioxidants. Tricalcium phosphate salts serve as humectants in table salts, sugar and baking powder.

12.2.3 Salts of magnesium

Magnesium and its salts are vital in our bodies as they take part in many metabolic and biochemical reaction processes, for example: in the energy metabolism reactions; synthesis of nucleic acids, amino acids or proteins; in cell reproduction; and in the transmission of nerve impulses. Magnesium salts are added to foodstuffs as supplements to prevent disorders such as nervousness, depression and anxiety.

Inorganic phosphate salts of magnesium (mono- and dimagnesium phosphate) play the role of anticaking in food industries.

12.2.4 Salts of ethylene diamine tetra-acetic acid (EDTA)

Synthetic salts of EDTA, mainly sodium and calcium EDTA salts, are also used as food additives. Disodium ethylenediamine tetra-acetate is used as food stabiliser as well as a remedy in cases of metal intoxication because it is a good chelating agent (hence can remove metals from the body). However, long-term exposure to high doses of disodium-EDTA may be detrimental as it may render the depletion of metal(s) such as iron from the body and cause metal deficiency disorders such as anaemia.

Calcium-EDTA is used as a preservative and flavouring, among other uses. However, its use as a food additive is discouraged from a health point of view as it has been associated with inhibition of enzyme and blood coagulation.

12.3   ORGANIC MINERAL SALTS

These are salts of mainly organic acids with metals such as sodium, potassium, calcium and magnesium. Like their inorganic counterparts, they play a variety of roles in the food industry either as food additives or supplements.

12.3.1   Magnesium

Examples of organic salts of magnesium that may be used as essential minerals, as well as food supplements in dairy products and fruit juices, include: dihydrate magnesium lactate $(Mg(CH_3CHOHCOO)_2.2H_2O)$; dimagnesium phosphate trihydrate $(MgHPO_4.3H_2O)$; and trimagnesium phosphate pentahydrate $(Mg_3(PO_4)_2.5H_2O)$. Other magnesium organic salts are: trimagnesium citrate nonahydrate $(Mg_3(C_6H_5O_7)_2.9H_2O)$; anhydrous magnesium carbonate $(MgCO_3$ or $4MgCO_3 Mg(OH)_2.4H_2O)$; magnesium chloride $(MgCl_2.6H_2O)$; magnesium sulphate heptahydrate $(MgSO_4.7H_2O)$; and magnesium gluconate $(Mg(CH_2OH(CHOH)_4COO)_2)$.

12.3.2   Citrate

These are mainly salts of citric acid. Those used in the food industry include ammonium ferric citrates which are added to foods as food acids, essential minerals and dietary iron supplements at recommended levels (if ingested at levels beyond prescribed thresholds, they become a health hazard). Tri-ammonium citrate salts also functions as food acids within the threshold limits stipulated; above these levels they are known to harm the functioning of the liver and the pancreas.

12.3.3   Fumaric acid

Organic salts of fumaric acid, such as calcium fumarate and potassium fumarate, are used as food acids and also function to regulate the acidity in foodstuffs. Sodium fumarate also finds use in the bakery industry to strengthen the bread dough.

12.3.4   Tartaric acid

Tartaric acids serve as food acids. They include sodium tartrates, calcium tatrates, potassium tartrates, potassium sodium tartrate and sodium potassium tartrate.

12.3.5   Malic acid

Malate salts of sodium, potassium and calcium serve as a flavouring as well as seasoning agent in soft drinks and dairy products.

12.3.6 Ascorbic acid

These include sodium, calcium, potassium, ascorbyl palmitate and ascorbyl stearate, and are used in the food industry as antioxidants, preservatives and as colouring agents. Ascorbyl palmitate and ascorbyl stearate are esterification products of palmitic acid and ascorbic acid and stearic acid and ascorbic acid, respectively. The two esters are used in food products as sources of vitamin C.

12.3.7 Other organic salts

There are many other salts of organic acids that are used in food industry to perform very similar functions to those mentioned here, for example, the organic salts of lactic acid which include sodium lactate, potassium lactate, calcium lactate, ammonium lactate and magnesium lactate. Others include the organic salts of citric acid such as the sodium citrates, potassium citrates and calcium citrates.

REFERENCES

Expert Panel on Food Safety and Nutrition and the IFT Committee on Public Information. (1980) Dietary salt; a scientific status summary. *Food Technology* 34 (1), 85–91.

Sebranek, J. G., Olson, D. G., Whiting, R. C., Benedict, R. C., Rush, R. E., Kraft, A. A. & Woychik, J. H. (1983) Physiological role of dietary sodium in human health and implications of sodium reduction in muscle foods. *Food Technology* 37 (7), 51–59.

Shank, F. R., Larsen, L., Scarbrough, F. F., Vanderveen, J. E. & Forbes, A. L. (1983) FDA perspective on sodium. *Food Technology* 37 (7), 73–77.

Thayer, D. W., Muller, W. S., Buchanan, R. L. & Phillips, J. G. (1987) Effect of NaCl, pH, temperature, and atmosphere on growth of salmonella typhimurium in glucose-mineral salts medium. *Applied and Environmental Microbiology* 53 (6), 1311–1315.

13 Dietary Supplements

Abstract: The deficiency of certain components in regular diets is known to cause some disease conditions or disorders. Human diets should be complete with all the ingredients to supply energy and support growth, nutrition and health. When a regular diet is deficient in some of the food/nutritious components necessary for the body, a common tendency is to artificially add them to diet either directly or indirectly. This introduces the wide area of what is generally known as dietary, food or nutritious supplements. These are normally added to either foodstuffs themselves, such as vitamins in baby formula milk or some minerals e.g. iron, selenium, etc. in maize meal (flour), or they may be contained in food additives, for example iodine in table salts. In this chapter, a detailed discussion of vitamins as food supplements if provided. Other types of supplements, such as iodised salts, probiotic dietary supplements, prebiotics and synbiotics, are covered in chapters 15, 18, 19 and 20, respectively.

Keywords: artificial/synthetic food supplements; complete diet; fat soluble vitamins; natural food supplements; water-soluble vitamins

13.1 INTRODUCTION TO DIETARY SUPPLEMENTS

As their name suggests, dietary supplements only augment a diet which may be deficient in some items. They are not whole or complete food by themselves, but are only food components. Examples of dietary supplements include vitamins, minerals, fibre, fatty acids and amino acids. Deficiencies arise when certain elements are completely missing or are only present in small amounts in a regular diet. Supplementation in the diet is therefore essential to increase the levels of a particular metabolite, simply to increase the total daily intake or both (Zeisel 1999; Kalra 2003).

13.1.1 Botanical and herbal dietary supplements

Botanical and herbal food supplements are those which are obtained mainly from plant or natural origin and include: soy extracts which are rich in isoflavones; tomato extracts which contain lycopene compounds; ginseng which is extracted from the plant species known as *Ginkgo biloba*; St John's wort which is extracted from *Hypericum perforatum*; and phytostanols which are a by-product of wood and are important in reducing the risk of high cholesterol levels. Herbal materials may also be a source of dietary supplements, for

Chemistry of Food Additives and Preservatives, First Edition. Titus A. M. Msagati.
© 2013 John Wiley & Sons, Ltd. Published 2013 by John Wiley & Sons, Ltd.

example extracts from some spices and herbs such garlic oil, rosemary and green tea are used as dietary supplements.

13.1.2 Vitamin dietary supplements

Human bodies are unable to synthesise a large enough quantity of vitamins; these therefore have to be supplemented through the diet. Vitamins are essential to human beings because of their involvement in a number of important biological and biochemical processes, for example vision and reproduction. They also act as mediators for cell regulation, tissue growth and differentiation (Olson 1984; Blomhoff 1994; Basu and Dickerson 1996; Napoli 1999). The function performed by each vitamin is unique; even vitamins in the same class do not necessarily have similarities in terms of their functions. In many other processes they play a crucial role as catalysts, coenzymes and transportation media in many metabolic pathways and also aid the metabolic activities taking place inside the body. Vitamin A is also crucial for proper gestation and offspring development (Clagett-Dame and DeLuca 2002) and for development and efficient performance of the body's immune system (Semba 1998).

   If vitamins are not supplied in sufficient quantities by the diet, deficiency disease syndromes can be instigated. The deficiency from water-soluble vitamins is always more serious than the deficiency due to fat-soluble vitamins, because water-soluble vitamins (C and D) are not stored in the body. Common vitamin deficiency diseases include scurvy (deficiency of vitamin C) or beriberi (deficiency of vitamin B1), although an excess of vitamins – especially the fat-soluble vitamins (A, D, E, K) – is also detrimental to the body. Vitamin A deficiency can lead to serious problems such as fetal resorption, stillbirth and malformation (Clagett-Dame and DeLuca 2002).

13.2 CLASSIFICATION OF VITAMINS

13.2.1 By class

There are 13 classes and subclasses of vitamins, including:

- vitamin A also known as retinol or retinoic acid;
- vitamin B complex, which is subdivided into the eight subclasses: vitamin B1 (thiamine-a thio containing vitamin); vitamin B2 (riboflavin); vitamin B3 (a nicotinic acid vitamin niacin); vitamin B5 (pantothenic acid or pantothenate); vitamin B6; vitamin B7 (biotin); vitamin B9 (folic acid or vitamin B_c or folacin); vitamin B12 (cobalamin);
- vitamins C (L-ascorbic acid or L-ascorbate);
- vitamin D, which exists in two forms: D2 (ergocalciferol) and D3 (cholecalciferol);
- vitamin E (tocopherols and tocotrienols); and
- vitamin K with its two natural forms: K_j (phylloquinone or phytomenadione or phytonadione) and vitamin K2 (menaquinone).

13.2.2 By solubility

Based on solubility, vitamins are classified as either fat soluble or water soluble. The naturally fat-soluble vitamins include vitamins A, D, E and K (Sections 13.3–13.6), while vitamins B,

C (Sections 13.7 and 13.8) and H are water soluble. The group to which a vitamin belongs reveals a lot about how it acts as well as its retention or elimination in the body.

Vitamins A, D, E and K are hydrophobic; they are normally concentrated and stored in fat tissues known as chylomicrons which are lipoprotein molecules composed mainly of triglycerides and a small percentage of phospholipids, cholesterol and other forms of proteins (Hussain 2000). Chylomicrons play a very important role in transporting dietary lipids from the intestines to other tissues and parts of the body. Fat-soluble vitamins can be stored in our bodies, and the advantage of this is that it eliminates the necessity for these vitamins to be supplied on a daily basis. The main sources of fat-soluble vitamins include vegetables and animal products.

Unlike fat-soluble vitamins and water, water-soluble vitamins are not stored in the body system. They are used immediately after being ingested and must be constantly supplemented through diet or other means. Since these vitamins cannot be stored, the excess is normally eliminated out of the body via urine. These water-soluble vitamins play different roles in the body, as discussed in Sections 3.7 and 3.8. The best source of such vitamins in our bodies is from a varied diet rich in fruits, vegetables and whole grains.

13.3 VITAMIN A (RETINOLS)

Vitamin A is the collective generic name for a number of chemical substances such as a diterpenoid and an alcohol retinol (2E,4E,6E,8E)-3,7-dimethyl-9-(2,6,6-trimethylcyclohex-1-enyl)nona-2,4,6,8-tetraen-1-ol), an aldehyde retinal (also known as retinaldehyde, retinoic acid and retinyl esters e.g. retinyl acetate and retinyl palmitate) and provitamin A carotenoids (e.g. β-carotene). These vitamin A derivatives have different sources; for example, retinol (Figure 13.1a) and its derivatives such as retinoic acid, retinyl esters etc. (Figures 13.1b–e) are of exclusively animal origin as opposed to β-carotene which is a plant derivative (Olson 1984; Blomhoff 1994; Basu and Dickerson 1996).

The functional dietary form of vitamin A normally takes the form of its β-carotene or retinyl esters. These undergo hydrolytic conversion in the lumen of intestines by the action of pancreatic and intestinal enzymes to yield retinol, which will then be absorbed through the intestinal lumen. The process of retinyl esters conversion to yield retinol is bidirectional, as retinol can also be converted to yield its aldehyde forms. This process is important in the body as it leads to the formation of the stable ester form of vitamin A from the unstable alcohol (retinol) form.

13.3.1 Structure-activity relationship

The chemical structure of vitamin A is characterised by the presence of two specific moieties: the ionone rings (Figure 13.2a–c) with an isoprenoid side (Figure 13.2d), commonly known as a retinyl group, attached to it.

The presence of these structural features is a prerequisite for vitamin activity. Studies have shown that food items known to be sources of vitamin A such as carrots (α and β-carotene) have these structural features associated with the vitamin activity; carotene without these features were unable to show any vitamin A activity (Berdanier 1997).

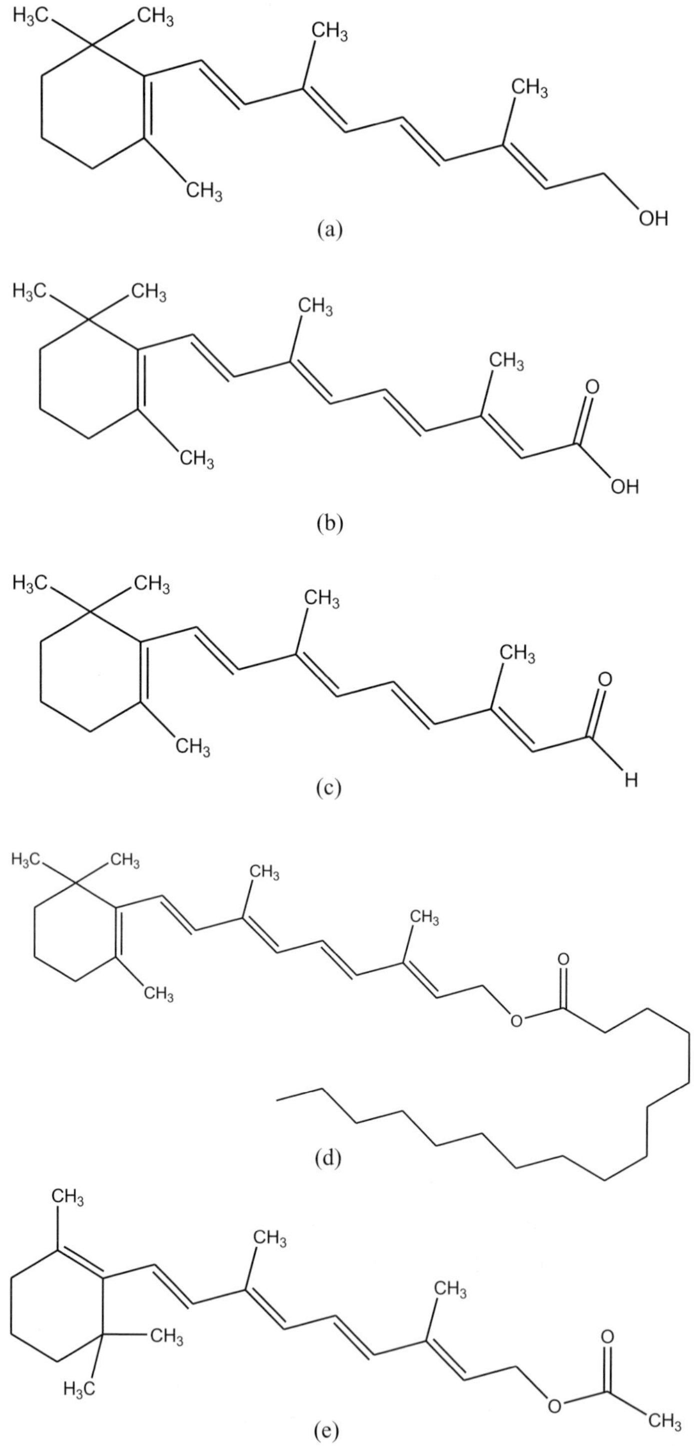

Fig. 13.1 Chemical structure of (a) retinol; (b) retinoic acid; (c) retinaldehyde; (d) retinyl palmitate; and (e) retinyl acetate (Olson 1984; Blomhoff 1994; Basu and Dickerson 1996).

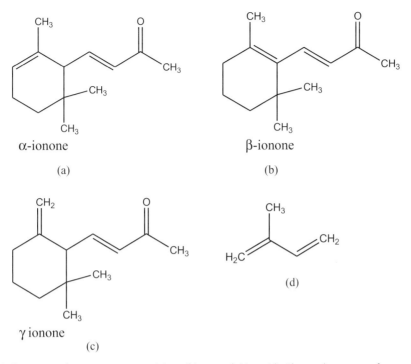

α-ionone

(a)

β-ionone

(b)

γ ionone

(c)

(d)

Fig. 13.2 Forms of ionone structures: (a) α; (b) β; and (c) γ. (d) Chemical structure of isoprene unit (Berdanier 1997).

13.3.2 Functions and mechanisms of vitamin A

The functions of vitamin A in the body fall within two classes: (1) vision and (2) systemic functions such as cellular differentiation, growth, bone development, immune system and reproduction. The mechanism of vitamin A in vision is different to that of its role in systemic functions.

13.3.2.1 Vitamin A mechanisms for vision

The retinol transported to the retina of the eye through the retinol binding protein–transthyretin (RBP-TTR) complex is converted to retinyl esters and stored (see Section 13.3.6). The stored retinyl esters will then be hydrolysed to form retinol, which will be oxidised in the photoreceptors cells (known as rod cells of the retina) to form all-trans retinaldehyde. The retinaldehyde in the retina exists in two main isomeric forms: the all-trans retinaldehyde and 11-cis retinaldehyde. The two isomers of retinaldehydes establish a state of equilibrium (Scheme 13.1).

The 11-cis retinaldehyde in the eye is incorporated into the visual pigment where it combines with the protein opsin to form a pigment known as rhodopsin (Figure 13.3). The photosensitive cells (rod cells) together with rhodopsin are capable of functioning even when there is only a small amount of light or at night.

As a result of all the compounds formed through this process, when light strikes the rhodopsin in the retina of the eye the retinaldehyde in the retina changes shape and there is an

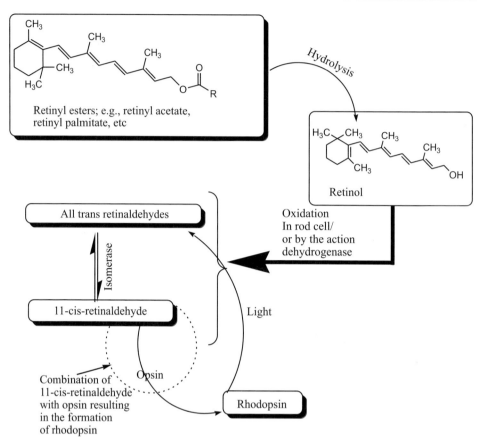

Scheme 13.1   The role of retinol and its derivatives in the vision process (Pozniakov 1986).

isomerisation reaction of 11-cis retinaldehyde with all-trans retinaldehyde. The isomerisation and conformational changes in the retina are catalysed by light-activated enzymes present in the retina. The processes will then trigger an electrical signal to the optic nerve, which causes light to be perceived. The all-trans retinaldehyde formed due to the light reaction will then be converted back to 11-cis retinaldehyde which combines with the protein opsin to form rhodopsin, hence completing the cycle.

The retinaldehyde form of vitamin A aids vision, while the 11-cis retinaldehyde binds the rods (rhodopsin) and cone (iodopsin) cells of the eye located at the residue of the amino acid lysine. When light strikes these cells, the 11-cis retinaldehyde is converted to its isomeric form of all-trans retinaldehyde. This then undergoes dissociation in a stepwise process known as photo-bleaching, resulting in the generation of stimuli from the optic nerve. These stimuli are sent to the vision centre located in the brain. This enzymatically controlled process will repeat itself over and over again.

13.3.2.2   Vitamin A mechanisms for systemic functions

The possible mechanism by which vitamin A controls systemic processes such as cell proliferation, differentiation and functioning may be through its influence on: the regulation

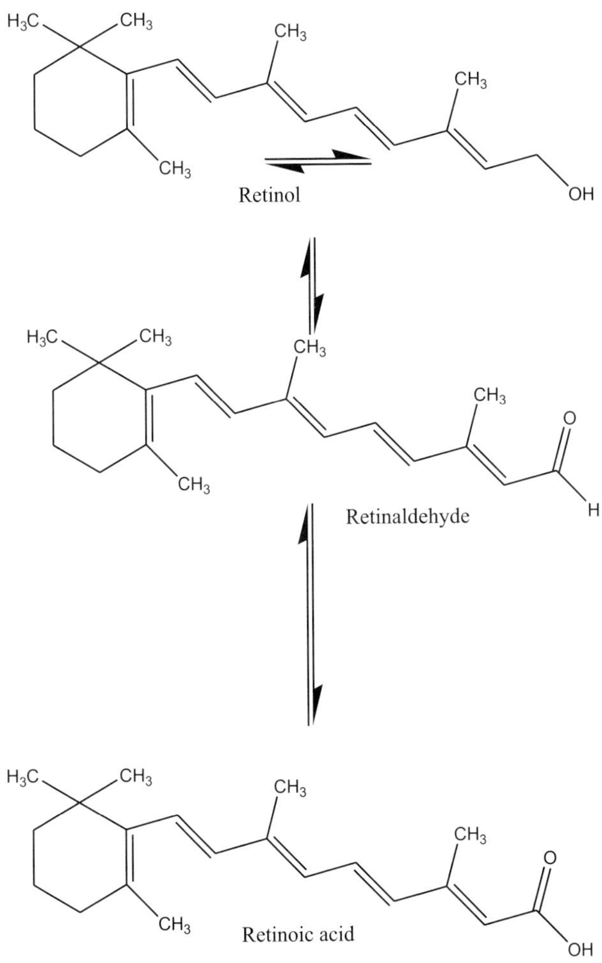

Fig. 13.3   Relationship between various forms of vitamin A (Pozniakov 1986).

of gene expression inside the nucleolus during the transcription process of some genes; processes which occur after the transcriptional modification of mRNA levels; or modifications in the membrane structural-functional organisation (Pozniakov 1986). Vitamin A is also known to influence the biosynthesis proteins, responsible for the regulation of development and cell functioning. Without vitamin A, the processes involving development and cell functioning are hindered. Moreover, vitamin A dictates the formation of proteins which functions like hormones. Vitamin A also has control over the sensitivity of cells to hormones and hormone-like factors (Pozniakov 1986).

Vitamin A has been linked to the process of cellular differentiation from the fact that retinoic acid has been reported to function as a hormone with the ability to induce gene expression and control of cell differentiation (Scheme 13.2). The mechanism by which vitamin A influences growth is not yet known.

The different classes of plant carotenoid compounds have been reported to have provitamin A activity as well as antioxidant activity, thus playing a very important role in immune

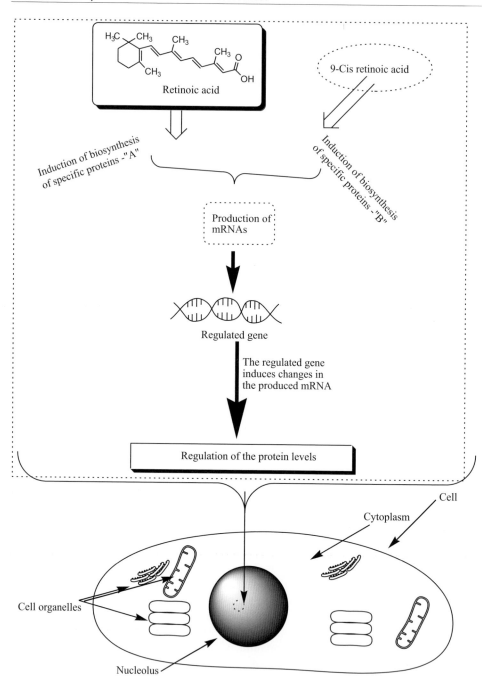

Scheme 13.2 The role of vitamin A in cell differention processes (Pozniakov 1986).

responses, gap junction communication and carcinogen-metabolising enzyme activity (Olson 1989; Wang 1994; Stahl *et al.* 1997).

13.3.3   Relationship of vitamin A to other dietary nutrients

Vitamin E is necessary for the cleavage of β-carotene to retinaldehyde, and is believed to play an important role in protecting the substrate as well as the products from undergoing oxidation reaction.

The metabolism of vitamin A is dependent on protein status, which is linked to the activity of the enzyme which cleaves β-carotene (carotenoid dioxygenase). The transport of vitamin A to other tissues is dependent upon a number of other vitamin-binding proteins. If these proteins are deficient, retinaldehyde in the liver cannot be distributed properly to other tissues and vitamin binding will not take place.

Zinc deficiency also affects the metabolism of vitamin A, and will lead to a general reduction in growth as well as a reduction in plasma protein synthesis (especially RBP). Zinc deficiency also results in a decreased mobilisation of free retinaldehyde which is stored in the liver, because the enzyme retinyl ester hydrolase becomes disturbed. Moreover, retinol requires a zinc-dependent enzyme known as alcohol dehydrogenase for its oxidation to retinaldehyde.

A decreased mobilisation of iron from the liver which is associated with microcytic anaemia has been linked to iron deficiency. The correlation between the amount of retinol in the serum and the concentration of blood haemoglobin means that individuals who suffer low levels of blood vitamin A have an increased risk of developing symptoms of iron deficiency or anaemia. There is also a direct relationship between vitamin A intake and concentration of blood haemoglobin.

13.3.4   Assessment of vitamin A status in individuals

Plasma retinol is normally used for the assessment of vitamin A status, even though no simple measurements of vitamin A nutriture exist. Plasma retinol levels are however constant in most cases, except in cases of severe deficiency of vitamin A, where the liver store is exhausted or (rare) where excessive intake of vitamin A means that it can no longer be stored in the liver.

A plasma level of <10 μg/dL is classed as a deficient condition, $10–20$ μg/dL implies marginal and >20 μg/dL is acceptable.

13.3.5   Chemistry of vitamin A and carotenoids

Vitamin A (sometimes referred to as preformed vitamin A) exists in a variety of chemical forms such as: alcoholic (retinol); aldehyde (retinal/retinaldehyde); retinoic acid (a metabolite of retinal); and as esters (retinyl esters). Retinol and retinal exist in equilibrium (Figure 13.3) and are very important as far as vision is concerned, while retinoic acid is not active in vision. Retinol is found mainly as fatty acid esters such as retinyl palmitate. Foodstuffs which are rich in retinyl esters include egg yolk, butter, whole milk products, liver and fish liver oils.

β-carotene and other classes of carotenoids found exclusively in plants and which can be converted to retinol are known as provitamin A. Such compounds are known as vitamin A precursors (Figure 13.4; Khachik 2006). Of the carotenoid compounds, β-carotene provides

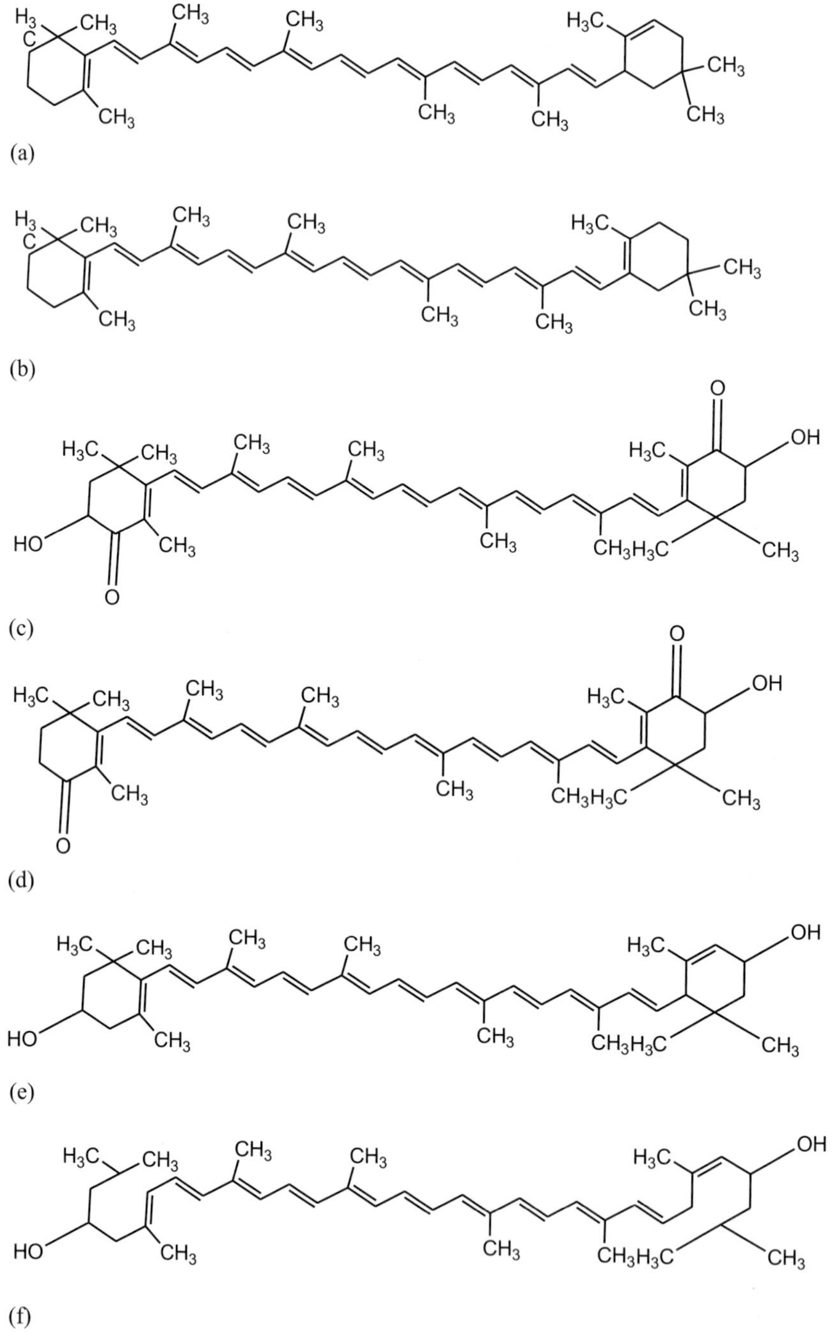

Fig. 13.4 Chemical structures of various carotenoid compounds found in common diets: (a) α-carotene; (b) β-carotene; (c) astaxanthin; (d) canthaxanthin; (e) lutein; and (f) lycopene (Khachik 2006).

the greatest amount of vitamin A activity. Other forms of carotene such as α-carotene and γ-carotene provide about 40–50% activity of that of the β-form.

13.3.6 Conversion of β-carotene to retinol

The process of β-carotene conversion to retinol takes place in the intestinal mucosa as well as in the liver. The β-carotene molecule is cleaved enzymatically at 15,15' position by the action of oxygenases to yield a retinaldehyde molecule (Figure 13.5). Some retinaldehyde molecules can then be reduced to form retinol, while others are oxidised to form retinoic acid. The kinetics of this enzymatic process is very slow (i.e. the activity of the enzyme is low) is such that some of the β-carotene may leave the intestine without being converted to retinol. Ongoing research is attempting to establish the actual amount of carotenoid needed to give a specified unit of vitamin A (retinol). Mathematically, it has been estimated that it requires 6 μg of β-carotene to produce 1 μg of retinol, which is taken as one retinol equivalent (1 RE; USDA 2008).

If a diet includes the carotenoid compounds provitamin A and vitamin A precursors, these will mix with other food ingredients and the retinyl esters (egg retinyl palmitate) will become bound to protein. They will then be hydrolysed enzymatically in the lumen of the intestines and then converted into 'free' retinal by the action of esterases and lipases.

The other dietary carotenoids (Figure 13.5), together with free retinol, lipids and bile, will be incorporated into micelles and diffuse into enterocytes. The retinol will then esterify the mucosal cellar into retinyl esters (e.g. retinyl palmitate) which, together with the unchanged carotenoids, will be incorporated into chylomicrons and carried via lymph fluids to the general circulation. In the general circulation, the chylomicrons will deliver retinyl esters and the unchanged carotenoids to the liver. Carotenoids which do not follow this route will be stored.

In the liver, retinyl esters are hydrolysed into retinol which combines with cellular retinol binding protein (CRBP), a form which can be transported via the circulation system to other tissues. It is the form which can be stored in specialised liver cells known as stellate cells. When the storage capacity of the stellate cells is exhausted, a condition known as hypervitaminosis A occurs; otherwise, the concentration of vitamin A in the plasma is constant for many different types of diet.

The rate of mobilisation and transportation of RBP from the liver to other body tissues is dependent on their synthesis and secretion by parenchymal cells located in the liver. The stored retinol in the liver reacts with RBP to form holo-RBP, which is then secreted into the blood stream. It should be noted that in plasma each molecule of holo-RBP is bound to a molecule of transthyretin (TTR), also known as prealbumin, which then forms a complex with retinol (RBP-TTR). During transportation of vitamin A, some tissues (e.g. adipose tissue, skeletal muscles, white blood cells and bone marrows) can take retinol from this complex.

13.3.7 Toxicity of vitamin A

The fact that vitamin A compounds such as retinol, retinaldehyde and retinoic acid are fat-soluble implies that their rate of withdrawal from the body is very low; getting rid of any surplus of these compounds ingested through diets rich in carotenoids is not easy. The body has a storage provision for excess vitamin A compounds, otherwise elevated concentrations

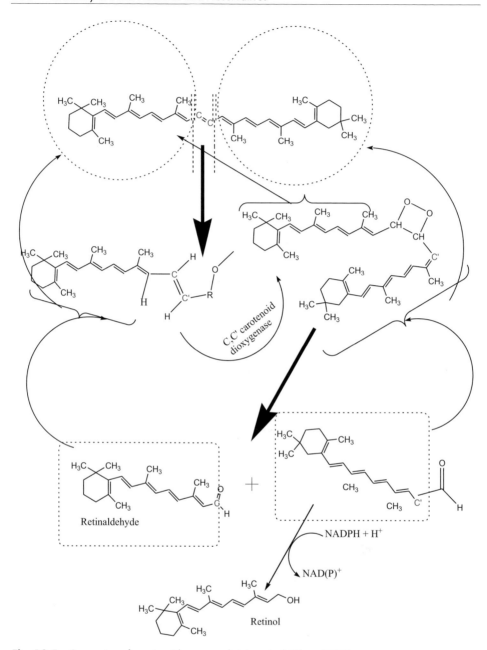

Fig. 13.5   Conversion of carotenoid compounds into retinol (Olson 1999).

will lead to toxicity (Rothman *et al.* 1995). Vitamin A compounds are normally converted by the action of enzymes (such as acyl transferase enzymes manufactured in the liver) into long-chain fatty acid esters of the corresponding vitamin A compound, which becomes primary storage mechanism of excess vitamin A. The long-chain fatty acid storage forms of vitamin A include retinyl palmitate, oleate, myristate, stearate and linoleate, and are normally found in the stellate cells in the liver, kidney, intestine and lung (Nagy *et al.* 1997). Whenever the

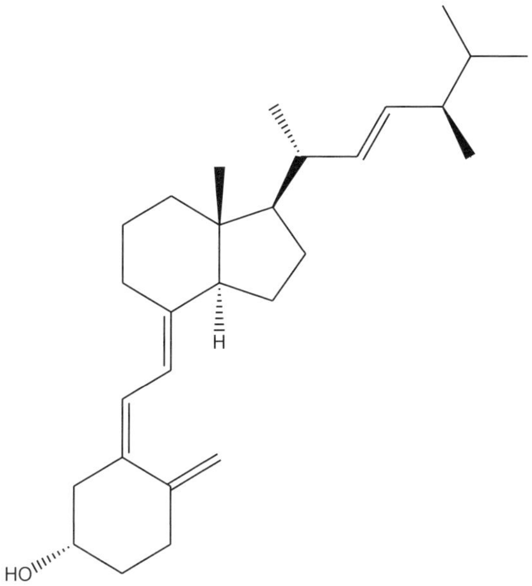

Fig. 13.6   Chemical structure of vitamin D_2 ergocalciferol (made from ergosterol) (Holick 2003, 2004; http://www.britannica.com/EBchecked/topic/631106/vitamin-D).

levels of vitamin A compounds decrease in the blood plasma, these long-chain fatty acid esters undergo enzymatic hydrolysis to yield retinol.

13.4   VITAMIN D (CALCIFEROL)

Chemically, vitamin D is a collective term which includes a number of steroidal hormones known as secosteroid hormones including ergocalciferol (vitamin D2) and cholecalciferol (vitamin D3). If not specified, the term 'vitamin D' refers to either of these two forms which are together known as calciferol. The structure of ergocalciferol (vitamin D2) and cholecalciferol (vitamin D3) are depicted in Figures 13.6–13.8. **Other structural forms of vitamin D, known as vitamins D1, D4 and D5, are depicted in** Figures 13.9–13.11.

The major source of vitamins D2 and D3 for humans is a photochemical process occurring within the 290–315 nm part of the electromagnetic spectrum. Ultraviolet B targets a pro-vitamin located in the epidermis (skin), chemically known as 7-dehydrocholesterol, as well as the eyes (Holick 2003, 2004).

13.4.1   The epidermal synthesis of vitamin D

The process of synthesizing calciferol (vitamin D) in humans begins at the epidermis and proceeds in other internal organs (mainly the liver and kidney), where a biologically active compound 1,25 dihydroxyvitamin D3 (1,25(OH)2D3) is produced through two hydroxy-lation steps of enzymatically controlled processes. The enzyme responsible for these two steps is 25 hydroxylase. The process takes place in the liver where another product, 25 hydroxyvitamin D (25(OH)2D), is produced. This undergoes further hydroxylation in the

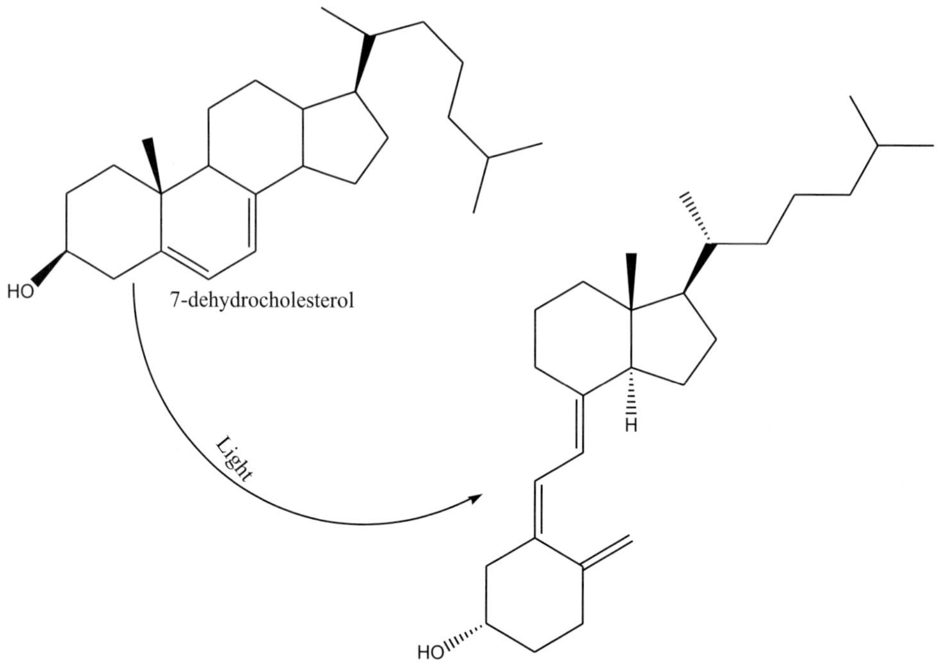

7-dehydrocholesterol

Light

Vitamin D3 (cholecaliferol)

Fig. 13.7 Formation of vitamin D$_3$ cholecalciferol from 7-dehydrocholesterol in the skin by light reaction (Holick 2003, 2004).

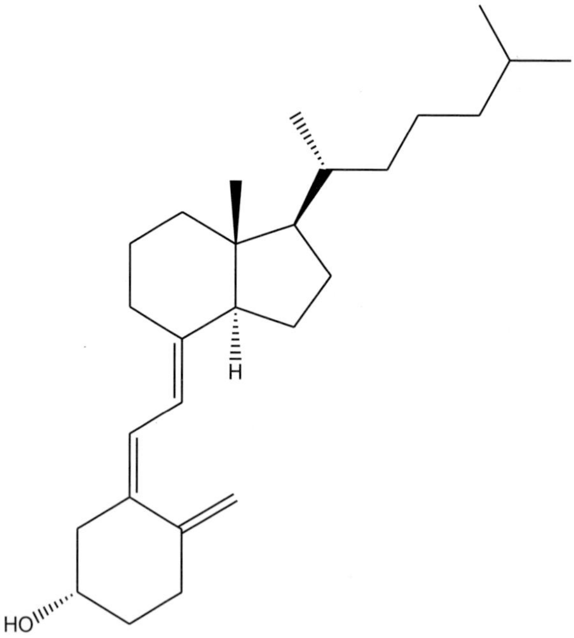

Fig. 13.8 Chemical structure of cholecalciferol (D$_3$)-calciol (Holick 2003, 2004; http://www.britannica.com/EBchecked/topic/631106/vitamin-D).

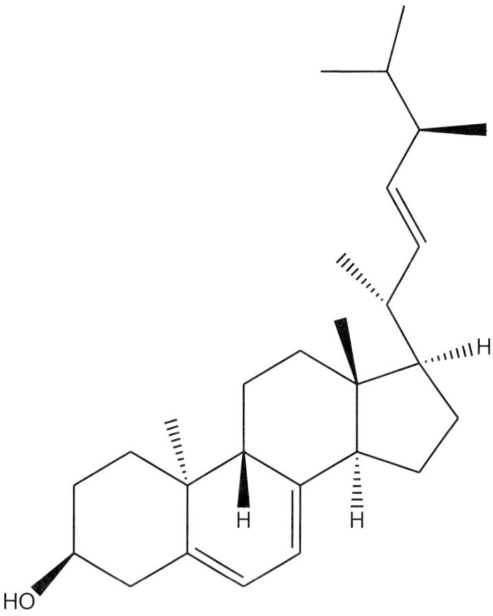

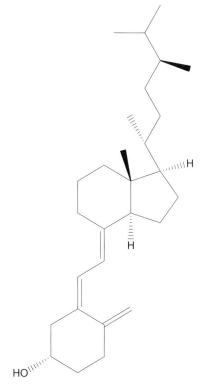

Fig. 13.9 Chemical structure of vitamin D1- molecular compound of ergocalciferol with lumisterol, 1:1 (Holick 2003, 2004; http://www.britannica.com/EBchecked/topic/631106/vitamin-D).

Fig. 13.10 Chemical structure of vitamin D4-22-dihydroergocalciferol (http://www.britannica.com/EBchecked/topic/631106/vitamin-D).

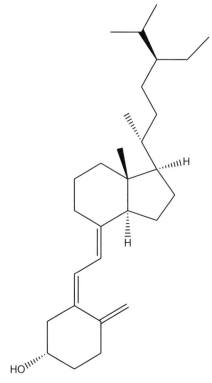

Fig. 13.11   Vitamin   D-5-sitocalciferol   (http://www.britannica.com/EBchecked/topic/631106/vitamin-D).

kidney by the action of enzyme 1-α hydroxylase (1α(OH)ase; CYP27B1) to form another compound 1,25(OH)2D3. In the kidney and liver, enzyme 24-hydroxylase (CYP24) converts 1,25(OH)2D3 to 24,25(OH)2D. The compound 1,25(OH)2D3 performs an important function in the physiology of the body as it binds with the high-affinity vitamin D receptor (VDR), a known ligand-dependent transcription factor (Dusso *et al.* 2005). VDRs take part in the regulation of gene transcription processes due to their ability to bind vitamin-D-responsive elements (VDREs). VDRs and 1,25(OH)2D3 are believed to play a role in the body's rapid response system (Dusso *et al.* 2005).

Despite the fact that vitamin D is not sourced from a natural diet (rather, the process of generating it takes place in the human skin), there is still the possibility of experiencing a vitamin D deficiency. Any deficiency is normally observed by determining the level of 25-hydroxy-vitamin D in the blood serum, which gives a representative measure of vitamin D produced in the epidermis. This also provides a measure of vitamin D obtained in foodstuffs which, in many cases, does not yield any quantitative information with regard to the in-system storage of vitamin D. A deficiency of vitamin D occurs when individuals are under-exposed to sunlight (Schoenmakers *et al.* 2008). This results in a deficiency of calcidiol with symptoms such as bone-softening diseases or impaired bone mineralisation (Grant and Holick 2005).

To treat diseases associated with the deficiency of Vitamin D, strategies include using either vitamin D supplements or commercially produced vitamin D.

13.4.2   Vitamin D supplementation in diets

The diet source of ergocalciferol (vitamin D2) is that of certain plants such as wild mushrooms and fungi, while the main diet source for vitamin D3 (cholecalciferol) is from animal (meat and meat products). It should be noted however that mushrooms and other fungi contain a provitamin ergosterol, which is converted to ergocalciferol (vitamin D2) (Lamberg-Allardt 2006). Types of foods which naturally contain vitamin D3 include fatty fish (such as eel, pike-perch, herring, salmon, tuna and cod), egg, egg yolk and internal organs of animals such as liver.

When diets rich in vitamins D3 and D2 are consumed, these vitamin compounds are assimilated and transported to the liver where they are hydroxylated by the action of 1-α-hydroxylase enzyme to form an intermediate secosteroid 25-hydroxyvitamin-D (25-D) (Albert *et al.* 2009). Under some modes of actions which are controlled by certain hormones, 1-α-hydroxylase enzyme catalyses the conversion of the formed 1,25-D product into other active secosteroidal homone species such as 1,25-dihydroxyvitamin-D3. Both 25-D and 1,25-D function as points to which vitamin D can be attached, and act as ligands for VDR (Cutolo *et al.* 2007).

13.4.3   Commercial production of vitamin D

Industrial production of vitamin D is dependent on 7-dehydrocholesterol or cholesterol as raw materials, obtained in most cases through organic solvent extraction from animal skins. For example, cholesterol is generally extracted from the lanolin of sheep wool using solvent extraction followed by purification and crystallisation. The purified product is then synthetically converted to 7-dehydrocholesterol (Holick 2005). The obtained crystalline product of 7-dehydrocholesterol is then solubilised in appropriate organic solvent and subjected to UV-B radiation. This mimics the transformation that takes place naturally in the epidermis and eye when irradiated with UV-B radiation to produce cholecalciferol (vitamin D). Vitamin D2 (ergocalciferol) is produced industrially using a similar strategy, but using ergosterol from yeast (Holick 2005). The produced D2 and D3 will then be formulated for application as milk/butter fortifications or even in animal feed formulations.

Other processes for extraction of 7-dehydrocholesterol from *Solanum glaucophyllum* cultures grown in the absence of light have been reported by Curino *et al.* (1998). *S. glaucophyllum* tissue (callus) and cell suspension cultures grown under strict conditions of darkness were extracted with chloroform/methanol (1:2, v/v). This is followed by purification of the lipid fraction by Sephadex LH-20. The extracts 7-dehydrocholesterol, vitamin D3, 25(OH)D3 and 1,25(OH)2D3 were obtained using high-performance liquid chromatography as well as mass spectrometry.

Similar work was reported by Aburjai *et al.* (1996), where they isolated 7-dehydrocholesterol from cell cultures of *Solanum malacoxylon* in a lengthy procedure. In their work, a number of solvent systems were used for extraction of the cholesterol derivatives. Extracts were then characterised using spectroscopic and chromatographic methods.

Man-Jeong *et al.* (2008) reported a method for the extraction of 7-dehydrocholesterol from rats. Cholesterol and its metabolic precursors (mainly lanosterol and 7-dehydrocholesterol) in rat plasma were saponified and extracted with hexane and diethyl ether in sequence. The product was derived using tert-butyldimethylsilyl (TBDMS) and determined using gas chromatography-mass spectrometry (GC-MS) in selected ion monitoring (SIM) mode (Man-Jeong *et al.* 2008).

13.4.4 Measuring vitamin D in foods

Methods of measuring vitamin D activity in food items include a number of bioassays such as rat line assays (which yield measurements in international units (IU) or μg); radioactive phosphorus assays; and in-house protein-binding assays (Hollis *et al.* 1981; Kunz *et al.* 1984; Parvianinen *et al.* 1984). Chromatographic methods such as reversed-phase high-performance liquid chromatography(RPHPLC) have also been reported in the determination of vitamin D contents as well as associated metabolites such as 25-(OH)D and 1,25-(OH)2-vitamin D3 (Mattila 1995; Ovesen *et al.* 2003a, 2003b). Vitamin D3 can be analysed by HPLC because the compound has UV-absorbing chromophores which can absorb UV light at 265 nm (Jakobsen and Saxholt 2009).

13.5 VITAMIN E

Vitamin E is another fat-soluble vitamin which is needed by all species of animal (McDowell 2000). Vitamin E plays many important roles in animal bodies including that of chain breaker in reactive species reactions (hence protecting tissues and cellular structures from oxidative damage due to oxygen free radicals as well as from reactive products of lipid peroxidation) and in exerting membrane stability (Burton and Ingold 1981). Moreover, as a biological antioxidant, vitamin E is important in the reaction mechanisms which prevent the oxidation of dietary polyunsaturated fatty acids (PUFAs; Weber *et al.* 1997).

Vitamin E occurs in nature in a number of chemical forms, such as α-tocopherol, β-tocopherol, γ-tocopherol, δ-tocopherol, DL-α-tocopheryl acetate (DL-α-TOA) and to-cotrienols (National Research Council 1983). Of the various forms of vitamin E, α-tocopherol has the highest vitamin E activity. DL-α-TOA form is known to be the most stable form of α-tocopherol and the most commonly used vitamin E supplement in animal feeds (Min-Hsien and Shi-Yen 2004).

In the same way as for tocopherols (Figure 13.12), tocotrienols (Figure 13.13) exist in four different forms of α, β, γ and δ; they have the same methyl group in the ring but the presence of unsaturations (three double bonds) in the side chain makes them structurally different from tocopherols. The presence of a single stereoisomeric carbon in tocotrienols gives it only 2 possible isomers per structural formula; tocopherols however have 3 chiral centres and 8 possible stereoisomers per structural formula. It has been reported that for stereoisomers which show vitamin E activity, a complete methylation in the α-form is advantageous as far as the enhancement of vitamin E is concerned since this is the form with the preference for the tocopherol binding protein (Sen *et al.* 2006).

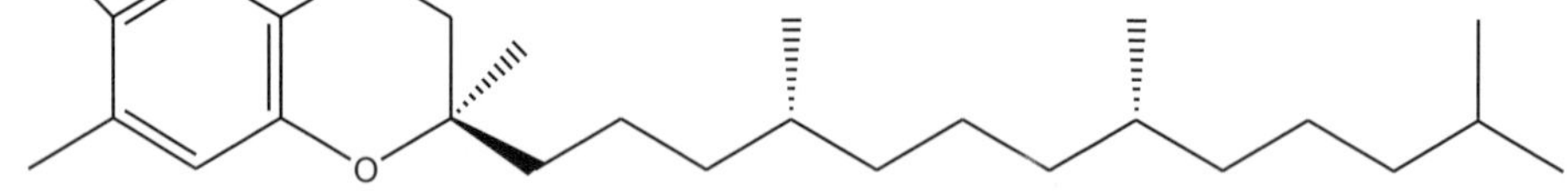

Fig. 13.12 The α-tocopherol form of vitamin E.

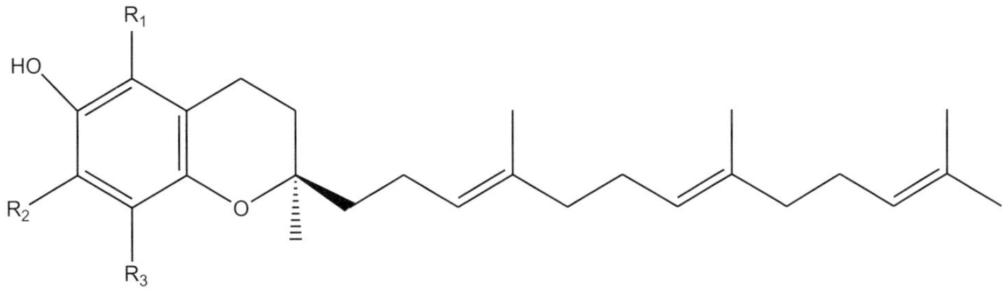

Type	R¹	R²	R³
α-tocotrienol	CH_3	CH_3	CH_3
β-tocotrienol	CH_3	H	CH_3
γ-tocotrienol	H	CH_3	CH_3
δ-tocotrienol	H	H	CH_3

Fig. 13.13 General structure of tocopherols and tocotrienols.

13.5.1 Role of vitamin E

Fish species such as salmon, catfish, bass, carp and tilapia, together other aquatic animal species, are known to be rich in vitamin E and unsaturated fatty acids. These play an important role in maintaining cell membranes when temperatures are low (Woodall *et al*. 1964; Wilson *et al*. 1984; Blazer 1992; Takeuchi *et al*. 1993; Kocabas and Gatlin 1999; Shiau and Shiau 2001). In mammals, vitamin E has an influence on the functioning of immunological responses (Panush and Delafuente 1985; Moriguchi *et al*. 1990; Beharka *et al*. 1997). In fish and other aquatic organisms, it enhances humoral and cellular defences (Hardie *et al*. 1984; Blazer and Wolke 1990; Ortuno *et al*. 2000).

13.5.2 Plant sources

A number of plant and plant-derived products are known to be good sources of vitamin E, including: sweet potatoes, pumpkins, mangoes, avocados, papayas, nuts and nut oils (e.g. almonds and hazelnuts), tomatoes, olives, sunflower oil, wheat germ oil and green leafy vegetables such as spinach and lettuce.

13.5.3 Industrial production of vitamin E

Despite the presence of many stereoisomers for vitamin E, only the α and γ forms show significant vitamin E activity. Acetate esters of RRR-α-tocopherol and all-rac-α-tocopherol (D and DL-α-tocopherol, respectively) are the main sources for industrial production of vitamin E (Machlin 1999). The major source of RRR-α-tocopherol is certain plant raw

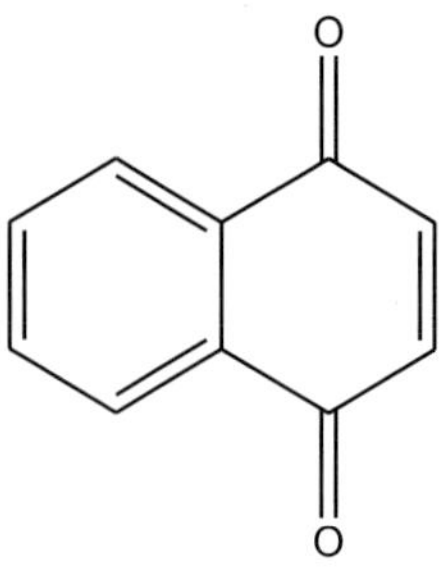

Fig. 13.14 Chemical structure of 1,4-naphthoquinone.

materials and is obtained through extraction procedures while the all-rac form is a synthetic product (Machlin 1999).

13.6 VITAMIN K

Fat-soluble vitamin K is a group of prenylated 2-methyl-1,4-naphtoquinone (Figure 13.14) derivatives required for blood coagulation processes as well as for metabolic pathways that exist in bone and certain body tissues (glutamates) (Craciun *et al.* 1998). For example, vitamin K plays an important role in the g-carboxylation of glutamyl residues needed for blood-clotting factors (Stenflo and Ganrot 1972; Shah and Suttie 1974; Stenflo *et al.* 1974). Vitamin K is also needed in extra hepatic proteins, for example the non-collagenous γ-carboxyglutamic acid-containing proteins present in bone-forming cells known as osteocalcin. Vitamin K therefore has a role in bone metabolism, assumed through carboxylation of osteocalcin (Hauschka *et al.* 1975; Price *et al.* 1976). Like other fat-soluble vitamins (A, D and E), vitamin K is also stored in the fat tissues.

13.6.1 Vitamin K structures

Structurally, vitamin K exists in two major forms: vitamin K1 (phylloquinone) and vitamin K2 (menaquinones) (Figures 13.14–13.16). Vitamin K1 exists singly as it has no isomers apart from its trans- (active biologically) and cis- (inactive biologically) forms (Hwang 1985; Indyk 1988). However, vitamin K2 is known to have about 14 isomers of vitamins (or vitamers), comprising several isoprene units at the 3-position of the naphthoquinone. These isomers of vitamin K2 are called menaquinones and are numbered 1–14 (MK-1 to MK-14).

Figure 13.16 shows that vitamins K1 and K2 have a similar methylated naphthoquinone ring structure. The difference in terms of their structure is observed from their hydrophobic aliphatic side chain, which is attached at the 3-position of the ring for vitamin K1. For vitamin K2 however, having four isoprene units and one unsaturation, the side chain is composed of several isoprene units which are unsaturated.

Despite this difference in their side-chain chemistry, the naphthoquinone functionality is believed to play a role in the activity of vitamin K. Some differences in terms of their bioavailability or absorption in the intestines might however be expected due to differences in the hydrophobicity of their side chain. Scientific reports have revealed that the extent to which the intestines (ileum and colon) absorb vitamin K is generally poor. Experiments

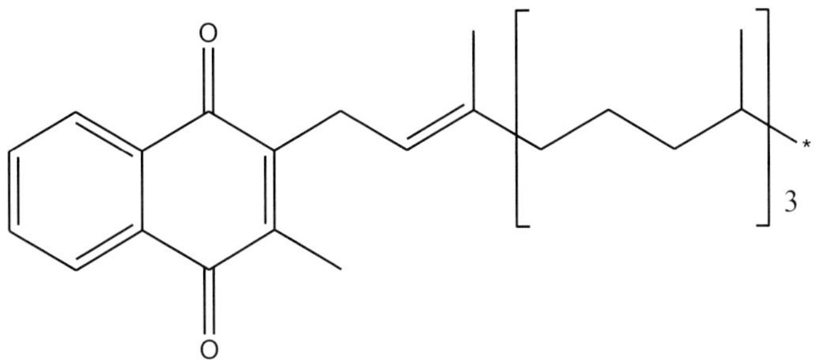

Fig. 13.15   Chemical structure of vitamin K1 (3 refers to the number of saturated isoprene units).

investigating the presence of incomplete g-carboxylated proteins in the blood, a test for vitamin K deficiency, found high levels. This implies that in most cases there is inefficient processing of vitamin K in the small and large intestines (Geleijnse *et al.* 2004; Komai and Shirakawa 2007; Beulens *et al.* 2009; Nimptsch *et al.* 2010).

These results explain how it is possible to suffer from vitamin K deficiency, despite being rare due to the variety of diets which are rich in vitamin K. Where there is a deficiency in vitamin K, symptoms such as anaemia, bleeding of the gums or nose and diseases such as coagulopathy, osteoporosis and coronary heart disease (Katsuyama *et al.* 2002; Ikeda *et al.* 2006).

13.6.2   Dietary sources of vitamin K

Major sources of vitamin K1 include green leafy vegetables such as spinach and cauliflower, avocados and grapes (Booth and Suttie 1998). Vitamin K2 vitamers such as menaquinone-4 and menaquinone-7 are found in meat, eggs and dairy products (Tsukamoto *et al.* 2000; Elder *et al.* 2006). Synthetic forms of vitamin K also exist, such as vitamins K3 (menadione-(2-methyl-1,4-naphthoquinone)), K4 (menadiol diphosphate) and K5 (2-methyl-4-amino-1-naphthol hydrochloride) (Figures 13.17 and 13.18). Vitamin K5 is believed to be a potential food preservative agent (Merrifield and Yang 1965).

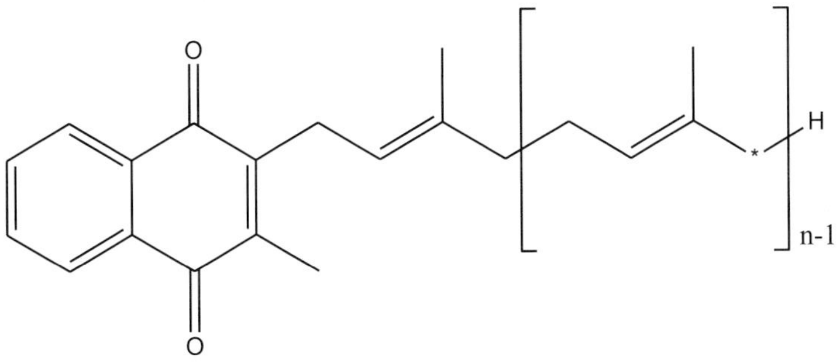

Fig. 13.16   General chemical structure of vitamin K_2 (where n = number of isoprene units; n−1 = number of unsaturations).

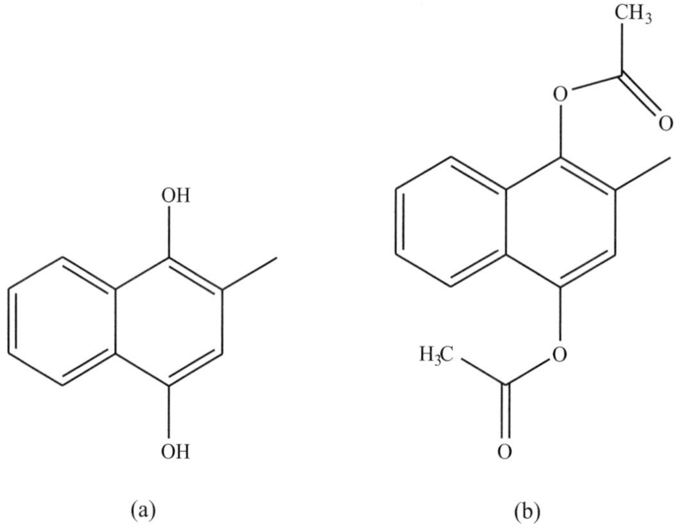

Fig. 13.17 Chemical structure of (a) menadiol (vitamin K4) and (b) menadiol diacetate (1,4-Diacetoxy-2-methylnaphthalene).

13.6.3   Methods of assessment

Methods of quantifying the bioactivity of different forms of vitamin K involve two different types of experimental animal systems: (1) vitamin-K-deficient rats and (2) anticoagulated rats method (Craciun *et al.* 1998). In the vitamin-K-deficient rats method, rats are treated with vitamin K antagonists (e.g. warfarin or brodifacoum) thus restricting the presence of glutamate due to a limited amount of thrombin present in the blood system. This effect is then reversed by a minimal supply of vitamin K, which allows prothrombin to be formed at higher levels. The extent of vitamin K utilisation is taken as the measurement of its bioactivity (Groenen-van Dooren *et al.* 1993, 1995).

In the anticoagulated method, an enzyme of vitamin K (vitamin KO or reductase; 2,3 epoxide reductase) is used to block the circulation of prothrombin. This implies that the stock of vitamin K will be completely depleted, so a higher concentration of vitamin K will be required to enable the formation of the carboxylation cofactors (Park and Leck 1982; Groenen-van Dooren *et al.* 1993; Reedstrom and Suttie 1995). The extent to which vitamin

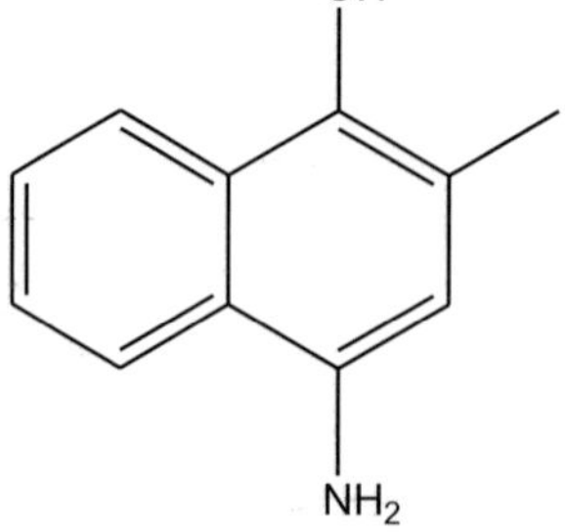

Fig. 13.18 Chemical structure of vitamin K5 (menadione).

K form is utilised gives a measure of its bioactivity (Groenen-van Dooren *et al.* 1993, 1995; Reedstrom and Suttie 1995).

Another bioactivity test for vitamin K is the prothrombin time test which measures the time required for blood to clot in an experimental set-up where blood samples are mixed with citric acid and then measured using a fibrometer. In this type of measurement, a delayed clot formation indicates a deficiency. The limitation of this method is that it lacks the sensitivity; the measurements may remain static until the prothrombin levels in the blood system drop considerably.

Another method for testing vitamin K bioactivity is known as plasma phylloquinone which correlates the phylloquinone intake. This test has however been reported to be gender biased (Thane *et al.* 2002).

Urinary γ-carboxyglutamic acid is another method for determining the bioactivity of vitamin K. It is used to indicate the glutamate response to the changes in dietary vitamin K intake in individuals. This method has however been reported to be age specific (Thane *et al.* 2002).

13.7 VITAMIN B

Vitamin B comprises a number of vitamins that play several important functions in the body. For example, vitamin B provides energy for glucose metabolism to form carbohydrates as well as other lipid and protein metabolism. Vitamin B is subdivided into 8 subclasses, namely: vitamin B1 (thiamine); vitamin B2 (riboflavin); vitamin B3 (niacin); vitamin B5 (pantothenic acid); vitamin B6 (pyridoxine or pyridoxal or pyridoxamine); vitamin B7 (or vitamin H or biotin); vitamin B9 (folic acid); and vitamin B12 (cyanocobalamin).

13.7.1 Vitamin B1 (thiamine)

This water-soluble vitamin contains a sulphur group, aminopyrine, thiazole rings and a side chain of methyl and hydroxyethyl joined by methylene molecules in its structure (Table 13.1). It has many other derivatives such as thiamine pyrophosphate, which is vital in sugar and amino acid metabolism as well as in alcoholic fermentation (Tanphaichitr 1999).

13.7.1.1 Deficiency of vitamin B1

The deficiency of vitamin B1 (thiamine) results in health problems such as beriberi (a neurological and cardiovascular disease), Wernicke's encephalopathy (also known as Wernicke–Korsakoff syndrome) and Korsakoff's psychosis, which is an alcohol amnestic disorder (Martin *et al.* 2003).

Under normal conditions, vitamin deficiency is caused by consumption of a diet lacking those particular vitamin ingredients. However, the deficiency of vitamin B1 may also result from malnutrition, for example consumption of diets containing high levels of the enzyme thiaminase or foodstuffs with high contents of anti-thiamine factors such as tea and coffee (Butterworth 2006).

Alcoholism can affect the uptake of vitamin B1 from the gastrointestinal tract, as well as the transport of the vitamin and its distribution into enterocytes (Butterworth 1993). The consumption of alcohol is also known to disrupt the enzymatic thiamine intermolecular

Table 13.1   Classification of B vitamins

Categorisation	Vitamin B	Chemical structure
B1	Thiamine or thio-thiamine	
B2	Riboflavin	
B3	Niacin or nicotinic acid	
B5	Pantothenic acid	
B6	Pyridoxine	

Table 13.1 (*Continued*)

Categorisation	Vitamin B	Chemical structure
B7 (H)	Biotin	
B9	Folic acid	
B12	Cyanocobalamin	

binding process which is facilitated by magnesium. This is because alcohol acts as an inhibitor of vitamin B1 transport in the gastrointestinal system, thus preventing phosphorylation of vitamin B1 which leads to the formation of its cofactor derivative known as thiamine diphosphate (ThDP; Rindi *et al.* 1986). These problems can be reversed with an appropriate diet or by abstaining from alcohol consumption.

13.7.1.2 Dietary sources of vitamin B1

There are many different types of diets which provide thiamine, including: yeast extract (e.g. marmite), milk, beans, lobsters, potatoes, brown rice, oranges, liver (chicken, beef, pork), eggs and cereal grains (both whole and refined, although wholegrain is known to be richer in terms of thiamine content). Most thiamine content is bound at the outer surfaces of the cereal grains as well as in the germ; refining processes tend to remove the husks together with the layer containing thiamine. In most cases, refined grains and flour are supplemented with thiamine in the form of thiamine mononitrate (as well as other nutritious ingredients such as ferrous iron, vitamin B2, iodised salt and folic acid which are all lost during the refining process). In some instances, vitamin B1 can be included in foodstuffs as an additive. The form of vitamin B1 which has been used in this category is thiamine hydrochloride, also known as betaxin which finds use as a flavouring in meat and soups (Mahan 2000).

13.7.1.3 Uptake, transport and distribution of thiamine

After digestion of vitamin B1, two types of enzymes are known to act to release the thiamine in the intestines: phosphatase and pyrophosphatase. These convert thiamine to thiamine diphosphate (Mahan and Escott-Stump 2000). The pyrophosphorylation process takes place in the intestine walls, where it has been shown to have high thiamine pyrophosphokinase activity (Mahan and Escott-Stump 2000).

During the transport of thiamine in the body, dephosphorylation takes place in the intestines by the action of sodium-dependent ATPase enzymes. These release free thiamine for absorption and uptake, mostly by free diffusion (Combs 1998). The metabolism of thiamine results in a number of acidic metabolites which include 2-methyl-4-amino-5-pyrimidine carboxylic acid, 4-methyl-thiazole-5-acetic acid and thiamine acetic acid, which the body expels with urine (Tanphaichitr 1999).

13.7.2 Vitamin B2 (riboflavin)

Vitamin B2, also known as riboflavin, is a water-soluble vitamin which performs important functions in a number of cellular processes dealing with energy metabolism, as well as the metabolism of major food components. The structural chemical formula for riboflavin shows that it is formed by two main components: (1) a ribitol (Figure 13.19a), which is a reduced ribose sugar molecule; and (2) a flavin (Figure 13.19b), which is a pteridine-based molecule made up of a tricyclic heteronuclear organic ring known as isoalloxazine. The two molecules fuse together to form riboflavin (Figure 13.19c).

13.7.2.1 Deficiency of vitamin B2

After digestion and absorption, the metabolites and undigested parent riboflavin molecules are excreted out of the body with urine (Brody 1999). This tendency to excrete riboflavin creates the possibility of riboflavin deficiency. In addition, some individuals possess abnormalities

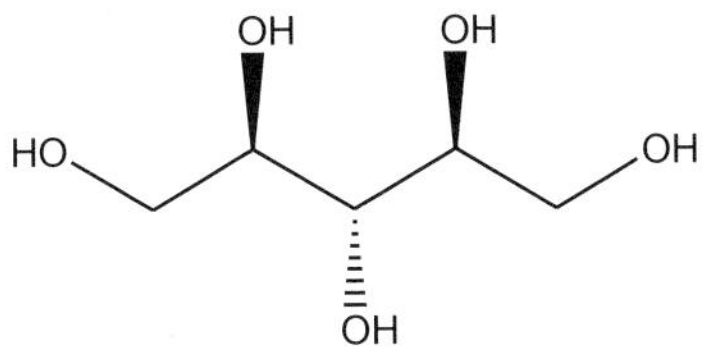

(a)

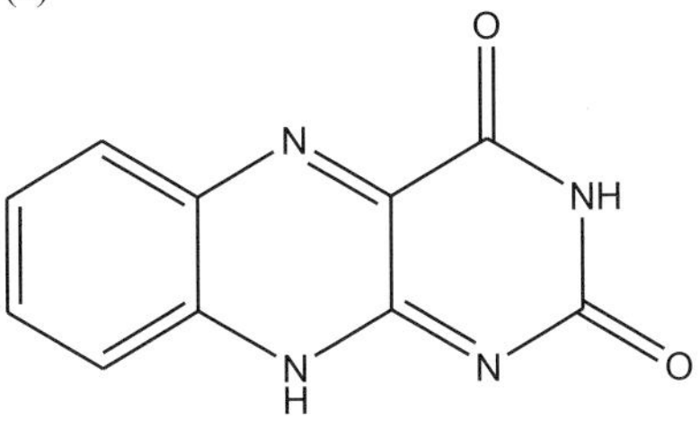

(b)

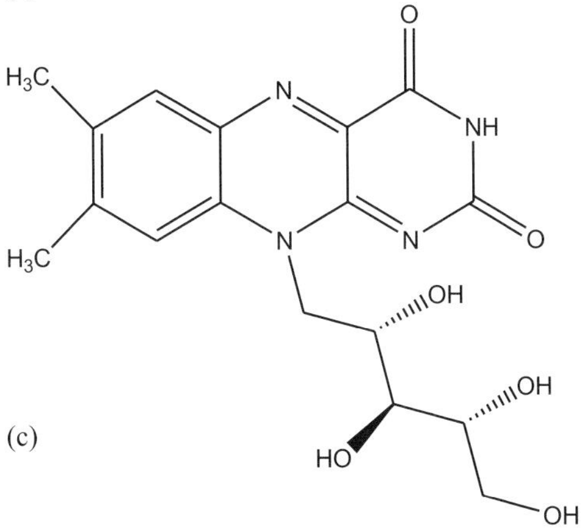

(c)

Fig. 13.19 Chemical formula of (a) ribitol; (b) flavin, showing the isoalloxazine tricyclic structure; (c) riboflavin (http://www.britannica.com/EBchecked/topic/631026/riboflavin).

which prevent the proper absorption of riboflavin in the intestines. Riboflavin deficiency syndromes and diseases such as ariboflavinosis or angular cheilitis (Brody 1999).

13.7.2.2 Dietary sources of vitamin B2

Foodstuffs known to be rich in vitamin B include leafy green vegetables, legumes, tomatoes, mushrooms, milk and animal organs (liver and kidney).

13.7.2.3 Measurements of riboflavin deficiency

Specific tests used to detect and confirm riboflavin deficiency in individuals are mostly biochemical in nature, and include erythrocyte glutathione reductase activity (Serrini *et al.* 1996) and urinary riboflavin excretion test (Sebrell and Butler 1939). The erythrocyte glutathione reductase activity works by measuring the activity coefficient of the flavin adenine dinucleotide (FAD) dependent enzyme, erythrocyte glutathione reductase (EGR), which is a nicotinamide adenine dinucleatide phosphate hydrogen (NADPH) (Powers 2003; Gibson 2005). The state of tissue saturation is measured both in the presence and in the absence of the FAD enzyme. If the ratio of activity coefficient in the presence of FAD to that in the absence of FAD is greater than 1:4, it implies that the body is deficient of riboflavin. If the ratio is less than 1:2, the measurement represents an acceptable level in the body (Gropper *et al.* 2009).

The urinary riboflavin excretion test is based on the fact that urinary riboflavin excretion rates increase slowly with increasing intakes to the point where the rate of saturation balances that of absorption (Bailey *et al.* 1997). If the urinary riboflavin excretion is less than 19 μg/g creatinine or 40 μg/day, riboflavin deficiency is implied (Gibson 2005).

13.7.3 Vitamin B3 (niacin, nicotinic acid or nicotinamide)

Vitamin B3, also referred to as niacin, nicotinic acid or vitamin PP (structure in Table 13.1), is one of the water-soluble vitamins essential processes such as DNA repair and the production of steroid hormones in the adrenal gland. Its chemical name is pyridine-3-carboxylic acid, implying that it is a derivative of pyridine and contains a carboxylic acid functionality side chain attached at 3-position. There are other chemical structures for niacin in which the carboxylic acid has been substituted by other functional groups such as carboxyamide group (CONH$_2$) or ester groups.

Niacin deficiency causes a disease known as pellagra. This is common in places where maize is used as a staple food, because the content of niacin in maize is low. To counter this, additives of niacin in the production of maize flour is important. A number of foodstuffs are known to be rich in niacin, such as liver, chicken, beef, fish, cereal, peanuts and legumes.

13.7.4 Vitamin B5 (pantothenic acid)

Vitamin B5 is another water-soluble vitamin with a chemical structure resembling both amide and b-alanine. This gives it the advantage of ubiquity in that it is an ingredient (in either CoA or acyl carrier protein forms) in almost all foodstuffs, especially in whole-grain cereals, legumes, eggs, meat, royal jelly, avocado and yogurt (Trumbo 2006). When foodstuffs rich in vitamin B5 are consumed, the pantothenic acid must be hydrolysed to 4'-phosphopantetheine which is further hydrolysed by pantotheinase enzyme into free pantothenic acid (Trumbo 2006; Gropper *et al.* 2009) before the absorption of the CoA or acyl carrier protein forms.

Due to the fact that pantothenic acid is found in almost every foodstuff, pantothenic acid deficiency is rare. Even in cases of starvation where the deficiency may be encountered, the symptoms (which are similar to other vitamin deficiencies discussed above) can easily be reversed with the supply of pantothenic acid (Gropper *et al.* 2009).

13.7.5 Vitamin B6 (pyridoxal phosphate or pyridoxal 5'-phosphate)

Water-soluble vitamin B6 is essential for a number of vital processes in the body including normal growth, development and metabolism (Tryfiates 1980; Merrill and Henderson 1987),

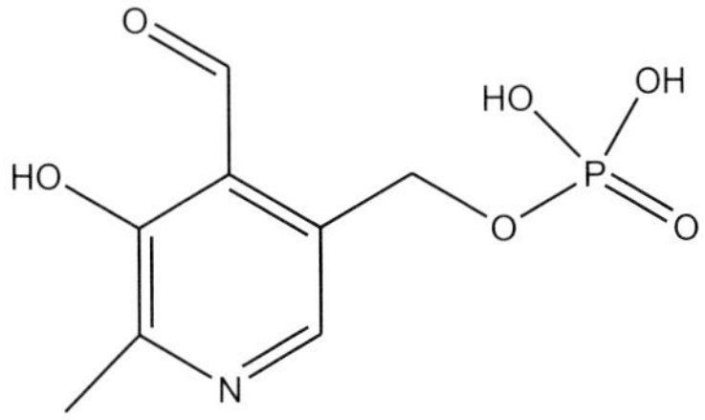

Fig. 13.20 Chemical structure of pyridoxal 5′-phosphate (PLP).

(a) Pyridoxine   (b) Pyridoxal   (c) Pyridoxamine

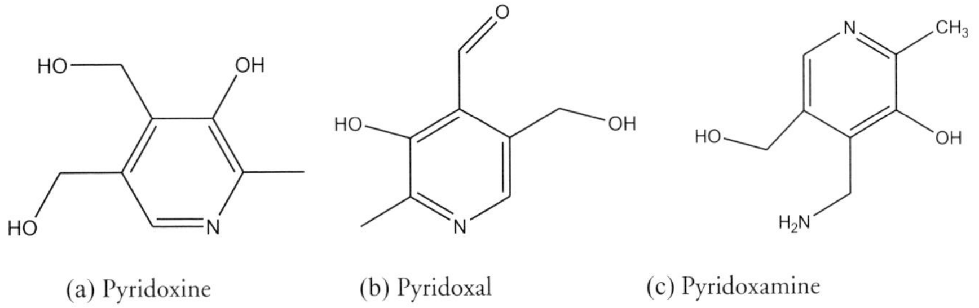

Fig. 13.21 Chemical structures of different forms of vitamin B6 (http://www.britannica.com/EBchecked/topic/631041/vitamin-B6).

the normal functioning of the central nervous system (Merrill and Burnham 1990) and as a cofactor in numerous enzyme reactions of amino acid metabolism (Merrill *et al.* 1984). There are a number of vitamin B6 forms, but is the most physiologically active and the most important form of the vitamin is derived from inactive dietary precursors, referred to as pyridoxal 5'-phosphate (PLP) (Figure 13.20; Merrill *et al.* 1984). Other forms of vitamin B6 include pyridoxal (PL), pyridoxamine, 4-pyridixic acid and pyridoxine (Figures 13.21a–c).

When vitamin B6 is consumed and digested, the active ingredient of vitamin B6 is absorbed by diffusion in the small intestines. The membrane-catalysed absorption process involves the respective dephosphorylation enzymes of the vitamin B6 forms, which are mostly the alkaline phosphatases and pyridoxial kinases (Combs 1998; McCormick 2006).

The main by-product of vitamin B6 is 4-pyridoxic acid (Figure 13.22), which is eliminated from the body with urine. In cases of high vitamin B6 consumption, some other extra by-products are also excreted with urine such as pyridoxal, pyridoxamine and pyridoxine and their respective phosphates (Combs 1998).

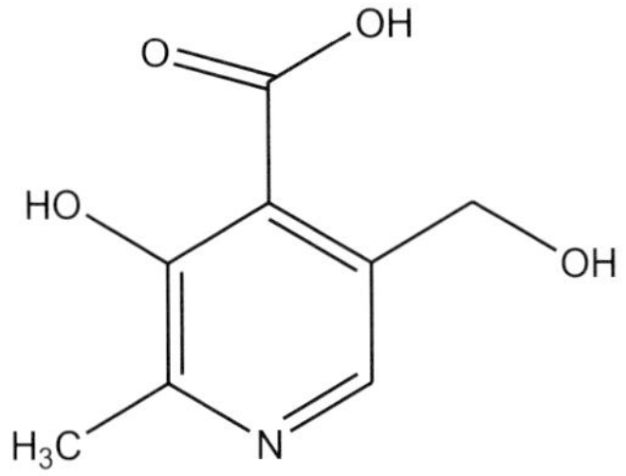

Fig. 13.22 Chemical structure of 4-pyridoxic acid.

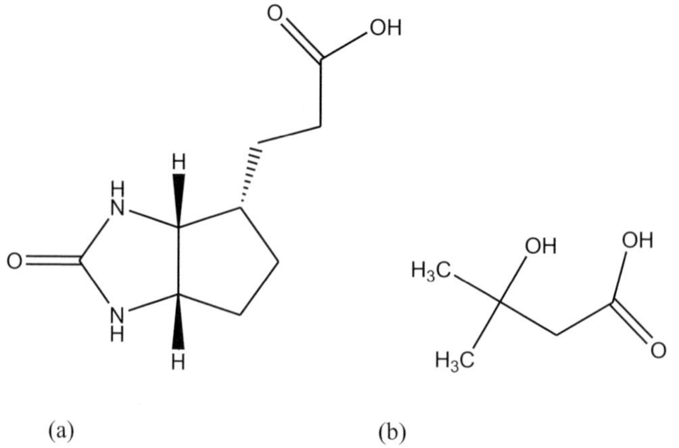

(a)   (b)

Fig. 13.23 Chemical structures of urinary metabolites for biotin: (a) bisnorbiotin; (b) 3-hydroxyisovaleric acid.

A deficiency of vitamin B6 leads to dermatological and neurological disorders (Rall and Meydani 1993; Bowman and Russell 2006). The deficiency can be corrected by consuming foodstuffs known to be rich in vitamin B6 such as meats, vegetables, whole grain and bananas.

13.7.6 Vitamin B7 or vitamin H (biotin)

Vitamin B7, also known as biotin or vitamin H, is another water-soluble vitamin. Its chemical structure comprises tetrahydroimidizalone and tetrahydrothiophene rings fused together with a side chain of valeric acid substituent attached to the tetrahydrothiophene ring part of the structure (see Table 13.1). As well as biotin being a coenzyme in the metabolism of fatty acids and amino acids such as leucine, it performs important functions in the gluconeogenesis processes and in ensuring the proper growth.

Vitamin B7 is an ingredient in many different types of foodstuffs, and those with a high content of biotin include raw egg yolk, liver, vegetables and peanuts. Synthetic biotin in the form of supplements is available, manufactured from the raw material of fumaric acid.

Low levels or decreased concentration of circulating biotin or bisnorbiotin (Figure 13.23a), as detected in urine or plasma samples, is an indicator of biotin deficiency. Another indicator of biotin deficiency is the increase in the urinary excretion of 3-hydroxyisovaleric acid (Figure 13.23b).

13.7.7 Vitamin B9 (folic acid)

Vitamin B9 or vitamin B_c (Figure 13.24), together with other B vitamins such as vitamins B6 and B12, play an important role in controlling the concentrations of the amino acid homocysteine in the blood. The action of vitamin B9 in this process is crucial; if the concentration of homocystein increases too much, it can cause heart and Alzheimer's disease. On the other hand, if vitamin B9 is deficient in the body system disorders such as degenerative growth, loss of appetite and mental retardation can develop.

Vitamin B9 deficiency is measured by performing a complete blood count (CBC) as well as measuring the level of vitamin B12 in the plasma. If the serum level of vitamin B9 is less

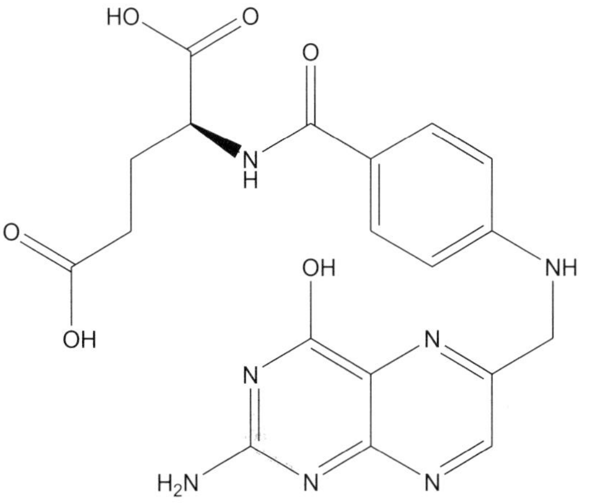

Fig. 13.24 Chemical structure of folic acid (vitamin B9).

than 3 μg/L, it is considered a deficiency condition. If the erythrocyte count is less than 140 μg/L in the serum, this also implies deficiency and reflects the realities in the tissues.

Other measurements that are used as indicators for vitamin B9 deficiency include the concentration of amino acid homocysteine; if there is an increase of this compound in the blood serum, a deficiency of folatic acid is implied. However, the homocysteine test is also used as an indicator of vitamin B6 and B12 deficiency, so a test for levels of methylmalonic acid (Figure 13.25) is used to distinguish between vitamin B9 and B12 deficiency cases. If the level of methylmalonic acid remains normal in the blood serum, this implies vitamin B9 deficiency. An elevated level of methylmalonic acid is indicative of vitamin B12 deficiency.

To counter the problems associated with vitamin B9 deficiency, either supplemental oral folate or the consumption of diets rich in folic acid/folate such as spinach, dark leafy greens, soybeans, beef liver, root vegetables, whole grains, wheat germ, kidney beans, white beans, mung beans, salmon, orange juice, avocado and milk is recommended. When foods rich in folate are consumed and digested, the active ingredient is converted to tetrahydrofolate by the action of dihydrofolate reductase enzymes (Scheme 13.3) present in the liver (Bailey and Ayling 2009).

13.7.8   Vitamin B12 (cyanocobalamine)

Vitamin B12 is a group of cobalt-containing vitamins which include cyanocobalamin, a compound which contains cyanide (Figure 13.26a); hydroxocobalamin (Figure 13.26b);

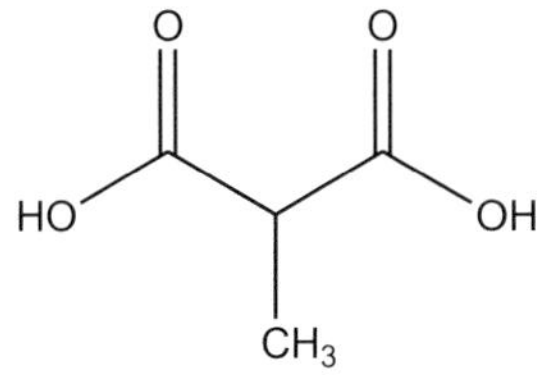

Fig. 13.25 Chemical structure of methylmalonic acid.

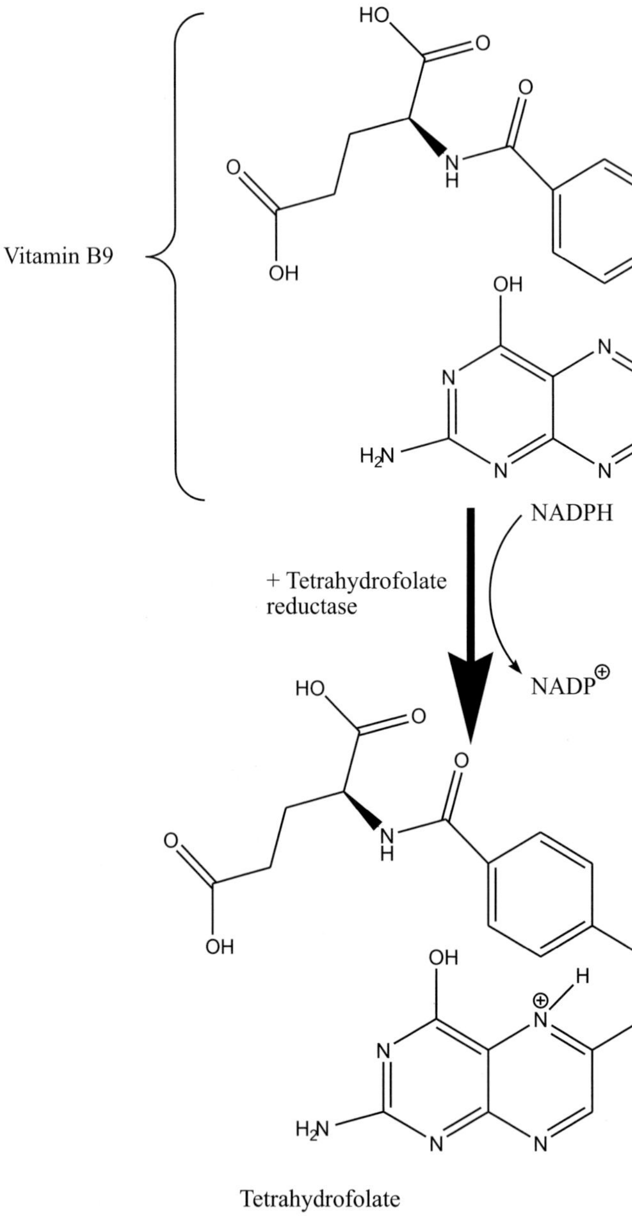

Vitamin B9

+ Tetrahydrofolate reductase

NADPH

NADP⊕

Tetrahydrofolate

Scheme 13.3   The reduction of folate (vitamin B9) to tetrahydrofolate by the action of dihydrofolate reductase enzyme (Bailey and Ayling 2009).

5'-deoxyadenosylcobalamin also known as adenosylcobalamin (Figure 13.26c); an enzyme known as methylmalonyl coenzyme A mutase; and 5'-methyltetrahydrofolate homocysteine methyltransferase, which is another enzyme.

The principle form of vitamin B12 is represented by cyanocobalamin which is actually the form used as food nutrition/supplement. Vitamin B12 performs important functions in the central nervous system as well as in the formation of blood cells. Together with sodium

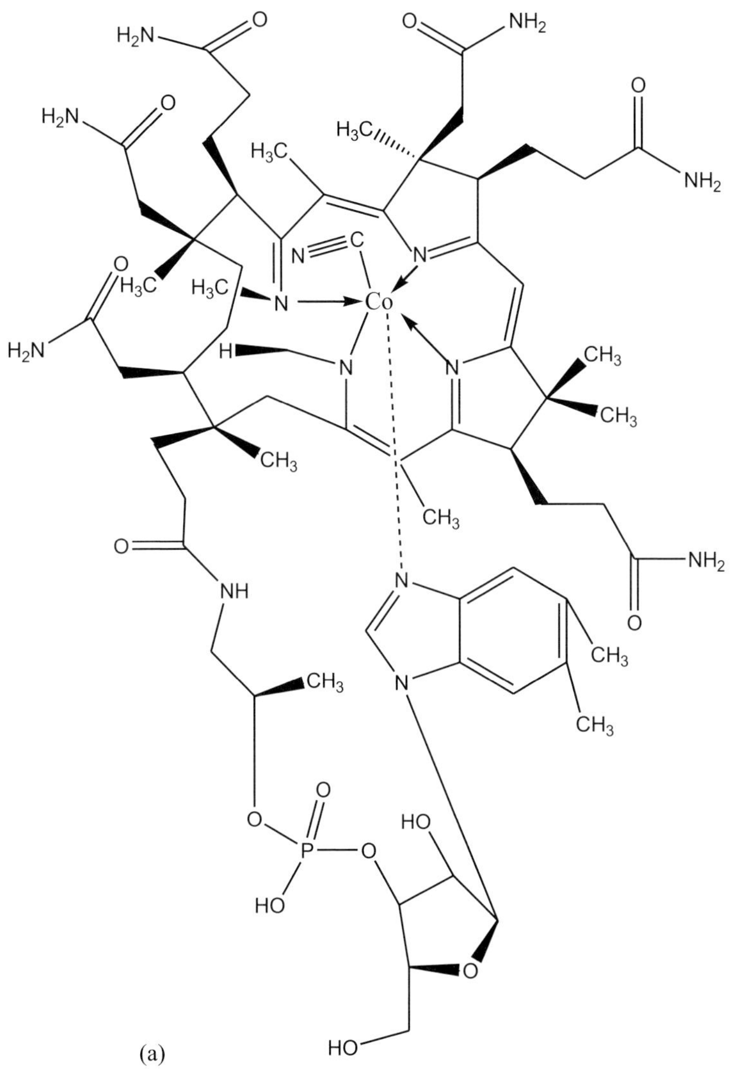

Fig. 13.26 Chemical structures for vitamin B12 groups: (a) cyanocobalamin; (b) hydroxocobalamine; and (c) 5'-deoxyadenosylcobalamin (http://www.britannica.com/EBchecked/topic/631051/vitamin-B12).

sulphate, cyanocobalamin is also used in the treatment of cyanide poisoning; hydroxycobal-amin hydroxide moiety becomes displaced by the cyanide toxin, forming a complex that is not toxic and is thus eliminated from the body with urine (Hall and Rumack 1987; Dart 2006).

Diet remains the major and most reliable source of vitamin B12. When foodstuffs con-taining cyanocobalamin, for example eggs, meat and chicken are consumed, they will be digested by the action of hydrochloric acid together with gastric protease enzyme to release vitamin B12, which is then absorbed by the body system. The process of absorption involves the interaction of vitamin B12 with gastric glycoprotein, which enables vitamin B12 to be

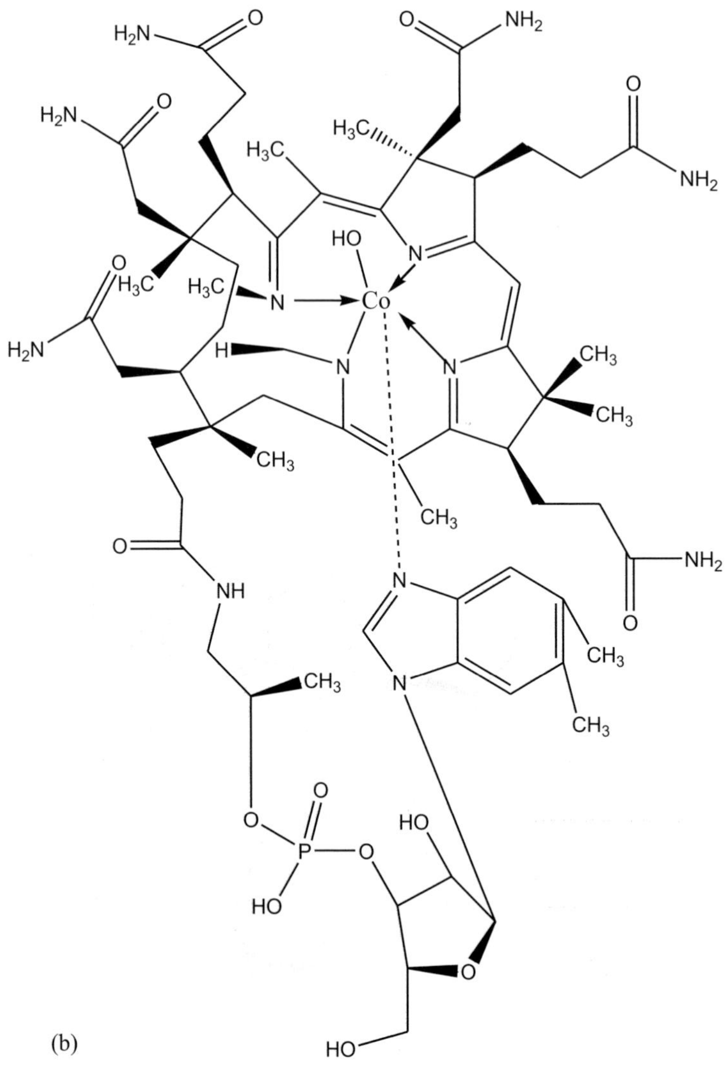

(b)

Fig. 13.26 (Continued)

absorbed by the intestines. Calcium ions in the intestines facilitate this process before vitamin B12 enters the blood stream. A deficiency of cyanocobalamin causes serious disease conditions associated with the central nervous system.

13.8 VITAMIN C (L-ASCORBIC ACID)

Vitamin C or L-ascorbic acid performs the important function of antioxidant, generating positive health effects as well as other benefits in the body as it protects against oxidative stress. L-ascorbic acid is also needed in a number of essential metabolic reactions in the body and is a cofactor in many enzymatic processes responsible in some metabolic

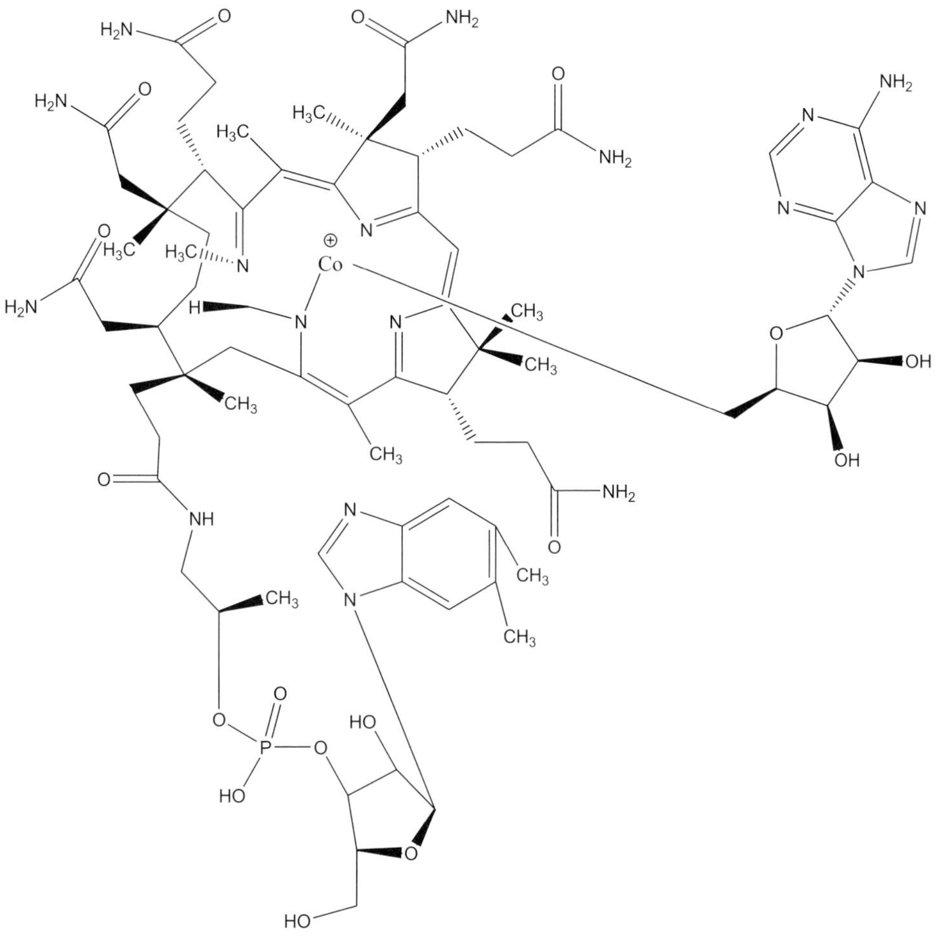

(c) 5'-deoxyadenosylcobalamin

Fig. 13.26   *(Continued)*

reactions (Padayatty *et al.* 2003). L-ascorbic acid is also known to play an important role as a natural antihistamine, where it prevents histamine release and enhances the detoxification of histamine (Johnston *et al.* 1992).

The structure of ascorbic acid (Figure 13.27a and b) reveals that there are two enantiomers: the L and D forms. The L-enantiomer is the form with physiological activity and the D-form is not active. L-ascorbic acid is a strong reducing agent and can easily be transformed to L-dehydroascobate (Figure 13.27c), which is its oxidised form of ascorbate, via an intermediate of a semidehydroascorbic acid radical. The reduction of L-ascorbic acid is reversible and is catalysed by the action of enzymes and glutathione.

The importance of the formed semihydroascorbate intermediate is that the ability of ascorbate free radical to react with oxygen is generally very low, hindering the formation of superoxide molecules. It therefore takes two semidehydroascorbate radicals to react to generate one ascorbate and one dehydroascorbate molecule. Moreover, the action of glutathione

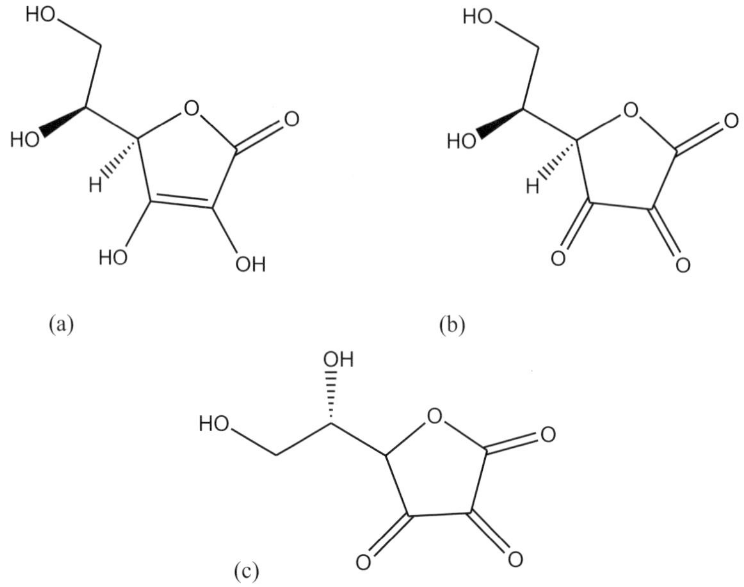

Fig. 13.27 Structures of ascorbic acid: (a) ascorbic acid (reduced form); (b) dehydroascorbic acid (oxidised form); (c) L-dehydroascorbate (http://www.britannica.com/EBchecked/topic/631079/vitamin-C).

is important as it protects ascorbate as well as enhancing the antioxidant capacity of blood L-ascobate (Nualart *et al.* 2003; Gropper *et al.* 2004).

In cases where L-ascorbic acid is deficient in humans, the body suffers a disease condition scurvy (Wilson 1975). To restore vitamin C levels in the body, either take nutritional supplements (commercially available in the form of tablets, capsules, drink mixes, etc.) or else consume a diet rich in vitamin C (both plant and animal origin). Natural sources which contain high amounts of vitamin C include fruits and vegetables, for example, citrus fruits, papaya and many others. Most animals are known to have the ability to synthesise their vitamin C and can therefore form good sources of L-ascorbic acid (Chatterjee 1973).

Upon consumption, L-ascorbic acid is transported all the way to the intestine using glucose-sensitive and glucose-insensitive mechanisms (Wilson 2005). The digested component is then absorbed through two active transport systems which involve (1) sodium ascorbate co-transporters and hexose transporters and (2) simple diffusion (Savini *et al.* 2007). The metabolism of L-ascorbic acid involves an enzymatic oxidation step by the action of L-ascorbate oxidase; by-products are then excreted together with urine.

13.9 CONCLUSIONS

Dietary supplements are needed for many purposes by people of all age groups in all geographical locations. Knowledge of how they work and what they can do is therefore vital, especially when a diet is deficient in some area. Ensuring that a diet contains all the required ingredients, with supplements if necessary, means that a number of deficiency diseases can be overcome with relative ease.

REFERENCES

Aburjai, T., Bernasconi, S., Manzocchi, L. & Pelizzoni, F. (1996) Isolation of 7-Dehydrocholesterol from cell cultures of *Solanum malacoxylon*. *Phytochemistry* 43, (4), 773–776.

Albert, P. J., Proal, A. D. & Marshall, T. G. (2009) Vitamin D: The alternative hypothesis. *Autoimmunity Reviews* 8, 639–644.

Bailey, A. L., Maisey, S., Southon, S., Wright, A. J., Finglas, P. M. & Fulcher, R. A. (1997) Relationships between micronutrient intake and biochemical indicators of nutrient adequacy in a free-living' elderly UK population. *British Journal of Nutrition* 77 (2), 225–242.

Bailey, S. W. & Ayling, J. E. (2009) The extremely slow and variable activity of dihydrofolate reductase in human liver and its implications for high folic acid intake. *Proceedings of National Academy of Science, USA*, 106 (36), 15424–15429.

Basu, T. K. & Dickerson, J. W. T. (1996) Vitamin A. In: *Vitamins in Human Health and Disease*, CAB International (ed.), Biddles Ltd, Guildford, UK, pp. 148–177.

Beharka, A., Redican, S., Leka, L. & Meydani, S. N. (1997) Vitamin E status and immune function. In: *Methods in Enzymology: Vitamins and Coenzymes*, Part L. McCormick, D. B., Suttie, J. W. & Wagner, C. (eds), Academic Press, New York, pp. 247–263.

Berdanier, C. (1997). *Advanced Nutrition: Micronutrients*. CRC Press, Boca Raton.

Beulens, J. W., Bots, M. L., Atsma, F., Bartelink, M. L., Prokop, M., Geleijnse, J. M., Witteman, J. C., Grobbee, D. E. & van der Schouw, Y. T. (2009) High dietary menaquinone intake is associated with reduced coronary calcification. *Atherosclerosis* 203 (2), 489–493.

Blazer, V. S. (1992) Nutrition and disease resistance in fish. *Annual Review of Fish Diseases* 2, 309–323.

Blazer, V. S. & Wolke, R. E. (1990) The effects of a-tocopherol on the immune response and non-specific resistance factors of rainbow trout (Salmo gairdneri Richardson). *Aquaculture* 37, 1–9.

Blomhoff, R. (1994) Overview of vitamin A metabolism and function. In: *Vitamin A in Health and Disease*, Blomhoff, R. (ed.), Marcel Dekker, New York, pp. 1–35.

Booth, S. L. & Suttie, J. W. (1998) Dietary intake and adequacy of vitamin K. *Journal of Nutrition* 128, 785–788.

Bowman, B. A. and Russell, R. (2006) *Present Knowledge in Nutrition*, 9th edition. ILSI Press, Washington, DC.

Brody, T. (1999) *Nutritional Biochemistry*. Academic Press, San Diego.

Burton, G. W. & Ingold, K. U. (1981) Autoxidation of biological molecules. 1. Antioxidant activity of vitamin E and related chain-breaking phenolic antioxidants in vitro. *Journal of American Chemical Society* 103, 6472–6477.

Butterworth, R. F. (1993) Pathophysiologic mechanisms responsible for the reversible (thiamine-responsive) and irreversible (thiamine non-responsive) neurological symptoms of Wernicke's encephalopathy. *Drug Alcohol Reviews* 12 (3), 315–322.

Butterworth, R. F. (2006) Thiamin. In: *Modern Nutrition in Health and Disease*, 10th edition. Shils, M. E., Shike, M., Ross, A. C., Caballero, B., Cousins, R. J. (eds), Lippincott Williams and Wilkins, Baltimore.

Chatterjee, I. B. (1973) Evolution and the biosynthesis of ascorbic acid. *Science* 182 (4118), 1271–1272.

Clagett-Dame, M. & DeLuca, H. F. (2002) The role of vitamin A in mammalian reproduction and embryonic development. *Annual Reviews in Nutrition* 22, 347–381.

Combs, G. F. (1998) *Vitamins: Fundamental Aspects in Nutrition and Health*. Academic Press, San Diego.

Craciun, A. M., Groenen-van Dooren, M. M. C. L., Thijssen, H. H. W. & Vermeer, C. (1998) Induction of prothrombin synthesis by K-vitamins compared in vitamin K-deficient and in brodifacoum-treated rats. *Biochimica et Biophysica Acta* 1380, 75–81.

Curino, A., Skliar, M. & Boland, R. (1998) Identication of 7-dehydrocholesterol, vitamin D3, 25(OH)-vitamin D3 and 1,25(OH)2-vitamin D3 in *Solanum glaucophyllum* cultures grown in absence of light. *Biochimica et Biophysica Acta* 1425, 485–492.

Cutolo, M., Otsa, K., Uprus, M., Paolino, S. & Seriolo, B. (2007) Vitamin D in rheumatoid arthritis. *Autoimmune Reviews* 7 (1), 59–64.

Dart, R. C. (2006) Hydroxocobalamin for acute cyanide poisoning: new data from preclinical and clinical studies; new results from the prehospital emergency setting. *Clinical Toxicology* 44 (Suppl 1), 1–3.

Dusso, A. S., Brown, A. J. & Slatopolsky, E. (2005) Vitamin D. *American Journal of Physiology: Renal Physiology* 289, F8–F28.

Elder, S. J., Haytowitz, D. B., Howe, J., Peterson, J. W. & Booth, S. L. (2006) Vitamin K contents of meat, dairy, and fast food in the U.S. Diet. *Journal of Agriculture and Food Chemistry* 54 (2), 463–467.

Geleijnse, J. M., Vermeer, C., Grobbee, D. E., Schurgers, L. J., Knapen, M. H. J., van der Meer, I. M., Hofman, A. & Witteman, J. C. M. (2004) Dietary intake of menaquinone is associated with a reduced risk of coronary heart disease: The Rotterdam Study. *American Society for Nutrition Journal of Nutrition* 134, 3100–3105.

Gibson, R. S. (2005) Riboflavin. In: *Principles of Nutritional Assessment*, 2nd edition. Oxford University Press, Oxford.

Grant, W. B. &. Holick, M. F. (2005) Benefits and requirements of vitamin D for optimal health: a review. *Alternative Medicine Review* 10 (2), 94–111.

Groenen-van Dooren, M. M. C. L., Soute, B. A. M., Jie, K.-S. G., Thijssen, H. H. W. & Vermeer, C. (1993) The relative effects of phylloquinone and menaquinone-4 on the blood coagulation factor synthesis in vitamin K-deficient rats. *Biochemical Pharmacology* 46 (3), 433–437.

Groenen-van Dooren, M. M. C. L., Ronden, J. E., Soute, B. A. M. & Vermeer, C. (1995) Bioavailability of phylloquinone and menaquinones after oral and colorectal administration in vitamin K-deficient rats. *Biochemical Pharmacology* 50 (6), 797–801.

Gropper, S. S., Smith, J. L. & Groff, J. L. (2004) *Advanced Nutrition and Human Metabolism*, 4th edition. Thomson Wadsworth, Belmont, CA, pp. 260–275.

Gropper, S. S., Smith, J. L. & Groff, J. L. (2009) Riboflavin. In: *Advanced Nutrition and Human Metabolism*, 5th edition. Thomson Wadsworth, Belmont, CA, Chapter 9, pp. 329–333.

Hall, A. H. & Rumack, B. H. (1987) Hydroxycobalamin/sodium thiosulfate as a cyanide antidote. *The Journal of Emergency Medicine* 5 (2), 115–121.

Hardie, L. J., Fletcher, T. C. & Secombes, C. J. (1984) The effect of vitamin E on the immune response of the Atlantic salmon (*Salmo salar* L.). *Aquaculture* 87, 1–13.

Hauschka, P. Y., Lian, J. B. & Gallop, P. M. (1975) Direct identification of the calcium-binding amino acid, gamma-carboxyglutamate, in mineralized tissue. *Proceedings of National Academy of Science* 72, 3925–3929.

Holick, M. F. (2003) Vitamin D: a millennium perspective. *Journal of Cellular Biochemistry* 88, 296–307.

Holick, M. F. (2004) Vitamin D: importance in the prevention of cancers, type 1 diabetes, heart disease, and osteoporosis. *American Journal of Clinical Nutrition* 79 (3), 362–371.

Holick, M. F. (2005) The vitamin D epidemic and its health consequences. *Journal of Nutrition* 135 (11), 2739S–2748S.

Hollis, B. W., Roos, B. A., Draper, H. H. & Lambert, P. W. (1981) Vitamin D and its metabolites in human and bovine-milk. *Journal of Nutrition* 111, 1240–1248.

Hussain, M. M. (2000) A proposed model for the assembly of chylomicrons. *Arteriosclerosis* 148, 1–15.

Hwang, S.-M. (1985) Liquid chromatographic determination of vitamin K1 trans- and cis-isomers in infant formula. *Journal of the Association of Official Analytical Chemists* 68, 684–689.

Ikeda, Y., Iki, M., Morita, A., Kajita, E., Kagamimori, S., Kagawa, Y. & Yoneshima, H. (2006) Intake of fermented soybeans, natto, is associated with reduced bone loss in postmenopausal women: Japanese Population-Based Osteoporosis (JPOS) Study. *Journal of Nutrition* 136 (5), 1323–1328.

Indyk, H. (1988) Liquid chromatographic study of vitamin K1 degradation: Possible nutritional implications in milk. *Milchwis-senschaft*, 43, 8.

Jakobsen, J. & Saxholt, E. (2009) Vitamin D metabolites in bovine milk and butter. *Journal of Food Composition and Analysis*, 22, 472–478.

Johnston, C. S., Martin, L. J. & Cai, X. (1992) Antihistamine effect of supplemental ascorbic acid and neutrophil chemotaxis. *Journal of the American College of Nutrition* 11 (2), 172–176.

Kalra, E. K. (2003) Nutraceutical: definition and introduction. *AAPS Pharmaceutical Sciences* 5 (3), 27–28.

Katsuyama, H., Ideguchi, S., Fukunaga, M., Saijoh, K. & Sunami, S. (2002) Usual dietary intake of fermented soybeans (Natto) is associated with bone mineral density in premenopausal women. *Journal of Nutritional Science and Vitaminology* 48 (3), 207–215.

Khachik, F. (2006) Distribution and metabolism of dietary carotenoids in humans as a criterion for development of nutritional supplements, *Pure Applied Chemistry* 78 (8), 1551–1557.

Kocabas, A. M. &. Gatlin D. M. (1999) Dietary vitamin E requirement of hybrid striped bass (*Morone chrysops* female X *M. saxatilis* male). *Aquaculture Nutrition* 5, 3–7.

Komai, M. & Shirakawa, H. (2007) Vitamin K metabolism. Menaquinone-4 (MK-4) formation from ingested VK analogues and its potent relation to bone function. *Clinical Calcium* 17 (11), 1663–1672.

Kunz, C., Niesen, M., Vonlilienfeldtoal, H. & Burmeister, W. (1984) Vitamin-D, 25-hydroxy-vitamin-D and 1,25-dihydroxy-vitamin-D in cow's milk infant Formulas and breast-milk during different stages of lactation. *International Journal of Vitamin & Nutrition Research* 54, 141–148.

Lamberg-Allardt, C. (2006) Vitamin D in foods and as supplements. *Progress in Biophysics and Molecular Biology* 92, 33–38.

Machlin, L. J. (1999) Vitamin E. In: *Handbook of Vitamins*, Machlin, L. J. (ed.) Marcel Dekker, Inc., New York.

Mahan, L. K. & Escott-Stump, S. (eds) (2000) *Krause's Food, Nutrition, and Diet Therapy*. 10th edition. W.B. Saunders Company, Philadelphia.

Man-Jeong, P., Jundong, Y., Man-Bae, H., Sung-Jean, K., Kyoung-Rae, K., Young-Hwan, A., Sangdun, C. & Gwang, L. (2008) Gas chromatographic-mass spectrometric analyses of cholesterol and its precursors in rat plasma as tert-butyldimethylsilyl derivatives. *Clinica Chimica Acta* 396, 62–65.

Martin, P. R., Singleton, C. K. & Hiller-Sturmhofel, S. (2003) The role of thiamine deficiency in alcoholic brain disease. *Alcohol Research and Health* 27 (2), 134–142.

Mattila, P. (1995) Analysis of cholecalciferol, ergocalciferol and their 25-hydroxylated metabolites in foods by HPLC. Dissertation, EKT Series 995. University of Helsinki.

McCormick, D. B. (2006) Vitamin B6. In: *Present Knowledge in Nutrition*, 9th edition, Bowman, B. A. and Russell, R. M. (eds), International Life Sciences Institute, Washington, DC, Vol. 2, p. 270.

McDowell, L. R. (2000) Vitamin E. In: *Vitamins in Animal Nutrition: Comparative Aspects to Human Nutrition*. Academic Press, San Diego, p. 93.

Merrifield, L. S. & Yang, H. Y. (1965) Vitamin K5 as a fungistatic agent. *Applied Microbiology* 13 (5), 660–662.

Merrill, A. H. & Henderson, J. M. (1987) Disease associated with deficiencies in vitamin B6 metabolism or utilization. *Annual Reviews in Nutrition* 7, 137–156.

Merrill, A. H. Jr. & Burnham, F. S. (1990) Vitamin B6. In: *Present Knowledge in Nutrition*, Brown, M.L. (ed.) International Life Science Institute, Washington, DC, pp. 155–162.

Merrill, A. H., Henderson, J. M., Wang, E., McDonald, B. W. & Millikan, W. J. (1984) Metabolism of vitamin B6 by human liver. *Journal of Nutrition* 114, 1664–1674.

Min-Hsien, L. & Shi-Yen, S. (2004) Vitamin E requirements of juvenile grass shrimp, Penaeus monodon, and effects on non-specific immune responses. *Fish and Shellfish Immunology* 16, 475–485.

Moriguchi, S., Kobayshi, N. & Kishino, Y. (1990) High dietary intakes of vitamin E and cellular immune function in rats. *Journal of Nutrition* 120, 1096–1102.

Nagy, N. E., Holven, K. B., Roos, N., Senoo, H., Kojima, N., Norum, K. R. & Blumhoff, R. (1997) Storage of vitamin A in extrahepatic stellate cells in normal rats. *Journal of Lipid Research* 38, 645–658.

Napoli, J. L. (1999) Interactions of retinoid binding proteins and enzymes in retinoid metabolism. *Biochimica Biophysica Acta* 1440, 139–162.

National Research Council. (1983) *Nutrient Requirements of Warmwater Fishes and Shellfishes*. National Academy Press, Washington DC.

Nimptsch, K., Rohrmann, S., Kaaks, R. & Linseisen, J. (2010) Dietary vitamin K intake in relation to cancer incidence and mortality: results from the Heidelberg cohort of the European Prospective Investigation into Cancer and Nutrition (EPIC-Heidelberg). *American Journal of Clinical Nutrition* 91 (5), 1348–1358.

Nualart, F. J., Rivas, C. I., Montecinos, V. P., Godoy, A. S., Guaiquil, V. H., Golde, D. W. & Vera, J. C. (2003) Recycling of vitamin C by a bystander effect. *Journal of Biological Chemistry* 278, 10128–10133.

Olson, J. A. (1984) Vitamin A. In: *Nutrition Review's Present Knowledge in Nutrition*, The Nutrition Foundation, Washington DC, pp. 176–191.

Olson, J. A. (1989) Provitamin A function of carotenoids: the conversion of b-carotene into vitamin A. *Journal of Nutrition* 119, 105–108.

Olson, J. A. (1999) Bioavailability of carotenoids. *Archives of Latinoamerican Nutrition*, 49 (3 Suppl 1), 21S–25S.

Ortuno, J., Esteban, M. A. & Meseguer, J. (2000) High dietary intake of a-tocopherol acetate enhances the non-specific immune response of gilthead seabream (*Sparus aurata* L.). *Fish Shellfish Immunology* 10, 293–307.

Ovesen, L., Andersen, R. &. Jakobsen, J. (2003a) Geographical differences in vitamin D status, with particular reference to European countries. *Proceedings of the Nutrition Society* 62, 813–821.

Ovesen, L., Brot, C. & Jakobsen, J. (2003b) Food contents and biological activity of 25-hydroxyvitamin D: a vitamin D metabolite to be reckoned with? *Annals of Nutrition and Metabolism* 47, 107–113.

Padayatty, S. J., Katz, A., Wang, Y., Eck, P., Kwon, O., Lee, J.-H., Chen, S., Corpe, C., Dutta, A., Dutta, S. K. & Levine, M. (2003) Vitamin C as an antioxidant: evaluation of its role in disease prevention. *Journal of the American College of Nutrition* 22 (1), 18–35.

Panush, M. E. & Delafuente, J. C. (1985) Vitamins and immuncompetence. *World Reviews in Nutrition and Diet* 45, 97–123.

Park, B. K. & Leck, J. B. (1982) A comparison of vitamin K antagonism by warfarin, difenacoum and brodifacoum in the rabbit. *Biochemical Pharmacology* 31 (22), 3635–3639.

Parvianinen, M. T., Koskinen, T., Ala-Houlala, M. & Visakorpi, J. K. (1984) A method for routine estimation of vitamin D activity in human and bovine milk. *Acta Vitaminology Enzymology* 6, 211–219.

Powers, H. J. (2003) Riboflavin (vitamin B-2) and health. *American Journal of Clinical Nutrition* 77, 1352–1360.

Pozniakov, S. P. (1986) Mechanism of action of vitamin A on cell differentiation and function. *Ontogenez* 17 (6), 578–586.

Price, P. A., Poser, J. W. & Raman, N. (1976) Primary structure of the γ-carboxyglutamic acid-containing protein from bovine bone. *Proceedings of National Academy of Science* 73, 3374–3375.

Rall, L. C. & Meydani, S. N. (1993) Vitamin B6 and immune competence. *Nutrition Review* 51, 217–225.

Reedstrom, C. K. & Suttie, J. W. (1995) Comperative distribution, metabolism, and utilization of phylloqui-none and menaquinone-9 in rat liver. *Proceedings of the Society for Experimental Biology and Medicine* 209, 403–409.

Rindi, G., Imarisio, L. & Patrini, C. (1986) Effects of acute and chronic ethanol administration on re-gional thiamin pyrophosphokinase activity of the rat brain. *Biochemical Pharmacology* 35 (22), 3903–3908.

Rothman, K. J., Moore, L. L., Singer, M. R., Ngyen, S., Mannino, S. & Milunsky, A. (1995) Teratogenicity of high vitamin A intake. *New England Journal of Medicine* 23, 1369–1373.

Savini, I., Rossi, A., Pierro, C., Avigliano, L. & Catani, M. V. (2007) SVCT2: Key proteins for vitamin C uptake. *Amino Acids* 34 (3), 347–355.

Schoenmakers, I., Goldberg, G. R. & Prentice, A. (2008) Abundant sunshine and vitamin D deficiency. *British Journal of Nutrition* 99 (6), 1171–1173.

Sebrell, W. H. & Butler, R. E. (1939) Riboflavin deficiency in man (ariboflavinosis). *Public Health Reports* 54, 2121.

Semba, R. D. (1998) The role of vitamin A and related retinoids in immune function. *Nutriton Reviews* 56, S38– S48.

Sen, C., Khanna, S. & Roy, S. (2006) Tocotrienols: Vitamin E beyond tocopherols. *Life Sciences* 78 (18), 2088–2098.

Serrini, G., Zhang, Z. & Wilson, R. P. (1996) Dietary riboflavin requirement of fingerling channel catfish (*Ictalurus punctatus*). *Aquaculture* 139 (3–4), 285–290.

Shah, D. V. & Suttie, J. W. (1974) The vitamin K dependent, in vitro production of prothrombin. *Biochemical and Biophysical Research Communications* 60, 1397–1402.

Shiau, S. Y. & Shiau, L. F. (2001) Reevaluation of the vitamin E requirements of juvenile tilapia (*Oreochromis niloticus* x *O. aureus*). *Animal Science* 72, 529–534.

Stahl, W., Nicolai, S., Briviba, K., Hanusch, M., Broszeit, G., Peters, M., Martin, H. D. & Sies, H. (1997) Biological activities of natural and synthetic carotenoids: induction of gap junctional communication and singlet oxygen quenching. *Carcinogenesis* 18, 89–92.

Stenflo, J. & Ganrot, P. O. (1972) Vitamin K and biosynthesis of prothrombin. *Journal of Biological Chemistry* 247, 8160–8166.

Stenflo, J., Fernlund, P., Egan, W. & Roepstorff, P. (1974) Vitamin K dependent modification of glutamic acid residues in prothrombin. *Proceedings of National Academy of Science USA* 71, 2730—2733.

Takeuchi, T., Watanabe, K., Satoh, S. & Watanabe, T. (1993) Requirement of grass carp fingerlings for alpha-tocopherol. *Nippon Suisan Gakkaishi* 58, 1743–1749.

Tanphaichitr, V. (1999) Thiamin. In: *Modern Nutrition in Health and Disease*, 9th edition. Shils, M. E., Olsen, J. A., Shike, M. & Ross., A. C. (eds), Lippincott Williams and Wilkins, Baltimore.

Thane, C. W., Bates, C. J., Shearer, M. J., Unadkat, N., Harrington, D. J., Paul, A. A., Prentice, A. & Bolton-Smith, C. (2002) Plasma phylloquinone (vitamin K1) concentration and its relationship to intake in a national sample of British elderly people. *British Journal of Nutrition* 87 (6), 615–622.

Trumbo, P. R. (2006) Pantothenic acid. In *Modern Nutrition in Health and Disease*, 10th edition. Shils, M. E., Shike, M., Ross, A. C., Caballero, B., Cousins, R. J. (eds), Lippincott Williams and Wilkins, Philadelphia, PA, pp. 462–467.

Tryfiates, G. P. (1980) *Vitamin B6 Metabolism and Role in Growth*. Food and Nutrition Press, Westport CT.

Tsukamoto, Y., Ichise, H., Kakuda, H. & Yamaguchi, M. (2000) Intake of fermented soybean (natto) increases circulating vitamin K_2 (menaquinone-7) and gamma-carboxylated osteocalcin concentration in normal individuals. *Journal of Bone and Mineral Metabolism* 18 (4), 216–222.

USDA. (2008) Composition of Foods Raw, Processed, Prepared. USDA National Nutrient Database for Standard Reference, Release 20 USDA, Feb. 2008.

Wang, X. D. (1994) Absorption and metabolism of β-carotene 1994. *Journal of American College of Nutrition* 13, 314–325.

Weber, P., Bendich, A. & Machlin, L. J. (1997) Vitamin E and human health: rationale for determining recommended intake levels. *Nutrition* 13 (5), 450–460.

Wilson, J. X. (2005) Regulation of vitamin C transport. *Annual Review of Nutrition* 25, 105–125.

Wilson, L. G. (1975) The clinical definition of scurvy and the discovery of vitamin C. *Journal of the History of Medicine and Allied Sciences* 30 (1), 40–60.

Wilson, R. P., Bowser, P. R. & Poe, W. E. (1984) Dietary vitamin E requirement of fingerling channel catfish. *Journal of Nutrition* 114, 2053–2058.

Woodall, A. N., Ashley, L. M., Halver, J. E., Olcott, H. S. & Van Der Veen, J. (1964) Nutrition of salmonid fishes. XII. The a-tocopherol requirement of chinook salmon. *Journal of Nutrition* 84, 125–135.

Zeisel, S. H. (1999) Regulation of nutraceuticals. *Science* 285, 185–186.

14   Glazing Agents

Abstract: There are many different types of food items which perform different functions. Some of these functions are directed at improving health, the nutritive value of foods, palatability, appearance, texture, storage or shelf life. Glazing agents, also called polishing agents, are however incorporated in foods for the purpose of imparting a shiny appearance as well as a protective coating to foods. In this chapter, the chemistry of various glazing agents (which are in most cases based on waxes) is discussed.

Keywords: cryoprotectives; food coating agents; mineral hydrocarbons; polishing agents; shiny appearance; waxes

14.1   INTRODUCTION TO GLAZING AGENTS

Several substances are known to play a role as cryoproctectives and glazes in the food industry. In many instances, these glazes are processing aids. These include mineral hydrocarbon oils, waxes (e.g. microcrystalline wax, crystalline wax, beeswax, polyethylene wax, candelilla wax, carnauba wax and paraffin wax), stearic acid, lanolin, shellac, colophonium esters, petrolatum, chitosan-based films, polyphosphates, tea extracts and many others.

Glazing agents perform the function of protective coating or sheen on the surface of foods, to give foods an improved appearance and also help to extending the shelf life of the food.

14.2   MINERAL HYDROCARBON GLAZES

Mineral hydrocarbons (MHC) are mixtures of alkane hydrocarbons (C15 to C40) derived from materials of mineral origin, particularly those a of non-vegetable (mineral) source and those which are obtained through distillation of petroleum (Heimbach *et al*. 2002). As well as being odourless, flavourless and colourless, mineral oils possess the ability to prevent water absorption or the sticking together of foods, making them suitable for use in foods.

Mineral oils have also been sprayed on some food items such as dried fruits in order to prevent them from sticking together. However, due to the possibility of their bioaccumulation in the body system, this practice is now discouraged. Mineral oils still find use in minute quantities in foods such as chewing gums and cheese as well as in the bakery industry where they prevent the sticking of bread to the baking trays. However, bioaccumulation can still occur with chewing gum; consumers are therefore warned not to swallow these food items

Chemistry of Food Additives and Preservatives, First Edition. Titus A. M. Msagati.
© 2013 John Wiley & Sons, Ltd. Published 2013 by John Wiley & Sons, Ltd.

to avoid the risks associated with mineral oils (Heimbach *et al.* 2002). For this reason, the use of mineral hydrocarbon oils in foods is regulated in many countries.

The term 'mineral hydrocarbons' generally refers to several hydrocarbon compounds such as white mineral oils, paraffin waxes, microcrystalline waxes and petrolatum. All of these compounds originate from petroleum products, after the separation of non-hydrocarbons by distillation processes. Typical hydrocarbons in MHC are mainly straight-chain alkanes (paraffins), branched alkanes (isoparaffins), saturated hydrocarbons and saturated hydrocarbons with rings and side chains (cycloparaffins, also known as naphthenics).

Mineral hydrocarbons are used in foods either directly – whereby MHC are intentionally added to foods to play a certain important function, e.g. as coatings, polishing agents or as dust control agents – or indirectly through physical contact with potential sources of MHC.

14.2.1 Direct application of MHC

Mineral hydrocarbons perform the important function as coating, for example wax coatings are used widely for cheeses. MHC coatings are also used for fruit and vegetable food items where they prevent moisture loss, protect from bruises and injuries and also give the food product a shiny appearance. MHC coats may sometimes be uni-coats, in the sense that the coating is only of one MHC (e.g. mineral oil), but the coating may also be a mixture of MHC compounds. Hetero-coating can also be formed, where MHC and other non-MHC compounds are combined to perform other functions such as inhibition of mould growth.

Other MHC compounds such as microcrystalline and paraffin waxes find use in the chewing gum industry where they are incorporated in the saliva-insoluble chewing gum base (note that the other part that forms chewing gum is saliva soluble). Mineral oil and petrolatum have important applications as releasing agents in moulding starch, and also as sealing and polishing agents in the processing of confectionery. In the bakery industry, MHC such as mineral oils are normally added to divide dough into portions for single loaves and rolls.

14.2.2 Indirect application of MHC

Indirect application of MHC refers to the intentional introduction of MHC in either packaging materials, seals in bottles or lubricants. For example, corrugated cartoons used as packaging materials for fruits and vegetables contain paraffin wax as well as microcrystalline wax which, through physical contact, migrate into these food items. The possibility of migration of MHC from containers to foods is not likely for a number of reasons, including the fact that these food items are normally kept at very low temperatures during transport or storage which discourages MHC such as paraffin wax or microcrystalline wax to migrate. The time span over which these foods are stored in such containers is normally very short, such that the amount migrating will be highly insignificant (if at all). Such foods are never in direct contact with the cartoon, as there is always a separating sheath that prevents mixing with the walls of the corrugated containers. Finally, some of the outer leaves of vegetables are removed in order to remove any threat of contamination (van Battum and Rijk 1979).

Some MHC are used as adhesives and others as food-grade lubricants with application in food-processing devices, where the danger of contamination is minimal. With domestic canning applications however, there may be possibilities of contact with foods and hence possible migration into food. This possibility of contamination is generally ruled out because of the aqueous compatibility nature of the food materials that are normally prepared

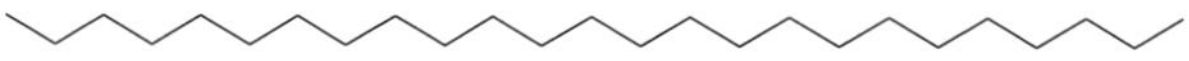

Fig. 14.1 Chemical structure of paraffin wax.

domestically (vegetables, fruits, etc.); paraffin waxes cannot mix with these foods as they are insoluble in such media (van Battum and Rijk 1979).

Mineral hydrocarbons are also incorporated in flexible wax packaging (wax paper, wax-coated cereal liners and wax-coated paperboard) which finds application in foods such as crackers and dry ready-to-eat cereals. For foods of high fat/oil content which are not kept at low temperatures and which do not have a separating medium to prevent any physical contact with the containers (wax coated papers), there may be the possibility of migration and hence contamination.

Wax-coated paper cups and other utensils may also allow migration of MHC; however, the chances are negligible since these drinks are very aqueous to allow dissolution of MHC and the time during which these are in use is very short, reducing any danger. Of more importance is the presence of MHC in both expanded and extruded foam polystyrene packaging and containers. Cups, plates and trays can be made of foamed polystyrene, and solid styrene can be found in kitchen knives as well as other kitchen products. Some reports have revealed the possibility of MHC contamination due to the use of polystyrene resin containers in foods (SPI 1996).

14.3 CHEMISTRY OF MHCs

14.3.1 Food-grade paraffin waxes

Members of paraffin wax, whose general molecular formula is $C_nH_{2n}+2$ where n=24–36, are mostly unbranched alkanes (Figure 14.1). Food-grade paraffin wax is used in the food industry to create a shiny coating on a number of food items such as confectionary. An important property of paraffin wax used in food products is that it is non-digestible.

14.3.2 Microcrystalline waxes

Unlike paraffin wax, microcrystalline wax is composed of high-molecular-weight saturated aliphatic hydrocarbons. These contain a higher percentage of isoparaffinic and naphthenic hydrocarbons (Figure 14.2a and b). Because of branching and cyclisation, microcrystalline waxes are more viscous and dense with a higher melting point and elastic than paraffin waxes.

14.3.3 Natural wax glazes

14.3.3.1 Beeswax

An example of a natural glaze used in the food industry is beeswax. For convenience, beeswax can be represented by the chemical formula $C_{15}H_{31}COOC_{30}H_{61}$ and is a natural wax product of honey bees (mainly of genus *Apis*). The composition of beeswax includes mainly esters of fatty acids and some long-chain alcohols. The major ester ingredients of beeswax are formed by palmitate, palmitoleate and hydroxypalmitate as well as long-chain oleate esters. Beeswaxes are widely used in the food industry as food coatings (e.g. cheese).

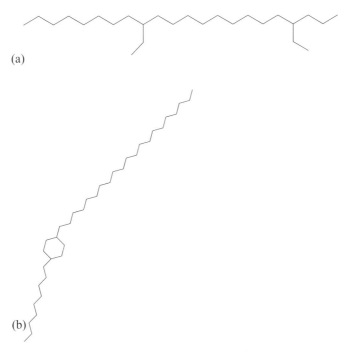

(a)

(b)

Fig. 14.2 Chemical structure of (a) isoparaffin; and (b) cycloparaffin.

14.3.3.2 Carnauba wax

Carnauba wax, of plant origin, is another natural wax glaze. Carnauba is obtained from the leaves of the palm plant *Copernicia prunifera* (Steinle 1936; Parish *et al.* 2002). It is insoluble in water and composed of aliphatic esters such as hydroxycinnamate, which are obtained through the reactions of hydroxycarboxylic acids or methoxycinnamic acid with fatty acid alcohols (Wolfmeier *et al.* 2002). Carnauba wax is normally used as a coating in sweets, chewing gums and other food products, and is also used as a food additive for the purposes of anticaking and as a lubricant.

14.3.3.3 Candelilla wax

Candelilla wax is another natural wax, originating from the leaves of candelilla plants. The composition of candelilla wax is mostly alkanes, esters, aliphatic acids and triterpenoid esters. Like carnauba wax, candelilla wax is insoluble in water, but differs by the presence of high hydrocarbon content (Wolfmeier *et al.* 2002).

14.3.3.4 Colophonium

Esters of colophonium such as rosin are one of the natural glazes used in foods such as chewing gums. Rosin is composed of resin acids, for example abietic acid (Figure 14.3; Palkin and Smith 1938; Fiebach and Grimm 2000).

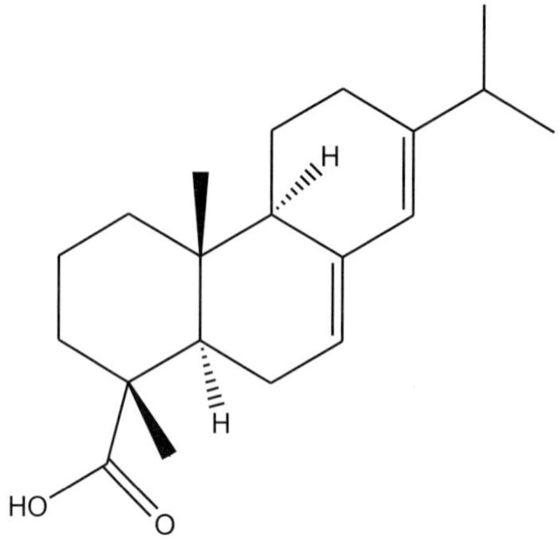

Fig. 14.3   Chemical structure of abietic acid.

14.3.3.5   Stearic acid glazes

Stearic acid (Figure 14.4) is a solid waxy compound with chemical formula $C_{18}H_{36}O_2$. It is one of the many natural waxes used in the food industry as glazers.

14.3.4   Commercial/synthetic food glazes

Of the synthetic food glazes currently in use in the food industry, casic methacrylate copolymer (Figure 14.5) is an example. The International Union of Pure and Applied Chemistry (IUPAC) nomenclature is poly(butylmethacrylateco-(2-dimethylaminoethyl)methacrylate-co-methyl methacrylate), whereby the three monomers are present in the ratio 1:2:1.

Basic methacrylate copolymer is a cationic copolymer composed of dimethylaminoethyl methacrylate, methacrylic esters, butyl methacrylate and methyl methacrylate (Figure 14.5). Basic methacrylate copolymer is normally synthesised bulk polymerisation, whereby the polymer chain increases through a radical polymerisation of the monomers, which takes place at elevated temperatures. Basic methacrylate copolymer finds application as a glazing agent or coating agent mostly in solid food supplements, and it has been proven to be safe for human consumption (Eisele *et al.* 2011).

14.4   CONCLUSION

Glazing agents have demonstrated many benefits in foods, including: imparting a shiny appearance to foods; acting as a preservative; and antimicrobial properties when used as a food packaging material (e.g. chitosan).

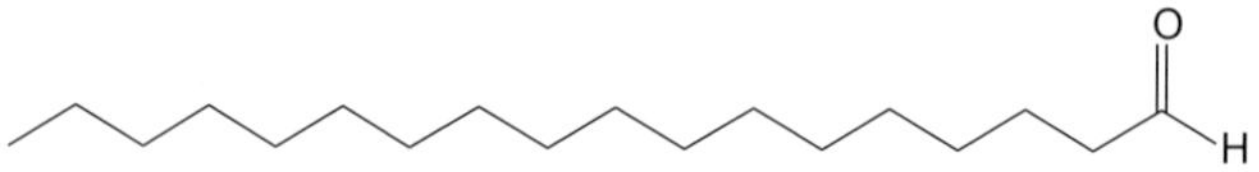

Fig. 14.4   Chemical structure of stearic acid.

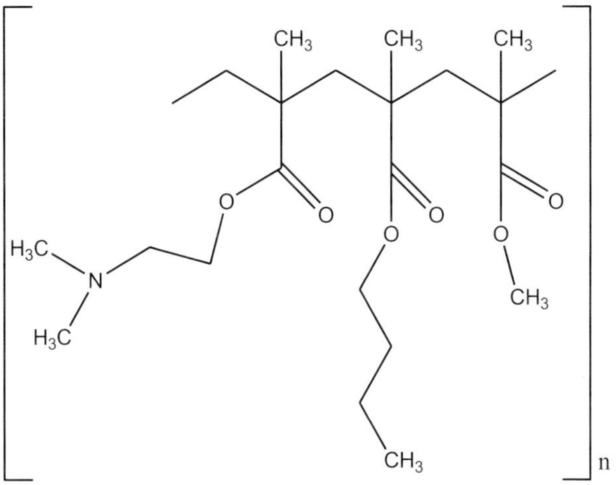

Fig. 14.5 General structure of basic methacrylate copolymer.

REFERENCES

Eiselea, J., Haynesb, G. & Rosamiliac, T. (2011) Characterisation and toxicological behaviour of basic methacrylate copolymer for GRAS evaluation. *Regulatory Toxicology and Pharmacology* 61 (1), 32–43.

Fiebach, K. & Grimm, D. (2000) Resins, Natural. In: *Ullmann's Encyclopedia of Industrial Chemistry*. Wiley-VCH, Weinheim.

Heimbach, J. T., Bodor, A. R., Douglass, J. S., Barraj, L. M., Cohen, S. C. Biles, R. W. & Fauste, H. R. (2002) Dietary exposures to mineral hydrocarbons from food-use applications in the United States. *Food and Chemical Toxicology* 40, 555–571.

Palkin, S. & Smith, W. C. (1938) A new non-crystallizing gum rosin. *Oil and Soap* 15 (5), 120–122.

Parish, E. J., Boos, T. L. & Li, S. (2002) The chemistry of waxes and sterols. In *Food Lipids: Chemistry, Nutrition, and Biochemistry*, 2nd edition, Casimir, C. & Akoh, D. B. (eds), Marcel Dekker, New York, p. 103.

SPI (Society of the Plastics Industry Inc.) (1996) *Report on Potential Exposure to Mineral Oil from Food-Contact Use of Polystyrene Resins*. Society of the Plastics Industry, Inc.

Steinle, J. V. (1936) Carnauba wax: an expedition to its source. *Industrial and Engineering Chemistry* 28 (9), 1004–1008.

van Battum, D. & Rijk, M. A. H. (1979) Migration Experiments with Waxed Papers. Fifth Report. Results Obtained with Various Foodstuffs. Report No. B77/2694A. *Central Institute for Nutrition and Food Research*.

Wolfmeier, U., Schmidt, H. & Heinrichs, F.-L. (2002) Waxes. In *Ullmann's Encyclopedia of Industrial Chemistry*. Michalczyk, G., Payer, W., Dietsche, W., Boehlke, K., Hohner, G. & Wildgruber, J. (eds), Wiley-VCH, Weinheim.

15 Preservatives

Abstract: Strategies to ensure quality, stability and food safety have been and always will be practised. To accomplish this, the food industry incorporates some compounds with the ability to form chelate complexes with certain polyvalent metal ions such as copper, iron and nickel (sequestrants) to improve the stability of foodstuffs. The strategy for the preservation of foods also requires the implementation of some steps whereby foodstuffs undergo some processing, for example they may be preserved at particular pH values, undergo pasteurisation or sterilisation or have additional ingredients such as salts to prevent microbial attack. In this chapter the different types of preservation techniques, as well as the chemistry of the preservative agents and how they work, is discussed.

Keywords: antimicrobial agents; food preservation; food sequestrants; food stability; microbial food attack; physical preservation techniques; thermal preservation techniques

15.1 PRESERVATIVES: PAST, PRESENT AND FUTURE

The idea of food preservation arises from the desire to maintain the food quality, its physico-chemical properties and maintain the functionality of its nutritious components without affecting the product itself (Farkas 1977). Throughout civilisation, there has always been a need to preserve food for future use for a number of reasons, including: to maintain the integrity of food products; for replenishment of food supply during famine (food security); or preservation of food products to be transported for use in a different locality (Gould 1995; Schellekens 1996; Peck 1997). Many methods and techniques including the use of classical or artificial preservative agents or naturally occurring antimicrobial biomolecules ('natural' preservatives) have been used. Artificial compounds include chemicals such weak organic acids. Natural preservative agents are derived from plants and other bio-preservatives are derived from the fermentation of lactic acid bacteria.

Preservation techniques are numerous and generally differ from one food product or food class to another. In practice, a combination of techniques is sometimes preferred in order to achieve the desired preservation without tampering with safety or the integrity of the food product.

15.1.1 Sequestrants as food additives

Sequestrants normally perform the function of preservative in foods. Most of the sequestrant compounds are either organic or inorganic salts such as calcium disodium ethylene diamine

Chemistry of Food Additives and Preservatives, First Edition. Titus A. M. Msagati.
© 2013 John Wiley & Sons, Ltd. Published 2013 by John Wiley & Sons, Ltd.

tetra-acetate (EDTA), glucono delta-lactone, sodium gluconate, potassium gluconate, sodium tripolyphosphate and sodium hexametaphosphate.

Salts such as EDTA, which is a synthetic chelating or sequestrating agent, are known to form strong complexes with cations thus making them good agents for use in food systems as stabilisers and sequestrants (Winter 1999). This salt also possesses other health benefits when incorporated in foods; it has antimicrobial activities due to the fact that it chelates with cations, which are essential elements needed for microbial growth, thus limiting their availability to pathogens. EDTA is also known to sequestrate with especially divalent cations which bridge lipopolysaccharides and other membrane biomolecules, destabilising the bacterial cytoplasmic membrane (Vaara 1992; Banin *et al.* 2006). In all these phenomena, EDTA also plays the role of food preservative.

Other sequestrants such as cholazol H, which has been reported to be a chemically functionalised insoluble fibre (Wilson *et al.* 1998) and GT16-239, an alkylated, cross-linked poly(allylamine) and cholestyramine (Wilson *et al.* 1998) are also useful because they are capable of forming chelation with bile acids in the intestine, thereby hindering their absorption. This triggers the liver to increase the production of bile in order to replace the bile that was not absorbed because it has been bound to sequestrants. One of the raw materials for the synthesis of bile inside the body is cholesterol; this means that cholesterol is taken from the blood circulation, hence providing a health benefit.

Cyclodextrins are also used in the food industry as important food additives which perform many functions such as: protection of lipophilic ingredients in foodstuffs which are oxygen sensitive and photosensitive; solubilisation of vitamins and food colouring agents; and stabilisation of fragrances and flavourings. Cyclodextrins are also sequestrants and used to complex food items such aspartame, glycyrrhizin and rubusoside (sweeteners), thereby stabilising them and enhancing their tastes and flavours (Singh *et al.* 2002). Moreover, β-cyclodextrins have been reported to bind cholesterol and eliminate it from the body (Kwak *et al.* 2004; Jung *et al.* 2008). As preservatives, cyclodextrins are useful in the food industry as packaging materials. They reduce the possibility of organic volatile contamination and are also good in terms of controlling the diffusion and transmission rates of the materials used for packaging without affecting the food quality (Wood 2001).

Further Thinking

Micro-organisms which bring about food spoilage are ubiquitous. Once they are exposed to favourable growth conditions of temperature, air and moisture, pH and nutrient supply, they can multiply very fast. Preservatives are included in foods to alter the environmental conditions such that growth of moulds and bacteria is discouraged, thus preventing food spoilage.

15.1.2 Food preservation processes

Food preservation during processing can be achieved through a number of operational manipulations such as the use of a mild heat stress, use of low levels of preservative agents or a combination of these approaches (Brul *et al.* 1997; Knorr 1998). The attractiveness of the synergy of preservation approaches is the flexibility of maintaining food safety without affecting the integrity, quality or the other physical attributes of the food product such as colour, flavour, texture and nutritious elements. Since heat treatment at a certain degree may destroy nutritious elements of food products, alternative physical treatments such as the use of ultra-high pressure (UHP) or the application of pulsed electric fields (PEF) (Brul and Coote 1999) have been considered.

Further Thinking

The same food which we eat is also a source of energy for a variety of microbes such as moulds. However, their action on foods causes spoilage due to the production of toxic compounds which endangers our health, making the food unfit for human consumption. Preservatives in foods are added to prevent the growth of harmful pathogens which may not only cause food spoilage but also lead to food poisoning, affecting the health of consumers. They are also intended for extending the shelf-life of foodstuffs, allowing them to be used long after processing and storage.

15.2 NATURAL FOOD PRESERVATIVES

Natural food preservatives are generally favoured by consumers due to their perceived safety as opposed to artificial preservatives. Living organisms (animals, plants as well as microorganisms) contain various molecules with antimicrobial properties which have evolved as host defence mechanisms and have potential application in the food industry as preservatives. Most if not all preservatives categorised as natural also perform a number of other important functions, for example, as antioxidants, flavourings or as antibacterials. Such preservatives tend to prolong the shelf life of many foods or foodstuffs such as meat and meat products.

Examples of natural preservative compounds include lactoperoxidase found in milk, lysozyme found in egg white, saponins and flavonoids which are extracts from herbs and spices, bacteriocins extracted from lactic acid bacteria, plant-derived antimicrobial compounds, microbial-derived compounds (mainly lytic enzymes), chitosan found in shrimp shells and chitosan-saccharide derivatives (e.g. chitosan-mint, chitosan-glucose and xylan-glucose).

15.2.1 Plant antimicrobial extracts

Plants, including herbs and spices, are known to contain compounds such as essential oils and other phenolic components with antimicrobial activity. For example, the phenolic glycoside oleuropein (Figure 15.1) from olives has been reported to have antimicrobial defence against *Staphylococcus aureus* and *Salmonella enteritidis* (Tranter *et al.* 1993; Tassou and Nychas 1994, 1995).

15.2.1.1 *Essential, volatile and ethereal oils*

From the chemistry point of view, essential oils (also known as volatile or ethereal oils; Guenther 1948) are aromatic oily liquids originating mainly from plant sources or plant parts

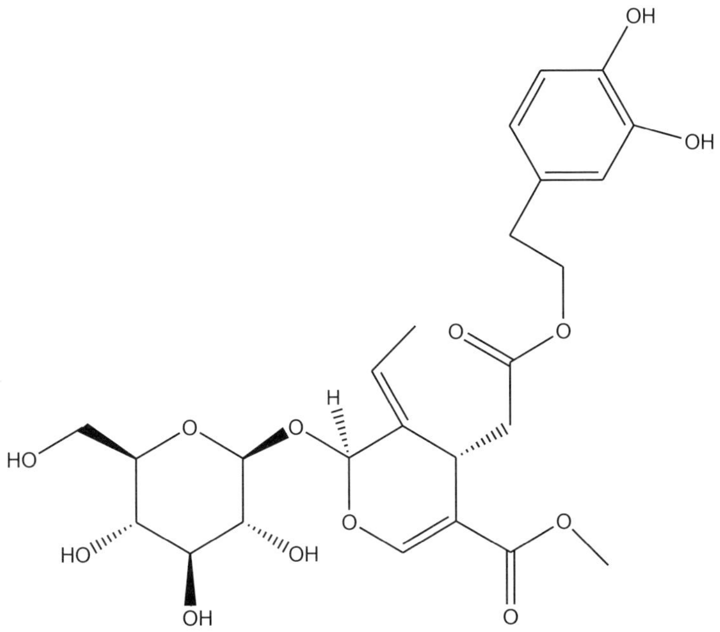

Fig. 15.1 Chemical structure of oleuropein.

(flowers, buds, seeds, leaves, twigs, bark, herbs, wood, fruits and roots). Essential oils are obtained from these plant parts through a number of chemical processes including fermentation, chemical extraction and, the most commonly used procedure, of steam distillation (Van de Braak and Leijten 1999). Essential oils have been reported to have antimicrobial properties with potential applications in the food industry (Burt 2004). For example, the essential oils extracted from mint and mastic gums were reported to be active against *Salmonella enteritidis* and *Listeria monocytogenes* (Tassou *et al.* 1995; Tassou and Nychas 1995).

Due to the diverse chemistries within essential oils, their antimicrobial properties may not follow a single mechanism or target in the microbial cell organelle (Skandamis and Nychas 2001; Carson *et al.* 2002). The key property of essential oils is their hydrophobicity, which enables them to cross the hydrophobic cytoplasmic membrane with a lipid bilayer structure (Figure 15.2) as well as membrane proteins and the mitochondria organelle. This makes them susceptible to permeability by biomolecules (cell contents) which are inside the cytoplasm (Knobloch *et al.* 1986; Sikkema *et al.* 1994).

15.2.1.2 Plant phenolic extracts

Plants contain phenolic compounds that play a role as preservatives in foodstuffs. Even the essential oils discussed in the previous section, especially those which possess the strongest antimicrobial activity, are composed of phenolic functionality such as carvacrol, eugenol (2-methoxy-4-(2-propenyl) phenol) and thymol (Farag *et al.* 1989; Thoroski *et al.* 1989; Cosentino *et al.* 1999; Dorman and Deans 2000; Juliano *et al.* 2000; Lambert *et al.* 2001). The mechanism of action of these phenolic compounds is believed to involve the disruption of the cytoplasmic membrane (Denyer and Hugo 1991; Sikkema *et al.* 1995; Davidson 1997). However, the chemical structures of the individual essential oil components affect

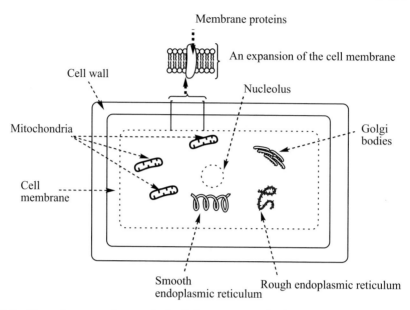

Fig. 15.2   General structure of a microbial cell showing targets for the microbial action of essential oils (Knobloch *et al.* 1986; Sikkema *et al.* 1994).

their respective mode of action as well as their antimicrobial properties (Dorman and Deans 2000).

Eugenol
The mechanism of action of eugenol (Figure 15.3) is attributed to the presence of the hydroxyl group on eugenol which plays an important role in binding to proteins, thus preventing the normal functioning of microbial enzymes (Wendakoon and Sakaguchi 1995).

Cinnamaldehyde
The mode of action of cinnamaldehyde (3-phenyl-2-propenal) (Figure 15.4) is attributed to the presence of the carbonyl group which binds to proteins and therefore disturbs the action of amino acid decarboxylases within the microorganism (Wendakoon and Sakaguchi 1995).

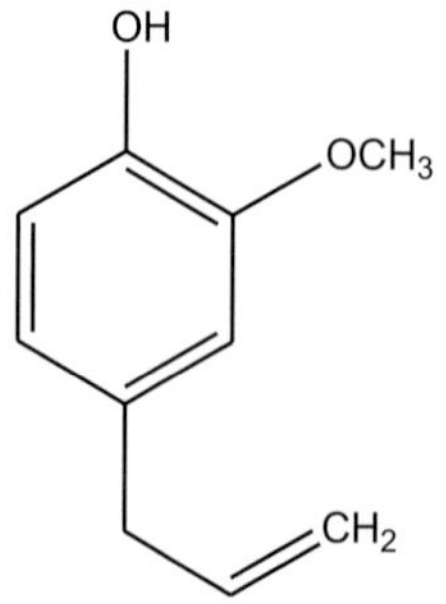

Fig. 15.3   Chemical structure of eugenol.

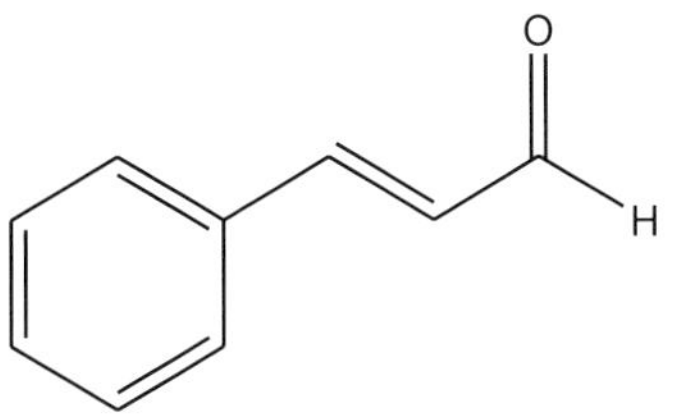

Fig. 15.4 Chemical structure of cinnamaldehyde.

p-cymene
p-cymene is known to be insoluble and thus contributes to the swelling of the cell membrane (Ultee *et al.* 2002), facilitating transport of its precursor compound (carvacrol) across the cytoplasmic membrane (Ultee *et al.* 2002) and disrupting the lipid bilayer structure of the microbial cell membrane.

Carvone
With chemical formula (2-methyl-5-(1-methylethenyl)-2-cyclohexen-1-one), carvone acts by disturbing the metabolic energy status of microbial cells (Oosterhaven *et al.* 1995).

Others
Other phenolic compounds extracted from spices (cinnamon), for example cinnamic acid (Figure 15.5) have also been used as preservatives as well as flavourings in a number of foodstuffs such as fish (Shimada *et al.* 1991).

Despite the abundance of plant extract with antimicrobial activities, there is one limitation to their effective use: they normally need to be used at a higher concentration to have a degrading effect on microbial cells, and this is undesirable in foodstuffs as it can affect other properties of the food such as flavour (Roller 1995).

15.2.2 Natural food preservatives derived from insects

The class insecta is also known to be a source of useful compounds that are used as additives in the food industry. One of these compounds is propolis, a resinous natural product produced by bees (*Apis mellifera*) and obtained from vegetable secretions (Tosi *et al.* 2007). Chemically, the varying composition of propolis is mainly dependent on the botanical nature of the plants from which the vegetable secretions are obtained, as well as their phytogeographical origin (Greenaway *et al.* 1991; Bonvehi *et al.* 1994; Bankova and Marcucci 2000). However, the chemical composition of propolis comprises a wide variety of substances including

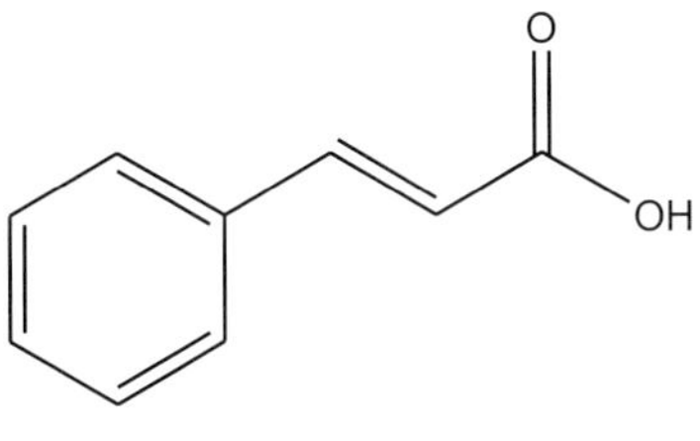

Fig. 15.5 Chemical structure of cinnamic acid.

polyphenols, quinones, coumarins, steroids, amino acids and inorganic compounds (Tosi *et al.* 2007). Most propolis components are of a phenolic nature, mainly flavonoids which are known to have strong antimicrobial activities (Cowan 1999). The three –OH group substitutions in ring B and a third –OH group in ring C in the flavonoid structures constitute the necessary structures for the flavonoid and myricetin phenolic compounds, present in propolis, to demonstrate antimicrobial activity (Mori *et al.* 1989; Puupponem-Pimia *et al.* 2001; Farre *et al.* 2004).

15.2.3 Microbial enzymatic-derived preservative agents (lytic enzymes)

Lytic enzymes are proteinous biomolecules capable of destroying/rupturing a biological cell or degrading any unwanted biomolecules such as polysaccharides. Studies have shown that they have the potential to be used as novel and 'natural' food preservatives (Roller 1995). Examples of such lytic enzymes include lysozyme from hen egg whites, which has antimicrobial activity against clostridium bacterial species and has been reported to be useful in preventing clostridial spoilage in hard-cooked cheeses in France (Fox and Grufferty 1991). A number of other lytic enzymes with activity against yeast spoilage in foods are also known, including mannanases and glucanases class of lytic enzymes which destroy the mannan and glucan components of yeasts, respectively.

15.2.4 Sulphur dioxide

Sulphur dioxide has been used as the main preservative, antiseptic and antioxidant in the winemaking industry to maintain the integrity of wine and to inhibit oxidation and the growth of harmful microbes such as wild yeast (Ough and Crowell 1987).

Further Thinking

There are a variety of microorganisms known to cause food spoilage, and they respond differently to different types of preservatives and in different mechanisms. This explains why there are a wide variety of food preservatives on the market.

15.2.5 Chitosan conjugates

Chitosan is a biopolymer which has been reported to have a potential use as an additive in the food industry (Rudrapatnam and Farooqahmed 2003). It has been shown to have strong antimicrobial properties against a range of foodborne microorganisms and is therefore a potential natural food preservative (Chen *et al.* 1998; El Ghaonth *et al.* 1992; Shahidi *et al.* 1999).

In conjunction with other polymeric compounds, chitosan (naturally from shrimp shells) has been reported in a number of articles as potential food preservers. The chitosan conjugates which have demonstrated the greatest potential to be used as food preservers include xylan-chitosan (Li *et al.* 2010); chitosan-mint mixture (Kanatt *et al.* 2008); and chitosan glucose complex (Kanatt *et al.* 2008).

Chitosan, a deacetylated derivative of chitin, forms complexes with some polymeric materials because it has amino groups in its chemical structure, which can offer many site-selective points for chemical modification using, for example, Schiff base, N-acylation and reductive alkylation (Kurita 2006).

15.3 TRADITIONAL FOOD PRESERVATION METHODS

Traditional food preservation techniques include intense heat treatments, salting, acidification, drying and chemical preservation. The use of preservative agents in foodstuffs is always considered hand in hand with the preservation methods and techniques for that particular food product. Preservation techniques tend to subject food items to particular physical conditions such as thermal treatment, so preserving agents must be able to withstand that particular heat treatment (they must be stable) and allow the food product to to maintain its texture, flavour and nutritional integrity.

There are numerous techniques which have been employed for food preservation for both short and long periods of time. For example, fresh food items of high moisture content such as meat may need low-temperature treatment to keep it safe from microbial spoilage, but this is only effective for a short period of time.

15.3.1 Wood smoke

Wood smoke has been used traditionally for preservation of food items such as meat. The smoke contains a number of antimicrobial compounds such as phenols, syringol and guaiacol and their derivatives as well as carbonyls, catechol as well as naphthalene derivatives. Wood smoke is also used in the food industry as a flavouring agent, and is discussed in more detail in Chapter 5.

15.3.2 Salting

In the same way as wood smoke, salt has been used as a preserver for longer than can be remembered. Apart from being used as a preservative agent, salts have also been used to impart characteristic flavour in many of the processed foodstuffs, thus maintaining food quality. In meat and meat processed products the amount of salt allowed ranges between 2.5% and 5.0% of the final processed product. For the salt to play a role as preservative, the level in the finished product should be about 17%.

Examples of salts that are used as preservatives include sodium chloride (NaCl), sodium nitrate ($NaNO_3$) and sodium nitrite ($NaNO_2$). Even at mild concentrations (up to 2%), sodium chloride (which is present in many food products) is capable of neutralising the antimicrobial character of natural compounds (Devlieghere *et al.* 2004). A similar observation has been reported in the case of lysozyme and for bacteriocins such as sakacin K (Leroy and De Vuyst 1999), curvacin (Verluyten *et al.* 2002) and chitosan (Devlieghere *et al.* 2002).

15.3.2.1 *Mode of action*

Microorganisms responsible for food spoilage have cell wall and cytoplasmic semi-permeable cell membranes in their outer structures. The presence of cytoplasmic semi-permeable membranes makes the microorganisms sensitive to changes in osmotic pressure and thus the concentrations of ions/salts. Most of these microbes live within a watery environment;

since the inside of a microbial cell is composed of cytoplasm (with relatively less water content than outside), this will encourage water movement from outside to the inside of the cell. If the outside environment becomes salty, this will change the water balance between the outside and the inside of the microbial cell such that the amount of water in the environment where the microbes are is less than inside the microbial cell; water will therefore move out of the cell (cytoplasm), causing dehydration of the cell, and the microbial cells will then stop growing and die. In this way, salt serves as a food preservative.

15.3.3 Canning

Canning involves putting foodstuffs in cans, sealing them and then applying heat to temperatures which can kill most pathogens. The sealing will keep the food safe from any further microbial attack.

15.3.4 Drying

Drying dehydrates food in the sense that it eliminates water/moisture from foods, excluding the important condition which encourages microbial growth which causes food spoilage.

15.3.5 Chilling and freezing

Temperature is one of the important factors which affect the ability of microbes to grow and multiply. Freezing or chilling slows down both the metabolic and enzymatic activities for the microbes, thus discouraging their growth and multiplication. Chilling and freezing only slows down the growth; the processes do not kill the pathogens. This approach needs to be coupled with other preservative means to effectively preserve foodstuffs.

15.3.6 Pickling

The acidity/alkalinity environment may have a large affect on microbial growth. Lowering of the pH environment where food is kept may discourage the multiplication of microbial cells, since this also suppresses the metabolic and enzymatic activity of the microbial pathogens.

15.3.7 Addition of sugars

Sugar tends to draw water from the microbes (plasmolysis). This process leaves the microbial cells dehydrated, thus killing them. In this way, the food will remain safe from microbial spoilage.

15.4 ARTIFICIAL PRESERVATIVE AGENTS

The most commonly used preservative agents include: caffeine (a trimethylated xanthine, also a psychoactive agent, which is also added as a flavouring agent in most soft drinks; Pena *et al.* 2005); saccharin; sorbic acid and its salts (antifungal preservative) is used for food products of higher pH (Wang *et al.* 2006); benzoic acid and its salts (inhibitor of bacterial development) is mainly used for acidic food products (Ochiai *et al.* 2002; Wang *et al.* 2006); parabens (p-hydroxybenzoic acid methyl ester (methylparaben), p-hydroxybenzoic acid ethyl ester(ethyl paraben), p-hydroxybenzoic acid propylester (propyl paraben), p-hydroxybenzoic

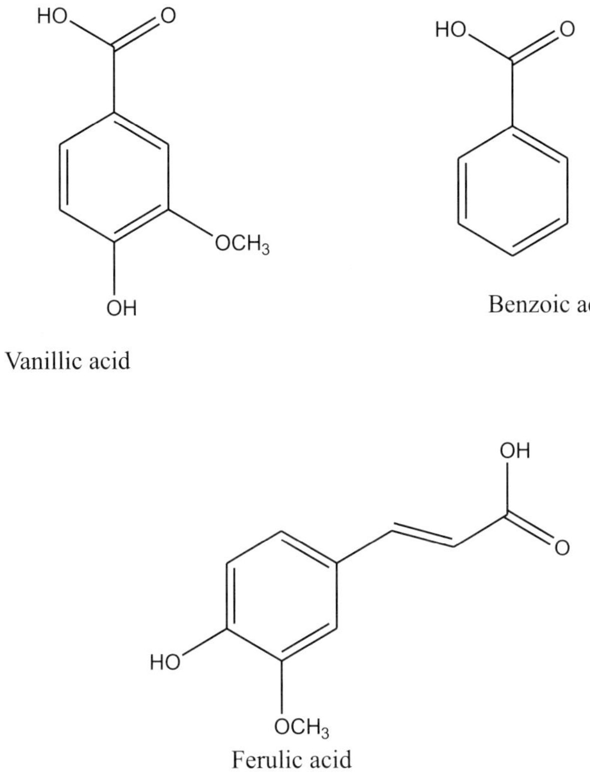

Vanillic acid

Benzoic acid

Ferulic acid

Fig. 15.6   Structures of some weak organic acid used as preservers in foods.

acid butyl ester (butyl paraben)); and salicylic acid (Boyce 1999). In most cases, these preservative agent are used in beverages such as soft drinks (Ochiai *et al.* 2002; Dong and Wang 2006; Techakriengkrai and Surakarnkul 2007; Wen *et al.* 2007).

15.4.1   Weak organic acids

A number of weak organic acids, such as benzoic acid and sorbic acid, are known to have antimicrobial properties (Figure 15.6). Weak organic acids are called weak simply because they do not ionise completely; at any time there will be a fraction which is undissociated and a fraction which is dissociated (charged species). The undissociated fractions of weak acids are readily soluble in lipids, while the dissociated charged fraction is lipid insoluble. Since the pH of the weak acid is low compared to the pH inside the microbial cell (cytoplasm has almost neutral pH) (Pilatus and Techel 1991; Legisa and Golic-Grdadolnik 2002; Jernejc and Legisa 2004; Plumridge *et al.* 2004), the uncharged species of the undissociated fraction of the weak acid will tend to diffuse through the microbial cell membrane to the cytoplasm. This will trigger the dissociation of the weak acid into protons and corresponding anions. Because the charged species are lipid insoluble, they will tend to accumulate inside the cell membrane (cytoplasm) with the resultant decrease in the pH of the cytosol. This is detrimental to the microbial cell, as this will hinder all the cellular metabolic activities (Krebs *et al.* 1983).

Benzoic acid and its salts such as sodium benzoate are widely used as additives in acidic foods such as fruit juices and soft drinks (International Programme on Chemical Safety

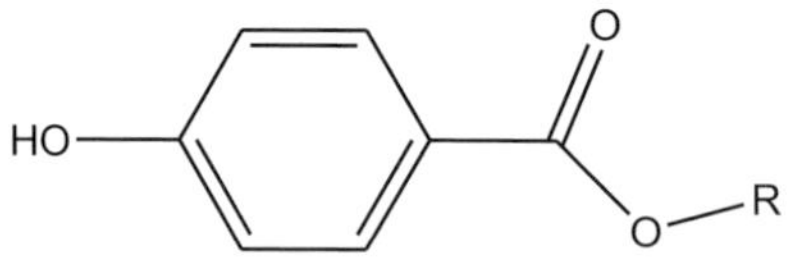

Fig. 15.7 General chemical structure of parabens, where R=alkyl group e.g. methyl (CH$_3$), ethyl (C$_2$H$_5$), propyl (C$_3$H$_7$), butyl (C$_4$H$_8$).

2000). When ingested, sodium benzoate and glycine undergoes a conjugation reaction to form a nitrogenous molecule, a hippurate, which is excreted via urine. Benzoic acid inhibits bacterial development and sorbic acid is an antifungal preservative against moulds and yeasts (Dong and Wang 2006). The pH at which benzoic and sorbic acids demonstrate effective antibacterial activity is different. Benzoic acid is mainly used for acidic food products, while sorbic acid is used for food products of higher pH (Ochiai *et al.* 2002).

The pH decline of the microbial cytoplasm due to the action of weak acids has been reported through experiments conducted with, for example: acetic acid (Neal *et al.* 1965); sulphite (Pilkington and Rose 1988); benzoic acid (Krebs *et al.* 1983); and a combination of sorbic and benzoic acids (Plumridge 2005).

The strength of weak acids generally tends to have an influence in their lowering of the pH of the cytoplasm. Those with the same dissociation capacity as measured by their pKa values tend to exert the same magnitude of lowering of the cytoplasmic pH.

Other food preservatives within this category include: calcium propionate (used bread, other baked goods, processed meat and in dairy products); sodium nitrate (obtained in leafy green vegetables), sodium nitrite which finds application as preservatives in meats and fish; and sulphites (found in wines where it protects against microbial spoilage and oxidation at several stages of winemaking; also used in dried fruits and potatoes). Small molecules such as sulphur dioxide (SO$_2$), hydrogen disulphide (H$_2$O$_2$) and certain chelators have also been used as preservatives.

15.4.2 Caffeine

Caffeine-based products (structurally related to uric acid) are among the most widely consumed foods in the world in the form of tea, coffee and cocoa. Caffeine (1, 3, 7-trimethyl xanthine) is a methylated xanthine alkaloid derivative found in numerous plant species at many locations (Ibrahim *et al.* 2006).

15.4.3 Parabens

Parabens are a group homologous series of hydroxybenzoic acid, which are esterified at the C-4 position (Figure 15.7). Examples of parabens include methyl-, ethyl-, propyl-, butyl-, heptyl- and benzyl-paraben. These compounds have been reported to have multiple bioactivities, including as active antimicrobial agents (Soni *et al.* 2005). They can either be used singly or in combination to exert an optimal antimicrobial effect. Their microbial inhibitory properties are believed to involve the disruption of cytoplasmic membrane and their effect on transport and mitochondrial functioning (Soni *et al.* 2005).

The use of parabens as food preservers is attractive because they meet several of the criteria of an ideal preservative in that they: possess a broad spectrum of antimicrobial activity; are stable at a broad pH range; are sufficiently soluble in water to produce the effective concentration in aqueous phase; and are heat stable such that they can safely be autoclaved and still maintain their antimicrobial activity (Maddox 1982). They are also known to have very low toxicity and are biodegradable (Maddox 1982).

Chemically, parabens are prepared by esterification of p-hydroxybenzoic acid with the corresponding alcohol in the presence of an acid catalyst.

15.5 MODERN FOOD PRESERVATION TECHNIQUES

New preservation technologies for food products include: non-thermal inactivation technologies such as high hydrostatic pressure (HHP) and pulsed electric fields (PEF); modern food packaging technologies such as modified atmosphere packaging (MAP); and active packaging, natural antimicrobial compounds and biopreservation (Devlieghere *et al.* 2004).

15.5.1 Non-thermal inactivation

Non-thermal inactivation techniques have been promoted as being associated with food attributes such as food nutritive integrity and an acceptable shelf life. There are a number of known inactivation technologies, including ionisation radiation, high hydrostatic pressure (HHP), pulsed electrical fields (PEF), high-pressure homogenisation (HPH), UV decontamination, pulsed high-intensity light (PHIL), high-intensity laser (HIL), pulsed white light (PWL), high-power ultrasound (HPU), oscillating magnetic fields (OMF), high-voltage arc discharge (HVID) and streamer plasma (SP). The most frequently used, described in the following sections, are HHP and PEF (Devlieghere *et al.* 2004).

15.5.1.1 *High hydrostatic pressure (HHP)*

This technique involves exposing food products to pressures of above 100 MPa which has an inactivation effect on microbes, thus extending microbial shelf life as well as improving the microbial safety of food products (Patterson 2000). The main drawback of this technology is that pressure treatment alone is often not effective enough to substantially eliminate microbial spores (Hoover 1993). To correct for this shortcoming, a combination of high-pressure and high-temperature treatments have been proposed. Some reports have indicated that there is a significant reduction of microbial spores (Kowalski *et al.* 1992; Seyderhelm and Knorr 1992). It has also been reported that the synergy of heat and pressure technologies works more effectively if there is a preheating treatment performed first then followed by pressurisation, as opposed to heating during pressurisation (Patterson 2000).

15.5.1.2 *Pulsed electric fields (PEF)*

Pulsed electric field is a non-thermal inactivation technology which is based on a pulsing power applied to the food product sandwiched between a set of electrodes, causing severe disruption of microbial cells (G'ongora-Nieto *et al.* 2002). However, as for HHP treatment, the inactivation effect of PEF alone is not sufficient. It can be enhanced by combination with other stressing technologies, for example, the addition of antimicrobial compounds such as nisin and organic acids. Other technologies to enhance the effect of PEF include increasing water activity, adjusting pH and the application of heat.

15.5.2 Modified atmosphere packaging (MAP)

MAP refers to the packaging/enclosure of food products in gas-barrier materials, in which the gaseous environment has been modified (Young *et al.* 1988). Despite the fact that this technology has contributed to the extension of the shelf life of many food products, the most effective limit for the gas atmosphere for a particular food product in a particular packaging system or design are not yet well established. This could be a difficult task as many other factors including pH, water activity, fat content, type of fat as well as the gas/product volume ratio in the chosen package type need to be considered. The optimisation of these factors is crucial as it will determine how the food product will react for specific microbial, chemical and enzymatic degradation.

For example, food items that are easily attacked by Gram-negative or yeasts would be suitably packaged in a CO_2-enriched atmosphere, since the growth of yeasts and Gram-negative microorganisms is significantly slowed down in such conditions. Oxygen is totally excluded from the gas mixture as it triggers oxidation. The limitation of CO_2 as a food preserver lies in the fact that it has appreciable solubility in water and fat; this may result in the collapsing of the package in cases where high levels of CO_2 are used, especially for food products containing high amounts of unsaturated fat. The use of too high CO_2 concentrations can bring a problem of pH drop during long periods of storage with the negative result on the decrease in the water binding capacity.

15.5.3 Biological food preservation technology

This technology makes use of lactic acid bacteria (LAB) and their antibacterial products such as lactic acid, bacteriocins and others to extend the storage life as well as improve the safety of food products (Hugas 1998). This technology introduces LAB as antagonistic cultures to inhibit microbes and also to prolong the shelf life of food products as the sensory properties of that particular foodstuff are only slightly modified . The LAB is referred to as antagonistic cultures because they inhibit the growth of other microbes in two ways: (1) by outcompeting microbes when competing for nutrients and (2) by producing some metabolic by-products such as organic acids (lactic and acetic, hydrogen peroxide, antimicrobial enzymes, bacteriocins and reuterin, which are known to have antimicrobial properties; Holzapfel *et al.* 1995). It should be noted that most of LAB (with the exception of enterococci) have been reported to be safe when present in food products (Adams and Marteau 1995).

15.5.4 Bacteriocinogenic cultures for food preservation

Bactericin are compounds which can be used to kill bacterial cells of other closely related bacterial species (bactericidal) or inhibit their growth (bacteriostatic). The mode of action of these bactericidal compounds involves rupturing the cell membrane of the microbe, thus destroying the lipid bilayer structure and forcing some cytoplasmic contents out (O'Keeffe and Hill 1999). Examples of bacteriocins include nisin extracted from Lactococcus lactis strains, pediocin produced by Pediococcus acidilactici strains and sakacins obtained from Lactobacillus sakei strains. All these strains acts as protective cultures (Jacobsen *et al.* 2002; Katla *et al.* 2002) in a variety of foods such as in fish (Duffes *et al.* 1999; Katla *et al.* 2001) and dairy products (Schwenninger *et al.* 2002), where they inhibit the growth of yeasts in foodstuffs containing selected *Propionibacterium/Lactobacillus* strains.

The performance of these bacterial species is however limited by the narrow activity spectrum in which they operate; they are known to be ineffective against Gram-negative pathogens and spoilage bacteria (Holzapfel *et al.* 1995; Rodriguez *et al.* 2002).

The possible alternative to using bacteriocinogenic cultures may be the use of non-bacteriocinogenic but very competitive cultures, such as *Lactobacillus alimentarius* BJ-33 (Andersen 1995); *Lactobacillus sakei* TH1 (Bredholt *et al.* 2001) and *Lactococcus lactis* (Vermeiren *et al.* 2004). These bacterial species operate effectively by producing lactic acid which lowers the pH thus inhibiting growth of spoilage and/or pathogenic bacteria (Juven *et al.* 1998). Moreover it has been hypothesised that the growth inhibition is brought about by a mixture of factors which may involve the production of antimicrobial compounds, competition for or depletion of the supply of nutrients (Gram and Melchiorsen 1996; Buchanan and Bagi 1997; Campo *et al.* 2001).

15.6 SAFETY CONCERNS OF FOOD PRESERVATIVES

A number of reports have highlighted that the many food preservatives, including sorbates, hydrogen peroxide, benzoic acid and sodium benzoate, have the potential to cause health problems, especially if used at higher concentrations in foodstuffs (Parke and Lewis 1990).

Propionic acid has been reported as carcinogenic, as experiments performed on rodents have demonstrated (Harrison 1992).

The sulphite, bisulphite, and metabisulphite preservatives have also been reported to cause some health-related problems, such as the sulphite-induced hypersensitivity, especially to asthmatic consumers (Taylor *et al.* 1988). For sulphite-sensitive patients who have the hepatic enzyme sulphite oxidase, which catalyses the conversion of sulphites to inorganic sulphates, the complications are more severe. The accumulation of sulphite will break down into sulphur dioxide, a chemical which causes pulmonary irritation (Gunnison and Jacobsen 1987).

Nitrates and nitrites as preservatives in food items such as meat and meat products, to prevent the growth of *Clostridium botulinum*, have also been reported to cause health problems. Once ingested, the nitrates undergo enzymatic reduction reactions to form nitrites. These have been implicated as oxidative stressors and also induce a disease condition known as methemoglobinemia (Chan 1996). Moreover, the nitrites may react with basic species containing amine functionality to produce *N*-nitrosamines, which may undergo decomposition reactions and form free radical species of alkyl carbonium. The alkyl carbonium radicals have been reported as potentially mutagenic and carcinogenic, especially in rodents (not in humans; Eichholzer and Gutzwiller 1998).

Further Thinking

The safety of preservatives used in foods is very important. Although preservatives are capable of introducing an environment that prohibits the growth and proliferation of harmful pathogenic microbes, this is not the case for body cells and tissues. Moreover, the guidelines which set the maximum allowed levels in food products must be observed and monitored.

15.7 ANALYTICAL METHODS FOR THE DETERMINATION OF PRESERVATIVE RESIDUES

The revelation that there is a potential threat to our health from the use of preservatives means that we must have methods to monitor the poisonous residue species. Methods involving high-performance liquid chromatography (HPLC) have been reported for the analysis of preservatives in food products (Tyler 1984; Veerabhadraro *et al.* 1987; Chen *et al.* 1997; Ferreira *et al.* 2000; Pylypiw and Grether 2000; Chen and Wang 2001; Tfouni and Toledo 2002; Garcia *et al.* 2003; Mota *et al.* 2003). Similarly, other chromatographic methods such as ion-pair HPLC and an electrically driven separation technique using a mixed micellar electrokinetic chromatography have been reported for the determination of preservatives (Chen and Fu 1995; Boyce 1999). All these methods are normally performed and validated in terms of limit of detection (LOD), limit of quantitation (LOQ), linearity, selectivity, accuracy and repeatability.

15.8 CONCLUSIONS

The importance of food preservation has been known since longer than can be remembered. Traditional and natural preservation methods and processes are still in use even today, and will be applicable for a long time. In addition to these traditional and natural techniques, artificial and modern methods have increased the effectiveness of food preservation. Foodstuffs can now be stored or transported without any fear of spoilage or of the natural properties of the foods being changed.

REFERENCES

Adams, M. R. & Marteau, P. (1995) On the safety of lactic acid bacteria. *International Journal of Food Microbiology* 27, 263–264.

Andersen, L. (1995) Biopreservation with FloraCarn L-2. *Fleischwirtschaft*, 75, 705–706, 711–712.

Banin, E., Brady, K. M. & Greenberg, E. P. (2006) Chelator-induced dispersal and killing of *Pseudomonas aeruginosa* cells in a biofilm. *Applied and Environmental Microbiology* 72, 2064–2069.

Bankova, V. & Marcucci, M. C. (2000) Standardization of propolis: Present status and perspectives. *Bee World* 8 (4), 182–188.

Bonvehi, J. S., Coll, F. V. & Jorda, R. E. (1994) The composition, active components and bacteriostatic activity of propolis in dietetics. *Journal of American Oil Chemists' Society* 71, 529–532.

Boyce, M. C. (1999) Simultaneous determination of antioxidants, preservatives and sweeteners permitted as additives in food by mixed micellar electrokinetic chromatography. *Journal of Chromatography A* 847, 369–375.

Bredholt, S., Nesbakken, T. & Holck, A. (2001) Industrial application of an antilisterial strain of Lactobacillus sakei as a protective culture and its effect on the sensory acceptability of cooked, sliced, vacuum-packaged meats. *International Journal of Food Microbiology* 66, 191–196.

Brul, S. & Coote, P. (1999) Preservative agents in foods: Mode of action and microbial resistance mechanisms. *International Journal of Food Microbiology* 50, 1–17.

Brul, S., Coote, P., Dielbandhoesing, S. K., Oomes, S., Naak-tgeboren, G., Stam, W. M. & Stratford, M. (1997) Natural composition for combating fungi. International Patent WO 1997/ 16973.

Buchanan, R. L. & Bagi, L. K. (1997) Microbial competition: Effect of culture conditions on the suppression of Listeria monocytogenes Scott A by *Carnobacterium piscicola*. *Journal of Food Protection* 60 (3), 254–261.

Burt, S. (2004) Essential oils: their antibacterial properties and potential applications in foods. *International Journal of Food Microbiology* 94, 223–253.

Campo, J. D., Carlin, F. & Nguyen, C. (2001) The effects of epiphytic enterobacteriacea and pseudomonads on the growth of Listeria monocytogenes in model media. *Journal of Food Protection* 64 (5), 721–724.

Carson, C. F., Mee, B. J. & Riley, T. V. (2002) Mechanism of action of Melaleuca alternifolia (tea tree) oil on Staphylococcus aureus determined by time-kill, lysis, leakage and salt tolerance assays and electron microscopy. *Antimicrobial Agents and Chemotherapy* 46 (6), 1914–1920.

Chan, T. Y. (1996) Food-borne nitrates and nitrites as a cause of methemoglobinemia. *Southeast Asian Journal of Tropical Medicine and Public Health* 27, 189–192.

Chen, B. & Fu, S. (1995) *Chromatographia*, 41, 43–50.

Chen, C., Lian, W. & Isai, G. (1998) Antibacterial effects of N-sulfonated and N-sulfobenzoyl chitosan and application to oyster preservation. *Journal of Food Protection* 61, 1124–1128.

Chen, Q. C. & Wang, J. (2001) Simultaneous determination of preservatives, sweeteners and antioxidants in foods by paired-ion liquid-chromatography. *Journal of Chromatography A*, 937, 57–64.

Chen, Q. C., Mou, S. F., Liu, K. N., Yang, Z. Y. & Ni, Z. M. (1997) Separation and determination of four artificial sweeteners and citric acid by high-performance anion-exchange chromatography. *Journal of Chromatography A*, 771, 135–143.

Cosentino, S., Tuberoso, C. I. G., Pisano, B., Satta, M., Mascia, V., Arzedi, E. & Palmas, F. (1999) In vitro antimicrobial activity and chemical composition of Sardinian Thymus essential oils. *Letters in Applied Microbiology* 29, 130–135.

Cowan, M. M. (1999) Plant products as antimicrobial agents. *Clinical Microbiology Reviews* 12 (4), 564–582.

Davidson, P. M. (1997) Chemical preservatives and natural antimicrobial compounds. In: *Food Microbiology: Fundamentals and Frontiers*, Doyle, M. P., Beuchat, L. R. & Montville, T. J. (eds), ASM, Washington, pp. 520–556.

Denyer, S. P. & Hugo, W. B. (1991) Mechanisms of antibacterial action - A summary. In: *Mechanisms of Action of Chemical Biocides*, Denyer, S. P. & Hugo, W. B. (eds), Blackwell Publishing Ltd., Oxford, pp. 331–334.

Devlieghere, F., Vermeulen, A. & Debevere, J. (2002) Chitosan as a food preservative: possibilities and limitations. *Proceedings of the 18th International ICFMH Symposium FOOD*, Lillehammer, Norway, August 18–23, pp. 83–86.

Devlieghere, F., Vermeiren, L. & Debevere, J. (2004) New preservation technologies: Possibilities and limitations. *International Dairy Journal* 14, 273–285.

Dong, C. & Wang, W. (2006) Headspace solid-phase microextraction applied to the simultaneous determination of sorbic and benzoic acids in beverages. *Analytica Chimica Acta* 562, 23–29.

Dorman, H. J. D. & Deans, S. G. (2000) Antimicrobial agents from plants: antibacterial activity of plant volatile oils. *Journal of Applied Microbiology* 88, 308–316.

Duffes, F., Corre, C., Leroi, F., Dousset, X. & Boyaval, P. (1999) Inhibition of Listeria monocytogenes by in situ produced and semipurified bacteriocins of Carnobacterium spp. On vacuum-packed refrigerated cold-smoked salmon. *Journal of Food Protection* 62, 1394–1403.

Eichholzer, M. & Gutzwiller, F. (1998) Dietary nitrates, nitrites, and N-nitroso compounds and cancer risk: a review of the epidemiologic evidence. *Nutrition Reviews* 56, 95–105.

El Ghaonth, A., Arul, J., Asselin, A. & Benhamon, N. (1992) Antifungal activity of chitosan on post-harvest pathogens: Induction of morphological and cytological variations on Rhizopur stolonfier. *Mycology Research* 96, 769–779.

Farag, R. S., Daw, Z. Y., Hewedi, F. M. & El-Baroty, G. S. A. (1989) Antimicrobial activity of some Egyptian spice essential oils. *Journal of Food Protection* 52 (9), 665–667.

Farkas, D. F. (1977) Unit operations concepts optimize operations. *Chem. Tech.* 7, 428–432.

Farre, R., Frasquet, I. & Sanchez, A. (2004) El propolis y la salud. *ARS Pharmaceutica* 45 (1), 21–43.

Ferreira, I. M. P. L. V. O., Mendes, E., Brito, P. & Ferreira, M. A. (2000) Simultaneous determination of benzoic and sorbic acids in quince jam by HPLC. *Food Research International* 33, 113–117.

Fox, P. & Grufferty, M. B. (1991) Exogenous enzymes in dairy technology. In: *Food Enzymology*, Fox. P. F. (ed.), Elsevier, London, pp. 219–269.

Garcia, I., Ortiz, M. C., Sarabia, L., Vilches, C. & Gredilla, E. (2003) Advances in methodology for the validation of methods according to the International Organization for Standardization Application to the determination of benzoic and sorbic acids in soft drinks by high-performance liquid chromatography. *Journal of Chromatography A* 992: 11–27.

G'ongora-Nieto, M. M., Sepulveda, D. R., Pedrow, P., Barbarosa-Canovas, G. V. & Swanson, B. G. (2002) Food processing by pulsed electric fields treatment delivery, inactivation level, and regulatory aspects. *Lebensmittel-Wissenschaft und Technologie* 35, 375–388.

Gould, G. W. (ed.) (1995) *New Methods of Food Preservation*, Chapman and Hall, London.

Gram, L. & Melchiorsen, J. (1996) Interaction between fish spoilage bacteria *Pseudomonas sp.* and Shewanella putrefaciens in fish extracts and on fish tissue. *Journal of Applied Bacteriology* 80, 589–595.

Greenaway, W., May, J., Scaysbrook, T. F. & Whatley, R. (1991) Identification by gas chromatography–mass spectrometry of 150 compounds in propolis. *Zeitschrift fur Naturforschung* C46, 111–121.

Guenther, E. (1948) *The Essential Oils*. D. Van Nostrand, New York.

Gunnison, A. F. & Jacobsen, D. W. (1987) Sulfite hypersensitivity. A critical review. *CRC Critical Reviews in Toxicology* 17, 185–214.

Harrison, P. T. (1992) Propionic acid and the phenomenon of rodent forestomach tumorigenesis: a review. *Food and Chemical Toxicology* 30, 333–340.

Holzapfel, W. H., Geisen, R. & Schillinger, U. (1995) Biological preservation of foods withreference to protective cultures, bacteriocins and food-grade enzymes. *International Journal of Food Microbiology* 24, 343–362.

Hoover, D. G. (1993) Pressure effects on biological systems. *Food Technology* 47 (6), 150–155.

Hugas, M. (1998) Bacteriocinogenic lactic acid bacteria for the biopreservation of meat and meat products. *Meat Science* 49 (S1), 139–150.

Ibrahim, S. A., Salameh, M. M., Phetsomphou, S., Yang, H. & Seo, C. W. (2006) Application of caffeine, 1, 3, 7-trimethylxanthine, to control *Escherichia coli* O157:H7. *Food Chemistry* 99, 645–650.

International Programme on Chemical Safety. (2000) *Concise international chemical assessment document No. 26 benzoic acid and sodium benzoate*. World Health Organization, Geneva.

Jacobsen, T., Koch, A. G., Gravesen, A. & Knochel, S. (2002) Biocontrol of class IIa bacteriocin sensitive and resistant Listeria monocytogenes on sliced meat products. Proceedings of the 18th International ICFMH Symposium FOOD MICRO 2002, Lillehammer, Norway, August 18–23, pp. 131–134.

Jernejc, K. & Legisa, M. (2004) A drop of intracellular pH stimulates citric acid accumulation by some strains of *Aspergillus niger*. *Journal of Biotechnology* 112, 289–297.

Juliano, C., Mattana, A. & Usai, M. (2000) Composition and in vitro antimicrobial activity of the essential oil of Thymus herba-barona Loisel growing wild in Sardinia. *Journal of Essential Oil Research* 12, 516–522.

Jung, T. H., Ha, H. J., Ahn, J. & Kwak, H. S. (2008) Development of cholesterolreduced mayonnaise with crosslinked b-cyclodextrin and added phytosterol. *Korean Journal of Food Science and Animal Resources* 28 (2), 211–217.

Juven, B. J., Barefoot, S. F., Pierson, M. D., McCaskill, L. H. & Smith, B. (1998) Growth and survival of Listeria monocytogenes in vacuum-packaged ground beef inoculated with Lactobacillus alimentarius FloraCarn L-2. *Journal of Food Protection* 61, 551–556.

Kanatt, S. R., Chander, R. & Sharma, A. (2008) Chitosan and mint mixture: A new preservative. *Food Chemistry* 107, 845–852.

Katla, T., Moretro, T., Holck, A., Axelsson, L. & Naterstad, K. (2001) Inhibition of Listeria monocytogenes in cold smoked salmon by addition of sakacin P and/or live Lactobacillus sakei cultures. *Food Microbiology* 18, 431–439.

Katla, T., Moretro, T., Sveen, I., Aasen, I. M., Axelsson, L., Rorvik, L. M. & Naterstad, K. (2002) Inhibition of Listeria monocytogenes in chicken cold cuts by addition of sakacin P and sakacin P producing *Lactobacillus sakei*. *Journal of Applied Microbiology* 93, 191–196.

Knobloch, K., Weigand, H., Weis, N., Schwarm, H.-M. & Vigenschow, H. (1986) Action of terpenoids on energy metabolism. In: *Progress in Essential Oil Research: 16th International Symposium on Essential Oils*. Brunke, E. J. (ed.), De Gruyter, Berlin, pp. 429–445.

Kurita, K. (2006) Chitin and chitosan: Functional biopolymers from marine crustaceans. *Marine Biotechnology* 8 (3), 203–226.

Knorr, D. (1998) Technology aspects related to microorganisms in functional foods. *Trends in Food Science Technology* 9, 295–306.

Kowalski, E., Ludwig, H. & Tauscher, B. (1992) Hydrostatic pressure to sterilize food. 1. Application to pepper (*Piper nigrum L.*). *Deutsche Lebensmittel Rundschau* 88, 74–75.

Krebs, H. A., Wiggins, D., Stubs, M., Sols, A. & Bedoya, F. (1983) Studies on the mechanism of the antifungal action of benzoate. *Biochemistry Journal* 214, 657–663.

Kwak, H. S., Kim, S. H., Kim, J. H., Choi, H. J. & Kang, J. (2004) Immobilized bcyclodextrin as a simple and recyclable method for cholesterol removal in milk. *Archives of Pharmacal Research* 27 (8), 873–877.

Lambert, R. J. W., Skandamis, P. N., Coote, P. & Nychas, G.-J. E. (2001) A study of the minimum inhibitory concentration and mode of action of oregano essential oil, thymol and carvacrol. *Journal of Applied Microbiology* 91, 453–462.

Legisa, M. & Golic-Grdadolnik, S. (2002) Influence of dissolved oxygen concentration on intracellular pH and consequently on growth rate of *Aspergillus niger*. *Food Technology and Biotechnology* 40, 27–32.

Leroy, F. & De Vuyst, L. (1999) The presence of salt and a curing agent reduces bacteriocin production by Lactobacillus sakei CTC 494, a potential starter culture for sausage fermentation. *Applied Environmental Microbiology* 65 (12), 5350–5356.

Li, X., Shi, X., Wang, M. & Du, Y. (2010) Xylan chitosan conjugate - A potential food preservative. *Food Chemistry* 126 (2), 520–525.

Maddox, D. N. (1982) The role of p-hydroxybenzoates in modern cosmetics. *Cosmetics and Toiletries* 97, 85–88.

Mori, A., Nishino, C., Enoki, N. & Tawata, S. (1989) Antibacterial activity and mode of action of plant flavonoids against Proteus vulgaris and Staphylococcus aureus. *Phytochemistry* 26 (8), 2231–2234.

Mota, F. J. M., Ferreira, I. M. P. L. V. O., Cunha, S. C., Beatriz, M. & Oliveira, P. P. (2003) *Food Chemistry* 82, 469–473.

Neal, A. L., Weinstock, J. O. & Lampen, J. O. (1965) Mechanisms of fatty acid toxicity for yeast. *Journal of Bacteriology* 90, 126–131.

Ochiai, N., Sasamoto, K., Takino, M., Yamashita, S., Daishima, S. & Hoffmann, A. (2002) Simultaneous determination of preservatives in beverages, vinegar, aqueous sauces, and quasi-drug drinks by stir-bar sorptive extraction (SBSE) and thermal desorption GC–MS. *Analytical and Bioanalytical Chemistry* 373, 56–63.

Oosterhaven, K., Poolman, B. & Smid, E. J. (1995) S-carvone as a natural potato sprout inhibiting, fungistatic and bacteriostatic compound. *Industrial Crops and Products* 4, 23–31.

O'Keeffe, T. O. & Hill, C. (1999) Bacteriocins. In: *Encyclopedia of Food Microbiology*, Robinson, R. K., Batt, C. A. & Patel P. D. (eds), Academic Press, San Diego, pp. 183–191.

Ough, C. S. & Crowell, E. A. (1987) Use of sulfur dioxide in winemaking. *Journal of Food Science* 52 (2), 386–389.

Parke, D. X. & Lewis, D. F. (1992) Safety aspects of food preservatives. *Food Additives and Contaminants* 9, 561–577.

Patterson, M. (2000) High pressure treatments of foods. In *Encyclopedia of Food Microbiology*, Robinson, R. K., Batt, C. A. & Patel P. D. (eds), Academic Press, San Diego, pp. 1059–1065.

Peck, M. W. (1997) *Clostridium botulinum* and the safety of refrigerated processed foods of extended durability. *Trends in Food Science and Technology* 8, 186–192.

Pena, A., Lino, C. & Silveira, M. I. N. (2005) Survey of caffeine levels in retail beverages in Portugal. *Food Additives and Contaminants* 22, 91–96.

Pilatus, U. & Techel, D. (1991) 31P-NMR-studies on intracellular pH and metabolite concentrations in relation to the circadian rhythm, temperature and nutrition in *Neurospora crassa*. *Biochimica et Biophysica Acta* 1091, 349–355.

Pilkington, B. J. & Rose, A. H. (1988) Reactions of Saccharomyces cerevisiae and Zygosaccharomyces bailii to sulphite. *Journal of General Microbiology* 134, 2823–2830.

Plumridge, A. (2005) *Sorbic acid stress in* Aspergillus niger. PhD thesis, University of Nottingham, UK.

Plumridge, A., Hesse, S. J. A., Watson, A. J., Lowe, K. C., Stratford, M. & Archer, D. B. (2004) The weak acid preservative sorbic acid inhibits conidial germination and mycelial growth of Aspergillus niger through intracellular acidification. *Applied Environmental Microbiology* 70, 3506–3511.

Puupponem-Pimia, R., Nohynek, L., Meier, C., Kahkonen, M., Heinonen, M., Hopia, A. & Oksman-Caldentey, K. M. (2001) Antimicrobial properties of phenolic compounds from berries. *Journal of Applied Microbiology* 90 (4), 494–507.

Pylypiw H. M., Jr. & Grether, M. T. (2000) Rapid high-performance liquid chromatography method for the analysis of sodium benzoate and potassium sorbate in foods. *Journal of Chromatography A*, 883, 299–304.

Rodriguez, J. M., Martinez, M. I., Horn, N. & Dodd, H. M. (2002) Heterologous production of bacteriocins by lactic acid bacteria. *International Journal of Food Microbiology* 80, 101–116.

Roller, S. (1995) The quest for natural antimicrobials as novel means of food preservation: status report on a european research project. *International Biodeterioration and Biodegradation* 36 (3–4), 333–345.

Rudrapatnam, N. T. & Farooqahmed, S. K. (2003) Chitin-the undisputed biomolecule of great potential. *Critical Reviews in Food Science and Nutrition* 43, 61–87.

Schellekens, M. (1996) New research in sous-vide cooking. *Trends in Food Science and Technology* 7, 256–262.

Schwenninger, M., von Ah, S., Teuber, M. & Meile, L. (2002) Application of propionibacteria and lactobacilli as protective cultures to prevent outgrowth of yeasts and bacterial pathogens in food. Proceedings of the 18th International ICFMH Symposium Food Microbiol. Lillehammer, Norway, August 18–23, pp. 123–125.

Seyderhelm, I. & Knorr, D. (1992) Reduction of Bacillus stearothermophilus spores by combined high pressure and temperature treatments. *Journal of the Food Industry* 43 (4), 17–20.

Shahidi, F., Arachchi, J. K. V. & Jeon, Y. J. (1999) Food applications of chitin and chitosans. *Trends in Food Science and Technology* 10, 37–51.

Shimada, K., Kimura, E., Yasui, Y., Tanaka, H., Matsushita, S., Nagakmura, M. & Kawahisa, M. (1991) Styrene formation by the decomposition of *Pichia carsonii* of trans-cinnamic acid added to a ground fish product. *Applied Environmental Microbiology* 58, 1577–1582.

Sikkema, J., De Bont, J. A. M. & Poolman, B. (1994) Interactions of cyclic hydrocarbons with biological membranes. *Journal of Biological Chemistry* 269 (11), 8022–8028.

Sikkema, J., De Bont, J. A. M. & Poolman, B. (1995) Mechanisms of membrane toxicity of hydrocarbons. *Microbiological Reviews* 59 (2), 201–222.

Singh, M., Sharma, R & Banerjee, U. C. (2002) Biotechnological applications of cyclodextrins. *Biotechnology Advances* 20, 341–359.

Skandamis, P. N. & Nychas, G.-J. E. (2001) Effect of oregano essential oil on microbiological and physico-chemical attributes of minced meat stored in air and modified atmospheres. *Journal of Applied Microbiology* 91, 101–1022.

Soni, M. G., Carabin, I. G. & Burdock, G. A. (2005) Safety assessment of esters of p-hydroxybenzoic acid (parabens) *Food Chemistry and Toxicology* 43, 985–1015.

Tassou, C. C. & Nychas, G. J. E. (1994). Inhibition of *Staphylococcus aureus* by olive phenolics in broth and in a model food system. *Journal of Food Protectection* 57 (2), 121–124.

Tassou, C. C. & Nychas, G. J. E. (1995) Inhibition of *Salmonella enteritidis* by oleuropein in broth and in a model food system. *Letters in Applied Microbiology* 20, 120–124.

Tassou, C., Drosinos, E. H. & Nychas, G.-J. E. (1995) Effects of essential oil from mint (Mentha piperita) on Salmonella enteritidis and Listeria monocytogenes in model food systems at 4 jC and 10 jC. *Journal of Applied Bacteriology* 78, 593–600.

Taylor, S. L., Bush, R. K., Selner, J. A., Nordlee, J. C., Wiener, M. B., Holden, K., Koepke, J. W., Busse, W. W. (1988) Sensitivity to sulfited foods among sulfite-sensitive subjects with asthma. *Journal of Allergy and Clinical Immunology* 81, 1159–1167.

Techakriengkrai, I. & Surakarnkul, R. (2007) Analysis of benzoic acid and sorbic acid in Thai rice wines and distillates by solid-phase sorbent extraction and high-performance liquid chromatography. *Journal of Food Composition and Analysis* 20, 220–225.

Tfouni, S. A. V. & Toledo, M. C. F. (2002) Determination of benzoic and sorbic acids in Brazilian food. *Food Control* 13, 117–123.

Thoroski, J., Blank, G. & Biliaderis, C. (1989). Eugenol induced inhibition of extracellular enzyme production by *Bacillus cereus*. *Journal of Food Protection* 52 (6), 399–403.

Tosi, E. A., Re, E., Ortega, M. E. & Cazzoli, A. F. (2007) Food preservative based on propolis: Bacteriostatic activity of propolis polyphenols and flavonoids upon Escherichia coli. *Food Chemistry* 104, 1025–1029.

Tranter, H. S., Tassou, S. C. & Nychas, G. J. (1993) The effect of the olive phenolic compound, oleuropein, on growth and enterotoxin B production by *Staphylococcus aureus*. *Journal of Applied Bacteriology* 74, 253–259.

Tyler, T. A. (1984) Liquid-chromatographic determination of sodium saccharin, caffeine, aspartame, and sodium benzoate in cola beverages. *Journal of Association of Official Analytical Chemists* 67, 745–747.

Ultee, A., Bennink, M. H. J., Moezelaar, R. (2002) The phenolic hydroxyl group of carvacrol is essential for action against the food-borne pathogen *Bacillus cereus*. *Applied Environmental Microbiology* 68 (4), 1561–1568.

Vaara, M. (1992) Agents that increase the permeability of the outer membrane. *Microbiology Review* 56, 395–411.

Van de Braak, S. A. A. J. & Leijten, G. C. J. J. (1999) *Essential Oils and Oleoresins: A Survey in the Netherlands and other Major Markets in the European Union.* CBI, Centre for the Promotion of Imports from Developing Countries, Rotterdam, p. 116.

Veerabhadraro, M., Narayan, M. S., Kapur, O. & Sastry, C. S. (1987) Reverse phase liquid chromatographic determination of some food additives. *Journal of the Association of Official Analytical Chemists* 70 (3), 578–582.

Verluyten, J., Schrijvers, V., Leroy, F. & De Vuyst, L. (2002) Modelling the behaviour of the potential meat starter culture *Lactobacillus curvatus* LTH 1174 as influenced by different environmental factors important for European sausage fermentations. Proceedings of the 18th International ICFMH Symposium Food Microbil. Lillehammer, Norway, August 18–23, pp. 167–172.

Vermeiren, L., Devlieghere, F. & Debevere, J. (2004) Evaluation of meat born lactic acid bacteria as protective cultures for the biopreservation of cooked meat products. *International Journal of Food Microbiology* 96, 149–164.

Wang, L., Zhang, X., Wang, Y. & Wang, W. (2006) Simultaneous determination of preservatives in soft drinks, yogurts and sauces by a novel solid-phase extraction element and thermal desorption-gas chromatography. *Analytica Chimica Acta* 577, 62–67.

Wen, Y., Wang, Y. & Feng, Y.-Q. (2007) A simple and rapid method for simultaneous determination of benzoic and sorbic acids in food using in-tube solid-phase microextraction coupled with high-performance liquid chromatography. *Analytical and Bioanalytical Chemistry* 388, 1779–1787.

Wendakoon, C. N. & Sakaguchi, M. (1995) Inhibition of amino acid decarboxylase activity of Enterobacter aerogenes by active components in spices. *Journal of Food Protection* 58 (3), 280–283.

Winter, R. (1999) *A Consumer's Dictionary of Food Additives.* Three Rivers Press, New York, NY.

Wilson, T. A., Romano, C., Liang, J. & Nicolosi, R. J. (1998) The hypocholesterolemic and antiatherogenic effects of cholazol H, a chemically functionalized insoluble fiber with bile acid sequestrant properties in hamsters. *Metabolism* 147 (8), 959–964.

Wood, W. E. (2001) Improved aroma barrier properties in food packaging with cyclodextrins. TAPPI: *Polymers, Laminations and Coatings Conference*, pp. 367–377.

Young, L. L., Reverie, R. D. & Cole, A. B. (1988) Fresh red meats: A place to apply modified atmospheres. *Food Technology* 42 (9), 64–66.

FURTHER READING

Walker, R. (1990) Toxicology of sorbic acid and sorbates. *Food Additives and Contaminants* 7, 671–676.

16 Nutraceuticals and Functional Foods

'Let food be your medicine and medicine be your food.'

Hippocrates, 460–370 BC

Abstract: There are components in foodstuffs which are useful for maintaining our body health and immune system. Scientifically, there is a strong relationship between food and health such that some food items contain medicinal constituents with the ability to keep the body healthy and fit. There has been much research activity recently on the types of food items and other natural products (mainly of plant origin) and their possibility of being sources of therapeutic and prophylactic agents. The types of foods in this regard are those functional foods that, when consumed regularly and appropriately, can result in beneficial health effects beyond their natural properties. Alternatively, dietary supplements can be prescribed in concentrated form to deliver an intended bioactive agent from some foodstuffs. These nutraceuticals are administered in higher doses than in normal food, but in a non-food composition to improve human health.

Keywords: dietary supplement; food and health; nutraceuticals; nutrition; pharmaceutical

16.1 WHAT ARE NUTRACEUTICALS?

The reasons for why we need food go beyond satisfying hunger and getting the energy we need to work or be active because of the presence of essential fuel-nutrient ingredients in food products. This introduces the concept of nutraceuticals, which is a term coined from the words 'nutrition' and 'pharmaceutical' and refers to a foodstuff or product with ingredients which can render health benefits to consumers. Nutraceutical can also refer to a product isolated or purified from foodstuffs, but which can be supplied to consumers in medicinal formulations that under normal circumstances may not be associated with foodstuffs. Another definition that is used for nutraceuticals is any non-toxic food component with health benefits that have been proven beyond reasonable doubt, including disease treatment and prevention. Another term used specifically in relation to animal husbandry is 'veterinary nutraceutical', defined as a substance produced in a purified or extracted form that is administered with the purpose of improving health.

In general, nutraceuticals perform the important functions of promoting wellness, preventing malignant processes and controling symptoms in humans. The exact difference between nutraceuticals and functional foods may be difficult to identify; however, while functional

Chemistry of Food Additives and Preservatives, First Edition. Titus A. M. Msagati.
© 2013 John Wiley & Sons, Ltd. Published 2013 by John Wiley & Sons, Ltd.

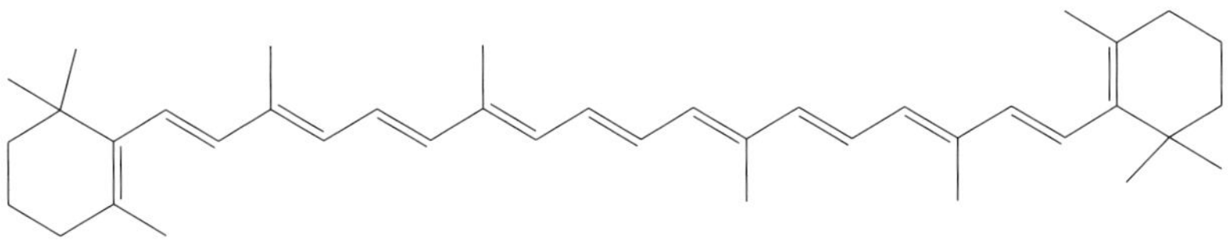

Fig. 16.1   Chemical structure of β-carotene.

foods are consumed normally like any other food, nutraceuticals are always consumed in the form of either tablets, capsules or other forms of medicinal formats.

For longer than can be remembered, it has been known that some types of food items or diet, such as varieties of fruits, vegetables or herbs, are generally composed of ingredients with medicinal activity that can pretreat or vent different types of diseases and ailments (e.g. diabetes or high blood pressure). Examples of food items known to contain medicinal ingredients include carrots, garlic, apples and honey, to mention a few. Carrots contains a terpenoid red pigment known as β-carotene (1,1'-(3,7,12,16-Tetramethyl-1,3,5,7,9,11,13,15,17-octadecanonaene-1,18-diyl)bis(2,6,6-trimethylcyclohexene; Figure 16.1). This isoprenoid pigment molecule in carrots has been linked to a reduced probability of developing lung cancer, stroke or heart attack, as well as keeping skin cells and hair fresh.

Garlic contains ingredients which are antibacterial and antiviral in nature, and has been associated with increasing immunity systems. Fruits such as apples contain fibres which help digestion functions and reduce the risks of cholesterol. Moreover, a flavonol chemical compound known as quercetin (2-(3,4-dihydroxyphenyl)-3,5,7-trihydroxy-4*H*-chromen-4-one; Figure 16.2) found in apple skin is known for its powerful antioxidant activities and is also suspected to play an important role in combating heart diseases, tumours and cancer.

16.2   CLASSIFICATION OF NUTRACEUTICALS

Nutraceuticals currently found in the market include both traditional or natural and non-traditional.

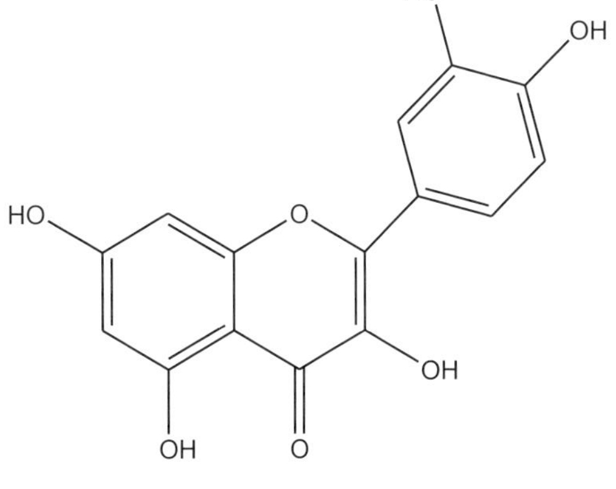

Fig. 16.2   Chemical structure of quercetin.

16.2.1 Natural nutraceuticals

The class of natural nutraceuticals includes whole foods with potential health qualities such as fish (contains active compounds known as ω-3 fatty acids which promotes health); vegetables; fruits, especially tomatoes which are known to contain lycopene; and soy beans which is rich in saponins. Other natural nutraceuticals which have been demonstrated to contain ingredients with health benefits include tea and chocolate.

16.2.2 Non-natural nutraceuticals

Non-natural nutraceuticals are scientifically produced in processes which tend to add some nutritional ingredients with health benefits to agricultural foods. For example, cereals can be enriched with some vitamins or β-carotene, qualifying these foods as nutraceuticals. In general practice, many food products are enriched with antioxidant and/or fat-soluble vitamins which are known to have many potential health benefits such as the prevention of cancer and cerebrovascular disease by the action of antioxidants.

Food items enriched with vitamins such as tocopherols (vitamin E) have been associated with the prevention of diseases such as Parkinson's disease, if taken constantly and consistently. The oxidised form of vitamin C (ascorbic acid) known as dehydroascorbic acid readily crosses the blood brain barrier, thus giving the potential to combat Alzheimer's disease. Some studies have shown that a synergy of vitamin E, C and β-carotene can play an important role in reducing low-density lipoprotein oxidation, thus reducing the risk of atherosclerosis.

16.3 MECHANISMS OF ACTION

The mechanism of certain food items in functioning as prophylactic and curative agents can be explained by the presence of chemicals known as secondary metabolites which have a wide range of biological activities. Although the activity of these secondary metabolites (present mainly in plant-derived foods) is lower than that of normal drugs prescribed for some diseases and ailments, a regular and consistent intake through diet can have a great effect on lasting physiological effects.

There are many different types of foods (both whole foods and processed foods) with a variety of ingredients and chemistries imparting health benefits (in treating or preventing different types of disease conditions), so a number of different biological mechanisms by which nutraceuticals work can therefore be expected. The following sections discussing the mechanisms of action of nutraceuticals is therefore based on the types of diseases in which they have been reported to have health benefits.

16.3.1 Obsessive compulsive disorder

The medical/clinical treatment for a mental condition known as obsessive–compulsive disorder (OCD) involves the prescription of tricyclic antidepressant clomipramine and other selective serotonin re-uptake inhibitors (SSRIs). Among the SSRIs usually prescribed include fluvoxamine, fluoxetine, sertraline, paroxetine, citalopram and escitalopram. Dual reuptake inhibitors, including either venlafaxine or duloxetine (Jenike 2004; Fineberg and Gale 2005; Dell'Osso et al. 2006; Denys et al. 2007; Camfield et al. 2011), are also prescribed.

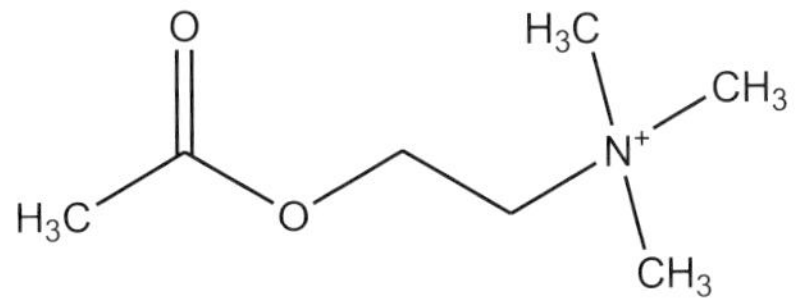

Fig. 16.3   Chemical structure of acetylcholine.

However, drugs used for OCD treatment have been reported to be associated with many side effects such as anxiety, nausea, dizziness, sedation, diarrhoea, headache and insomnia (Jenike 2004; Camfield *et al.* 2011). Apart from the side effects, the treatment itself is said to be temporal in the sense that once patients stop using the medication, they immediately become sick again. This situation forces them to continue with the dose for a period after OCD symptoms have stopped (Pato *et al.* 1988; Steiner 1995; Jenike 2004). In addition to all these drawbacks, the OCD treatment takes a considerable period of time before curative effects are realised by patients. Finally, OCD treatment is not universally effective in the sense that only 40–60% of patients treated with these medications experience improvements in their condition (Pallanti *et al.* 2002; Jenike 2004; Camfield *et al.* 2011). Because of these disadvantages in the treatment of OCD, research on the use of nutraceuticals for OCD treatment is underway.

16.3.1.1   *Nutraceuticals for OCDs*

Among the foodstuffs that are recommended as nutraceuticals for OCD treatment are those in which N-acetylcysteine (NAC; Figure 16.3), which is also a neurotransmitter, has been added as a supplement. NAC is a derivative of cysteine which, unlike cysteine, is not usually obtained naturally from dietary sources and therefore must be added as a supplement (Bonanomi and Gazzaniga 1980). When a diet rich in NAC is consumed, the NAC is metabolised in the liver system to form cysteine which has the ability to cross and enter the brain system (Smith 2000). There, it plays the very important role of inhibiting the synaptic release of glutamate, bringing positive effects as far as the treatment of OCD and many other psychiatric disorders are concerned.

16.3.1.2   *Mechanisms of action in OCD*

To understand the mechanism of action of NAC in the treatment of OCD, it is necessary to introduce metabotropic glutamate receptors (mGluRs) which bind glutamate, an amino acid compound which performs the function of an excitatory neurotransmitter (Bonsi *et al.* 2005). The mGluRs are a class of receptors divided into three main groups (group I, II and III). Group I is mainly located in postsynaptic sites while groups II and III are mainly located in presynaptic site on all neurons (Wright *et al.* 2001). The mGluRII and mGluRIII receptors play a crucial role in regulation of the synaptic release of glutamate; when stimulated by the extracellular glutamate, they inhibit the synaptic release of glutamate (Moran *et al.* 2001). The extracellular glutamate levels are known to be controlled by cysteine (a disulphide derivative of cysteine)-glutamate antiporter, found on the neuroglia cells in the neurons (Baker *et al.* 2002). When there is an increase of extracellular glutamate in the neurons, the cysteine-glutamate antiporter located in the neuroglia cells facilitates the exchange of extracellular cysteine for the intracellular glutamate, which in turn stimulates the activity of mGluRII receptors and, in the process, inhibits the release of synaptic glutamate (Moran *et al.* 2003).

An investigation involving a randomised controlled trial of NAC 3000 mg/day has been proposed (Camfield *et al.* 2011) to test the potential efficacy of NAC in the treatment strategy of OCD.

16.3.1.3   Glycine, myo-inositol, tryptophan and 5-hydroxytryptophan

A number of other amino acids and glucose-based nutrients, which are also used for the treatment of psychiatric disorders, are being investigated as potential nutraceuticals for OCD treatment (Camfield *et al.* 2011). They operate via many different types of mechanisms involving various receptors located in the brain cells. However, the N acetylcholine has received more attention than glycine. This is due to the fact that the administration of the amino acid glycine is problematic, as large doses are associated with the side effect of nausea (Greenberg 2009).

16.3.2   Mental health problems

Myo-inositol is an isomer of glucose which has been used for the treatment of mental-related problems for a long time, and is the most abundant chemically active compound (Frey *et al.* 1998; Sarris *et al.* 2010). Myo-inositol is an ingredient found abundantly in many foodstuffs including fruits, beans, grains and nuts (Clements and Darnell 1980). It has been observed that the exogenous administration of myo-inositol is directly related to the increase in the levels of myo-inositol in the cerebrospinal fluid as well as in the brain (Einat and Belmaker 2001). The mechanisms by which myo-inositol exerts its therapeutic effects are yet to be determined; there are many theories which are yet to be proven (Marazziti *et al.* 2000, 2002; Harvey *et al.* 2001, 2002).

Myo-inositol administration is known to have some drawbacks in the form of gastro-intestinal side effects. Normally categorised as mild side effects, these include diarrhoea, flatulence, bloating and nausea (Levine 1997; Carey *et al.* 2004). There are a number of food products which are known to be rich in the amino acid nutrients of tryptophan and 5-hydroxytryptophan, including egg white, cheese, Atlantic cod, spirulina, soybeans and pumpkin seed (Camfield *et al.* 2011). Unlike tryptophan, which can be obtained directly from diet, 5-hydroxytryptophan has to be synthesised from tryptophan in the body system. Prescribing a diet high in tryptophan with the intention of raising the levels of 5-hydroxytryptophan is an option which does not work however, due to the minute fraction of tryptophan which is converted to 5-hydroxytryptophan (Brown 1994). Moreover, the amount of tryptophan that crosses the blood–brain barrier is still dependent on the partitioning ratio due to the presence of other large neutral amino acids (LNAA) which compete for the same transporter (Birdsall 1998).

These limitations mean that tryptophan and 5-hydroxytryptophan are suitable as nutraceuticals for OCD because, under normal circumstances, OCD patients have been reported to have lower tryptophan as well as tryptophan/LNAA ratios in their blood plasma. On the other hand, a tryptophan-rich diet may not add much value to consumers free of OCD disorders, since research has shown that the depletion of tryptophan does not result in OCD symptoms in a healthy set of testants. It can therefore be interpreted that tryptophan may not be the only parameter that has an influence in the expression of OCD symptoms (Smeraldi *et al.* 1996).

16.3.3 Obesity and diabetes mellitus

There are two main basic types of diabetes mellitus, referred to as types 1 and 2. Excessive elevation of blood glucose in the body is the major cause of diabetic complications, affecting many organs and body systems in both type 1 and 2 diabetes mellitus. Type 1 diabetes mellitus is a condition of absolute insulin deficiency, and type 2 diabetes mellitus is a relative insulin deficiency condition caused by a beta cell failure as well as varying degrees of insulin resistance (Kaiser *et al.* 2003).

Diabetes diseases normally occur in people who tend to consume foodstuffs with a high content of trans-fats and saturated fatty acids. Controlling the consumption of saturated and trans-fats is therefore essential for the improvement of blood lipid levels, diabetic parameters and reversing of the progression of diabetes. For this reason, diabetes treatment involves the modification of behaviour, lifestyle and eating habits (Dubnov *et al.* 2003).

Among the nutraceuticals that have been reported to treat diabetes is R-lipoic acid (R-LA), which is both a natural coenzyme and an antioxidant known to lower levels of serum lactate and pyruvate in sufferers of diabetes mellitus. The decrease of serum lactate and pyruvate has the effect of enhancing the process of glucose uptake by activating insulin-signalling pathways (Korotchkina *et al.* 2004). Polar extract from some plants, for example a plant species from Gramin family known as *Triticum repens*, has demonstrated success in treating diabetes as it has shown potential in lowering the blood glucose levels (Maghrani *et al.* 2004).

Quercetin flavonoids are among those compounds which are known to have antidiabetic activity. A study by Vessal *et al.* (2003) has indicated that quercetin flavonoid was able to exert regeneration of pancreatic islets, and this compound was suspected of having an effect on increasing the release of insulin. Hif and Howell (1984, 1985) published findings which showed that quercetin flavonoid was able to induce insulin release together with an increase in the uptake of calcium ions from isolated islets.

16.3.4 Cardiovascular diseases

Diet rich in flavonoids, especially plant-derived phenolic compounds such as those from fruits, vegetables and grains, have been reported to have anticancer and anti-inflammatory activity with beneficial health effects on cardiovascular and neurodegenerative diseases (see Chapter 1). Pomegranate juice (POM) is known to be the richest source of polyphenolic antioxidants with health benefits related to heart conditions. Pomegranate is rich in a number of flavonoid compounds, including tannins and anthocyanins. The health benefits of diets rich in polyphenolic compounds is due to the ability of polyphenols to transfer electrons, chelate ferrous ions and scavenge reactive oxygen species (Kandaswami and Middleton 1994).

It has been reported that a significant and constant intake of diet containing the long-chain polyunsaturated fatty acids of eicosapentaenoic acid (EPA) and docosahexaenoic acid (DHA), also known as ω-3 fatty acid (mainly found in fish such as salmon, rainbow trout, mackerel and sardines), has benefits to health (Mandel *et al.* 2005). These ingredients are especially beneficial to individuals who are overweight, suffering from hypertension and on weight loss diets. The compounds decrease triglycerides, thus lowering the heart rhythms which will tend to lower the blood pressure and improve the blood-clotting regulation mechanisms (Leaf *et al.* 1999).

It has been reported that a diet rich in flavonoid compounds has a positive effect on the prevention of endothelial dysfunction. This is possible as these compounds tend to enhance the vasorelaxant process, which cause the reduction of arterial pressure (Iijima and

Aviram 2001; Bernatova *et al.* 2002). The reduction of arterial pressure which is important in the reduction of endothelial dysfunction is important because otherwise the process will encourage the initiation of cardiovascular diseases (Jayakody *et al.* 1985).

16.3.5 Cancer

Fresh fruits and vegetables are among the functional nutrition that has been reported to have health benefits including that of decreasing cancer risk (Donaldson 2004). The presence of antioxidant flavonoids in these foodstuffs contributes greatly to their anticancer properties. As well as acting as radical scavengers these flavonoid antioxidants play a very important role as inducers of enzymes, decreasing the cancer risk. These enzymes include catalases and glutathione peroxidase (Frei and Higdon 2003).

The mechanisms of action of these anticancer compounds are believed to be acting on multiple target-signalling pathways, for example activator protein-1 (AP-1) and/or nuclear factor κ B (NF-nB). All of these signalling pathways play crucial roles in tumour promoter-induced cell transformation and tumour promotion (Sarkar and Li 2004). In some other reports from animal studies, it has been demonstrated that flavonoid antioxidants from plants such as green tea catechins, curcumin, anthocyanins, quercetin and silibinin have high chemopreventive activity (Kresty *et al.* 2001). In many other studies, the anti-inflammatory activity of flavonoids and other related class of compounds and derivatives have been numerously reported. For instance, flavone glycosides, flavonol glycosides, flavonoid aglycons, apigenin, luteolin and quercetin have been shown to have health benefits as far as anti-inflammatory activity is concerned (Lee *et al.* 1993; Farmica and Regelson 1995; Hang *et al.* 2002). Other flavonoid compounds such as hesperidin, an extract from citrus plants, has been reported to exert anti-inflammatory as well as analgesic effects (Shahid *et al.* 1998).

16.3.5.1 *Mechanism of flavonoid antioxidants in reducing cancer risk*

Nutraceuticals of interest in treating cancer diseases include the isoflavones (Mason 2001). The major source of dietary isoflavones which occur naturally in the form of glycosides is soybeans (Messina 1999). Apart from being nutraceuticals for cancer diseases, isoflavones are also suspected to perform the protective function against osteoporosis, menopausal symptoms and high levels of cholesterol in the blood (Mason 2001; Awaisheh *et al.* 2005).

Huang *et al.* (2002) have reported that extracts of raspberries inhibited benzo(a)pyrene diol-epoxide-induced transactivation of activated protein 1 and nuclear factor kappa B. The mechanism of the inhibition has been suggested to be via both the down-regulation of protein (enzyme) cyclooxygenase-2 (COX-2) as well as the down-regulation of the transcription activators, AP-1 and NF-nB in JB-6 epidermal cells (Huang *et al.* 2002). A similar mechanistic pathway has been suggested for the phenolic antioxidants from green tea and soybeans, i.e. a polyphenol epigallocatechin-3-gallate and isoflavone genistein, respectively. These polyphenols play important roles in the inhibition of the activation of NF-nB and AP-1 transcription activators, thus preventing the COX-2 induction in the epidermal/epithelial cells (Park and Surh 2004).

Extracts from some plants such as *Curcuma longa* L. from the family Zingiberaceae, which also yields the yellow colouring agent curcumin, have been reported to inhibit the expression of COX-2 in epidermal cells. They also suppress carcinogenesis in the oral cavity, skin, liver, lung, colon, stomach and breast. The mechanism for the inhibition of the extracts from *Curcuma longa* has been suggested to be via the inactivation of transcription factor NF-nB, the regulation of the transcription factors NF-nB, signal transducer and activator of transcription 3 (STAT-3) and suppression of InB kinase (Surh 2002).

Apart from phytoantioxidants, phenolic extracts from dietary supplement milk-thistle, known as silibinin and silymarin, have been reported to have antihepatotoxic activity as well as activity against skin and prostate cancer (Bhatia and Agarwal 2001). The reported mechanism of action of silibinin has been linked to its ability to induce transcription factors which then inhibits the kinase activity of CDKs and cyclins, resulting in the arrest of G1 as well as the control of cell growth (Bhatia and Agarwal 2001).

16.3.6 Brain development

Scientific studies have shown that there is an association between the build up of iron/iron gradient together with reactive oxygen species/reactive nitrogen species (ROS/RNS) (superoxide, hydroxyl radical and nitric oxide) with the death of brain cells (neurons) (Mandel *et al.* 2005). Moreover, other studies have linked the sulphoxide oxidation of some methionine amino acid residues with the occurrence of certain brain diseases. This is because the build up of cellular toxicity leads to the accumulation of some peptides including amyloid-beta peptides, a-synuclein and cytochrome C (Moskovitz 2002). To reverse this oxidation process, the presence of methionine sulphoxide reductase enzymes is required (Moskovitz 2002). Together with this enzyme, studies have shown that the presence of effective brain-permeable iron-chelating and antioxidant-containing bioactive compounds play a very important role in the prevention of neurons death (Mandel *et al.* 2005).

There exist a number of natural dietary antioxidant polyphenolic flavonoids, present in fruits and vegetables such as green tea, pomegranate juice, raspberries, blueberries and red wine (all of which are rich in polyphenol (-) -epigallocatechin-3-gallate), which show activity towards neurodegeneration disorders (Mandel *et al.* 2003). The activity by the polyphenolic flavonoids in fruits and vegetables is believed to be due to their radical scavenging property, transition metal chelating ability and their anti-inflammatory activities (Mandel and Youdim 2004; Weinreb *et al.* 2004; Joseph *et al.* 2005). Polyphenol epigallocatechin-3-gallate are beneficial not only as antioxidant and iron chelaters, but also as molecules which play an important role in the processes which govern cell survival or death genes and signal transduction pathways (Schroeter *et al.* 2002). Other antioxidant flavonoid compounds, for example EGb 761 which is extracted from *Ginkgo biloba* leaves, have been reported to have activity towards cognitive function, in memory impairment cases and in extending life span (Luo 2001).

Other nutraceuticals known to be beneficial in brain development are polyunsaturated fatty acids (PUFAs) which include ω-3-fatty acids such as linolenic, ecosapentaenoic and docosahexaenoic acids as well as ω-6 fatty acids (linoleic and arachidonic acids). These compounds are precursors of long-chain fatty acids such as eicosapentaenoic acid (EPA) and docosahexaenoic acid (DHA), which human bodies are unable to synthesise. These same fatty acids have been reported to be important in reducing plaque formation in the arteries (Milner and Alison 1999). The best dietary sources of ω-3-fatty acids include vegetable oils or fatty fish such as herrings or sardines (Childs and King 1993).

The mode of action of PUFAs which enable them to provide health benefits involves the formation of compounds such as prostacyclins and thromboxanes, as well as pro-inflammatory cytokine production which include: tumour necrosis factor alpha and interleukin-1; modulation of the hypothalamic–pituitary–adrenal anti-inflammatory responses; and induction of the release of acetylcholine (Das 2000).

As well as playing an important role in brain development, diets rich in natural PUFAs have been reported to have health benefits in controlling immune function, blood pressure, bloodstream cholesterol and triglycerides levels (Russo *et al.* 1995; Kawasaki *et al.* 2000; Straniero *et al.* 2008; Chen *et al.* 2009).

16.3.7 Age-related diseases

When the functioning of body tissues and organs slows down due to miscellaneous mutations, especially the mutations associated with neurons as well as the general damage to proteins caused by oxidative stress, the body become prone to age-related diseases and disorders. As discussed in Chapter 1, the generation of oxidative stress due to the reactive species (ROS/RNS) is considered responsible for accelerating the aging process which is characterised by, among other symptoms, brain-related diseases. These diseases/disorders include age-associated cognitive decline and neuronal loss in neurodegenerative diseases such as Alzheimer's, Parkinson's and Huntington's diseases (Markesbery 1997; Jenner 1998).

There exist compounds which are important constituents of nutritional dietary supplements, for example choline, folic acid and their metabolites (e.g. betaine, a choline metabolite), which are crucial for the normal and smooth functioning of body cells. These compounds are known to play important roles in the structural integrity and signalling functions of cell membranes (Zhu *et al.* 2004).

16.3.8 Viral diseases

There are a number of naturally occurring flavonoids which have been reported to have a promising activity against certain virus particles, include quercetin, morin, rutin, dihydroquercetin (taxifolin), apigenin, catechin and hesperidine (Selway 1986). The antiviral activity of the flavonoids seems to be dependent on their structural chemistries, which demonstrate features such as the absence of sugar/glycoside moiety and the presence of hydroxyl group at the 3-position in the flavonoid structure.

Other functional groups related to flavonoids such as flavonols and flavons have also shown to have antiviral activity; flavonols have been observed to be more active than flavones against *Herpes simplex* virus type 1 (Tapas *et al.* 2008). In terms of comparison of antiviral activity, compounds can be ranked (in order of effectiveness): galangin, kaempferol and quercetin (Thomas *et al.* 1988). Activity towards HIV types I and II or other similar immunodeficiency viruses have found flavans to be more active than flavones and flavonones (Gerdin and Srensso 1983).

16.3.9 Anti-gastrointestinal ulcers and anti-acids

Alarcon *et al.* (1994) and Izzo *et al.* (1994) reported the activity of flavonoids extracted from *Ocimum basilicum* plant species, which belong to a family Labiatae, against ulcers. The flavonoid extracts (quercetin, rutin and kaempferol) were not only active in decreasing the ulcer index, but also inhibited the effect of gastric acid and pepsin secretions.

16.3.10 Liver cell protection

A number of flavonoid compounds such as silymarin, apigenin, quercetin and naringenin have shown health benefits when they protect liver cells against cyanobacterial toxins (microcrystin LR) (Carlo *et al.* 1993). Other flavonoid compounds, including rutin and venoruton, have also been reported to play a protective role in cases of cirrhosis (Lorenz *et al.* 1994).

16.3.11 Blood coagulation and thrombotic diseases

Diseases associated with the formation of a blood clot within blood vessels (thrombosis), causing the obstruction of the flow of blood in the body, are known to affect many people worldwide (Furie and Furie 2008). The formation of a blood clot is known to have advantages, for instance when there is an injury on the blood vessels which may cause bleeding, the thrombocytes (also known as platelets) and fibrin aggregate to form a blood clot to stop bleeding. However, there are instances where blood clots may form in the vessels even when a blood vessel is not injured, for instance if there is a disturbed blood flow or a genetical disorder known as hypercoagulability. In cases where clotting is very severe, the situation may lead to thrombotic diseases.

Some older publications have reported that extracts from tea have health benefits in combating thrombotic diseases by reducing blood coagulability, increasing fibrinolysis and prevention of thrombocytes adhesion and aggregation (Lou *et al.* 1989). Some flavonoid derivatives such quercetin, kempferol and myricetin have also been reported as being able to prevent the formation of blood clots in animals (Osman *et al.* 1998). Flavonoids in the prevention of blood clot formation have been linked to their ability to inhibit the activity of cyclooxegenase (COX) and lipoxigenase pathways (Alcaraz and Ferrandiz 1987). Flavonol derivatives also have been reported to be antithrombotic due to their ability to scavenge free radicals. This is advantageous because it tends to maintain the proper concentration of endothelial prostacyclin and nitric oxide (Gryglewski *et al.* 1987). Thrombocytes themselves are known to have the self-mechanism of releasing lipid peroxides and oxygen free radicals. Like flavonols, these have the ability to inhibit the endothelial formation of prostacyclin and nitrous oxide (Gryglewski *et al.* 1987).

16.3.12 Blood cholesterol cases

Diets containing natural phytosterols have been reported to have benefits in maintaining the level of cholesterol in the bloodstream. Phytosterols work by interfering with the uptake of cholesterol from the intestinal tract (Heinemann *et al.* 1991; Jones *et al.* 1999), keeping it at acceptable lower levels.

16.4 CONCLUSION

This chapter provides evidence of the promising trend for nutraceuticals in combating or lessening the risks of some chronic diseases as well as other physiological health benefits. Since the active ingredients of these nutraceuticals are found in many foodstuffs, future research activities may further strengthen their reliability as remedies of the health problems faced by many people.

REFERENCES

Alarcon, D. L., Martin, M. J., Locasa, C. & Motilva, V. (1994) Antiulcerogenic activity of flavonoids and gastric protection. *Journal of Ethnopharmacology* 42, 161–170.

Alcaraz, M. J. & Ferrandiz, M. L. (1987) Modification of arachidonic metabolism by flavonoids. *Journal of Ethnopharmacology* 21, 209–229.

Awaisheh, S. S., Haddadin, M. S. Y. & Robinson, R. K. (2005) Incorporation of selected nutraceuticals and probiotic bacteria into a fermented milk. *International Dairy Journal* 15, 1184–1190.

Baker, D. A., Xi, Z. X., Shen, H., Swanson, C. J. & Kalivas, P. W. (2002) The origin and neuronal function of in vivo nonsynaptic glutamate. *Journal of Neuroscience* 22, 9134–141.

Bernatova, I., Pechanova, O. & Balal, P. (2002) Wine polyphenols improve cardiovascular remodeling and vascular function in NO-deficient hypertension. *American Journal of Physiology, Heart Circulatory Physiology* 282, 942–948.

Bhatia, N. & Agarwal, R. (2001) Detrimental effect of cancer preventive phytochemicals silymarin, genistein and epigallocatechin 3-gallate on epigenetic events in human prostate carcinoma DU145 cells. *Prostate* 46, 98–107.

Birdsall, T. C. (1998) 5-Hydroxytryptophan: a clinically-effective serotonin precursor. *Alternnative Medicinal Reviews* 3, 271–280.

Bonanomi, L. & Gazzaniga, A. (1980) Toxicology, pharmacokinetics and metabolism of acetylcysteine. *European Journal Respiratory Diseases* 111, 45–51.

Bonsi, P., Cuomo, D., De Persis, C., Centonze, D., Bernardi, G., Calabresi, P. & Pisani, A. (2005) Modulatory action of metabotropic glutamate receptor (mGluR) 5 on mGluR1 function in striatal cholinergic interneurons. *Neuropharmacology* 49 (Suppl 1), 104–113.

Brown, R. R. (1994) Tryptophan metabolism: In: *L-Tryptophan: Current Prospects in Medicine and Drug Safety*. Kochen, W. & Steinhart, H. (eds), Walter de Gruyter, Berlin, pp. 17–30.

Camfield, D. A., Sarris, J. & Berk, M. (2011) Nutraceuticals in the treatment of Obsessive Compulsive Disorder (OCD): A review of mechanistic and clinical evidence, *Progress in Neuro-Psychopharmacology and Biological Psychiatry* 35, 887–895.

Carey, P. D., Warwick, J., Harvey, B. H., Stein, D. J. & Seedat, S. (2004) Single Photon Emission Computed Tomography (SPECT) in obsessive–compulsive disorder before and after treatment with inositol. *Metabolism and Brain Disorders* 19, 125–134.

Carlo, G. D., Autore, G. & Izzo A. A. (1993) Inhibition of intestinal motility and secretion by flavonoids in mice and rats; structure activity relationships. *Journal of Pharmacology* 45, 1045–1059.

Chen, Z. Y., Peng, C., Jiao, R., Wong, Y. M., Yang, N. & Huang, Y. (2009) Anti-hypertensive nutraceuticals and functional foods. *Journal of Agriculture and Food Chemistry* 57, 4485–4489.

Childs, M. T. & King, I. B. (1993) Dietary importance of fish and shellfish. In *Encyclopaedia of Food Science, Food Technology and Nutrition*, Macrae, R., Robinson, R. K. & Sadler M. S. (eds), Academic Press, London, pp. 1877–1881.

Clements R. S., Jr, & Darnell, B. (1980) Myo-inositol content of common foods: development of a high-myo-inositol diet. *American Journal of Clinical Nutrition* 33 1954–1967.

Das, U. N. (2000) Beneficial effect(s) of n-3 fatty acids in cardiovascular diseases: but, why and how? *Prostaglandins Leukot Essent Fatty Acids* 63, 351–362.

Dell'Osso, B., Nestadt, G., Allen, A. & Hollander, E (2006) Serotonin-norepinephrine reuptake inhibitors in the treatment of obsessive–compulsive disorder. *Journal of Clinical Psychiatry* 67, 600–610.

Denys, D., Nieuwerburgh, F., Deforce, D. & Westenberg, H. G. M. (2007) Prediction of response to paroxetine and venlafaxine by serotonin-related genes in obsessive–compulsive disorder in a randomized, double-blind trial. *Journal of Clinical Psychiatry* 68, 747–253.

Donaldson, M. S. (2004) Nutrition and cancer: a review of the evidence for an anti-cancer diet. *Nutrition Journal* 3, 19.

Dubnov, G., Brzezinski, A. & Berry, E. M. (2003) Weight control and the management of obesity after menopause: the role of physical activity. *Maturitas* 44, 89–101.

Einat, H. & Belmaker, R. H. (2001) The effects of inositol treatment in animal models of psychiatric disorders. *Journal of Affective Disorders* 62, 113–121.

Farmica, J. V. & Regelson, W. (1995) Review of the biology of quercetin and related bioflavonoids. *Food Chemistry and Toxicology* 33 (12), 1061–1080.

Fineberg, N. A. & Gale, T. M. (2005) Evidence-based pharmacotherapy of obsessive–compulsive disorder. *International Journal of Neuropsychopharmacology* 8, 107–129.

Frei, B. & Higdon, J. V. (2003) Antioxidant activity of tea polyphenols in vivo: evidence from animal studies. *Journal of Nutrition* 33, 3275S–3284S.

Frey, R., Metzler, D., Fischer, P., Heiden, A., Scharfetter, J. & Moser, E. (1998) Myo-inositol in depressive and healthy subjects determined by frontal 1H-magnetic resonance spectroscopy at 1.5 tesla. *Journal of Psychiatric Research* 32, 411–420.

Furie, B. & Furie, B. C. (2008) Mechanisms of thrombus formation. *New England Journal of Medicine* 359 (9), 938–949.

Gerdin, B. & Srensso, E. (1983) Inhibitory effect of flavonoids on increased microvascular permeability induced by various agents in rat skin. *International Journal of Microcirculation – Clinical and Experimental* 2, 39–46.

Greenberg, W. M., Benedict, M. M., Doerfer, J., Perrin, M., Panek, L. & Cleveland, W. L. (2009) Adjunctive glycine in the treatment of obsessive–compulsive disorder in adults. *Journal of Psychiatric Research* 43, 664–670.

Gryglewski, R. J., Korbut, R., Robak, J. & Swips J. (1987). On the mechanism of anti thrombotic action of flavonoids. *Biochemical Pharmacology* 36, 317–322.

Hang, T., Jin, J. B., Cho, S., Cyang J. C (2002). Evaluation of anti-inflammatory effects of baicalein on dextran sulfate sodium-induced colitis in mice. *Planta Medica* 68 (2), 268–271.

Harvey, B. H., Scheepers, A., Brand, L. & Stein, D. J. (2001) Chronic inositol increases striatal D2 receptors but does not modify dexamphetamine-induced motor behavior –relevance to obsessive–compulsive disorder. *Pharmacology and Biochemical Behaviors* 68, 245–253.

Harvey, B. H., Brink, C. B., Seedat, S. & Stein, D. J. (2002) Defining the neuromolecular action of myoinositol: application to obsessive– compulsive disorder. *Progress in Neuropsychopharmacology, Biology and Psychiatry* 26, 21–32.

Heinemann, T., Kullak- Ublick, G. A., Pietruck, B. & von Bergmann, K. (1991) Mechanisms of action of plants sterols on inhibition of cholesterol absorption. *European Journal of Clinical Pharmacology* 40, 859–863.

Hif, C. S. & Howell, S. L. (1984) Effects of epicatechin on rat islets of langerhans. *Diabetes* 33, 291–296.

Hif, C. S. & Howell, S. L. (1985) Effects of flavonoids on insulin secretion and $45Ca^{+2}$ handling in rat islets of langerhans. *Journal of Endocrinology* 107, 1–8.

Huang, C., Huang, Y., Li, J., Hu, W., Aziz, R. & Tang, M. S. (2002) Inhibition of benzo(a)pyrene diolepoxide-induced transactivation of activated protein 1 and nuclear factor kappaB by black raspberry extracts. *Cancer Research* 62, 6857–6863.

Iijima, K. & Aviram, M. (2001) Flavonoids protect LDL from oxidation and attenuate atherosclerosis. *Current Opinion Lipidology* 12, 41–48.

Izzo, A. A., Carlo, G. D., Mascolo, N., Capasso, F. & Autore, G. (1994) Effect of quercetin on gastrointestinal tract in rats and mice. *Phytotherapy Research* 8, 179–185.

Jayakody, T. L., Senaratne, M. P. J., Thompson, A. B. R. & Kappagoda, C. T. (1985) Cholesterol feeding impairs endothelium-dependent relaxation of rabbit aorta. *Canadian Journal of Physiology and Pharmacology* 63, 1206–1209.

Jenike, M. A. (2004) Obsessive–compulsive disorder. *The New England Journal of Medicine* 350, 259–265.

Jenner, P. (1998) Oxidative mechanisms in nigral cell death in Parkinson's disease. *Movements and Disorders* 13, 24–34.

Jones, P. J., Ntanios, F. Y., Raeini-Sarjaz, M. & Vanstone, C. A. (1999) Cholesterol-lowering efficacy of a sitostanol-containing phytosterol mixture with a prudent diet in hyperlipidemic men. *American Journal of Clinical Nutrition* 69, 1144–1150.

Joseph, J. A., Shukitt-Hale, B. & Casadesus, G. (2005) Reversing the deleterious effects of aging on neuronal communication and behavior: beneficial properties of fruit polyphenolic compounds. *American Journal of Clinical Nutrition* 81, 313S– 316S.

Kaiser, N., Leibowitz, G. & Nesher, R. (2003) Glucotoxicity and beta-cell failure in type 2 diabetes mellitus. *Journal of Pediatrics, Endocrinology and Metabolism* 16, 5–22.

Kandaswami, C. & Middleton, E. (1994) Free radical scavenging and antioxidant activity of plant flavonoids. In: *Free Radicals in Diagnostic Medicine*, 2nd edition. Armstrong, D. (ed.) Plenum Press, New York, pp. 351–376.

Kawasaki, T., Seki, E., Osajima, K., Yoshida, M., Asada, K. & Matsui, T. (2000) Antihypertensive effect of valyl-tyrosine, a short chain peptide derived from sardine muscle hydrolyzate, on mild hypertensive subjects. *Journal of Human Hypertensions* 14, 519–523.

Korotchkina, L. G., Sidhu, S. & Patel, M. S. (2004) R-lipoic acid inhibits mammalian pyruvate dehydrogenase kinase. *Free Radical Research* 38, 1083–1092.

Kresty, L. A., Morse, M. A., Morgan, C., Carlton, P. S., Lu, J. & Gupta, A. (2001) Chemoprevention of esophageal tumorigenesis by dietary administration of lyophilized black raspberries. *Cancer Research* 61, 6112–6119.

Leaf, A., Kang, J. X., Xiao, Y. F., Billman, G. E. & Voskuyl, R. A. (1999) The antiarrhythmic and anticonvulsant effects of dietary n-3 fatty acids. *Journal of Membrane Biology* 172 (1), 1–11.

Lee, S. J., Son, K. H. & Chang, H. W. (1993) Anti-inflammatory activity of naturally occurring flavone and flavonol glycosides. *Archives of Pharmaceutical Research* 16 (1), 25–28.

Levine, J. (1997) Controlled trials of inositol in psychiatry. *European Neuropsychopharmacology* 7, 147–155.

Lorenz, W., Kusche, J., Barth, H. & Mathias, C. H. (1994) Action of several flavonoids on enzyme of histidine metabolism in vivo. In: *Histamine*, Maslinski, C. Z. (ed.) Hutchinson and Ross, Pennsylvania, pp. 265–269.

Lou, F. Q., Zhang, M. F., Zhang, X. G., Liu, J. M. & Yuan, W. L. (1989) A study on tea pigment in prevention of antherosclerosis. *Chinese Medicinal Journal (Engl)* 102, 579–583.

Luo, Y. (2001) Ginkgo biloba neuroprotection: therapeutic implications in Alzheimer's disease. *Journal of Alzheimers Diseases* 3, 401–407.

Maghrani, M., Lemhadri, A., Zeggwagh, N. A., El Amraoui, M., Haloui, M. & Jouad, H. (2004) Effects of an aqueous extract of Triticum repens on lipid metabolism in normal and recent-onset diabetic rats. *Journal of Ethnopharmacology* 90, 331–337.

Mandel, S. & Youdim, M. B. H. (2004) Catechin polyphenols: neurodegeneration and neuroprotection in neurodegenerative diseases. *Free Radical Biology and Medicine* 37, 304–317.

Mandel, S., Reznichenko, L., Amit, T. & Youdim, M. B. H. (2003) Green tea polyphenol (-)-epigallocatechin-3-gallate protects rat PC12 cells from apoptosis induced by serum withdrawal independent of P13-Akt pathway. *Neurotoxicology Research* 5, 419–424.

Mandel, S., Packer, L., Youdim, M. B. H. & Weinreb, O. (2005) Proceedings from the Third International Conference on Mechanism of Action of Nutraceuticals. *Journal of Nutritional Biochemistry* 16, 513–520.

Marazziti, D., Masala, I., Rossi, A., Hollander, E., Presta, S. & Giannaccini, G. (2000) Increased inhibitory activity of protein kinase C on the serotonin transporter in OCD. *Neuropsychobiology* 41, 171–177.

Marazziti, D., Dell'Osso, L., Masala, I., Baroni, S., Presta, S. & Giannaccini, G. (2002) Decreased inhibitory activity of PKC in OCD patients after six months of treatment. *Psychoneuroendocrinology* 27, 769–776.

Markesbery, W. R. (1997) Oxidative stress hypothesis in Alzheimer's disease. *Free Radical Biology and Medicine* 23, 134–147.

Mason, P. (2001) Isoflavones. *Pharmaceutical Journal* 266, 16–19.

Messina, M. J. (1999) Legumes and soybeans: overview of their nutritional profiles and health effects. *American Journal of Clinical Nutrition* 70, 439S–450S.

Milner, J. A. & Alison, R. G. (1999) The role of dietary fat in child nutrition and development. *Journal of Nutrition* 129, 2094–2105.

Moran, M. M., Stein, T. P. & Wade, C. E. (2001) Hormonal modulation of food intake in response to low leptin levels induced by hypergravity. *Experimental Biology and Medicine (Maywood)* 226 (8), 740–745.

Moran, M. M., Melendez, R., Baker, D., Kalivas, P. W. & Seamans, J. K. (2003) Cystine/glutamate antiporter regulation of vesicular glutamate release. *Annals of the New York Academy of Science* 1003, 445–447.

Moskovitz, J., Singh, V. K., Requena, J., Wilkinson, B. J., Jayaswal, R. K. & Stadtman, E. R. (2002) Purification and characterization of methionine sulfoxide reductases from mouse and Staphylococcus aureus and their substrate stereospecificity. *Biochemistry and Biophysics Research Communications* 290, 62–65.

Osman, H. E., Maalej, N., Shanmuganayagam, D., Folts. J. D. (1998) Grape juice but not orange or grapefruit juice inhibits plate activity in dogs and monkeys. *Journal of Nutrition* 128, 2307–2312.

Pallanti, S., Hollander, E., Bienstock, C., Koran, L., Leckman, J. & Marazziti, D. (2002) Treatment non-response in OCD: methodological issues and operational definitions. *International Journal of Neuropsychopharmacology* 5, 181–191.

Park, O. J. & Surh, Y. J. (2004) Chemopreventive potential of epigallocatechin gallate and genistein: evidence from epidemiological and laboratory studies. *Toxicology Letters* 150, 43–56.

Pato, M. T., Zohar-Kadouch, R., Zohar, J. & Murphy, D. L. (1988) Return of symptoms after discontinuation of clomipramine in patients with obsessive–compulsive disorder. *American Journal of Psychiatry* 145, 1521–1525.

Russo, C., Olivieri, O., Girelli, D., Azzini, M., Stanzial, A. M. & Guarini, P. (1995) Omega-3 polyunsaturated fatty acid supplements and ambulatory blood pressure monitoring parameters in patients with mild essential hypertension. *Journal of Hypertension* 13, 1823–1826.

Sarkar, F. H. & Li, Y. (2004) Cell signaling pathways altered by natural chemopreventive agents. *Mutation Research* 555, 53–64.

Sarris, J., Kavanagh, D. & Byrne, G. (2010) Adjuvant use of nutritional and herbal medicines with antidepressants, mood stabilizers and benzodiazepines. *Journal of Psychiatric Research* 44, 32–41.

Schroeter, H., Boyd, C., Spencer, J. P., Williams, R. J., Cadenas, E. & Rice-Evans, C. (2002) MAPK signaling in neurodegeneration: influences of flavonoids and of nitric oxide. *Neurobiology and Aging* 23, 861–880.

Selway, J. W. T. (1986) Antiviral activity of flavones and flavans. In: *Plant Flavonoids in Biology and Medicine: Biochemical, Pharmacological and Structure Activity Relationships*. Cody, V., Middleton, E. & Harborne, J. B. (eds) Alan R Liss Inc, New York, pp. 521–536.

Shahid, F., Yang, Z. & Saleemi, Z. O. (1998) Natural flavonoids as stabilizers. *Journal of Food Lipids* 1, 69–75.

Smeraldi, E., Diaferia, G., Erzegovesi, S., Lucca, A., Bellodi, L. & Moja, E. A. (1996) Tryptophan depletion in obsessive–compulsive patients. *Biology and Psychiatry* 40, 398–402.

Smith, Q. R. (2000) Transport of glutamate and other amino acids at the blood-brain barrier. *Journal of Nutrition* 130 (4S Suppl), 1016S–22S.

Steiner, M. (1995) Long-term treatment and prevention of relapse of OCD with paroxetine. *148th* Annual Meeting of the American Psychiatric Association, Miami, Florida, p. 150.

Straniero, S., Cavallini, G., Donati, A., Metelli, M. R., Tamburini, I. & Pietrini, P. (2008) Deficiency and supplementation of PUFA in the diet have similar effects on the age-associated changes in rat-plasma cholesterol levels. *Mechanisms of Ageing Development* 129, 759–762.

Surh, Y. J. (2002) Anti-tumor promoting potential of selected spice ingredients with antioxidative and anti-inflammatory activities: a short review. *Food Chemistry and Toxicology* 40, 1091–1097.

Tapas, A. R., Sakarkar, D. M. & Kakde, R. B. (2008) Flavonoids as Nutraceuticals. *Tropical Journal of Pharmaceutical Research* 7 (3), 1089–1099.

Thomas, P. R. S., Nash, G. B. & Dormandly, J. A. (1988) White cells accumulation in dependent legs of patients with venous hypertension: A possible mechanism for trophic changes in the skin. *British Medical Journal* 296, 1673–1695.

Vessal, M., Hemmati, M. & Vasei, M. (2003) Antidiabetic effects of quercetin in streptozocin induced diabetic rats. *Comperative Biochemistry and Physiology C* 135, 357–364.

Weinreb, O., Mandel, S., Amit, T. & Youdim, M. B. H. (2004) Neurological mechanism of green tea polyphenols in Alzheimer's and Parkinson's diseases. *Journal of Nutritional Biochemistry* 15 (9), 506–516.

Wright, R. A., Arnold, M. B., Wheeler, W. J., Ornstein, P. L. & Schoepp, D. D. (2001) (3H)LY341495 binding to group II metabotropic glutamate receptors in rat brain. *Journal of Pharmacology and Experimental Therapeutics* 298, 453–460.

Zhu, X., Mar, M. H., Song, J. & Zeisel, S. H. (2004) Deletion of the Pemt gene increases progenitor cell mitosis, DNA and protein methylation and decreases calretinin expression in embryonic day 17 mouse hippocampus. *Brain Research and Development* 149, 121–129.

FURTHER READING

Moran, M. M., McFarland, K., Melendez, R. I., Kalivas, P. W. & Seamans, J. K. (2005) Cystine/glutamate exchange regulates metabotropic glutamate receptor presynaptic inhibition of excitatory transmission and vulnerability to cocaine seeking. *Journal of Neurosciences* 25, 6389–6393.

Schoepp, D. D. (2001) Unveiling the functions of presynaptic metabotropic glutamate receptors in the central nervous system. *Journal of Pharmacology and Experimental Therapeutics* 299, 12–20.

17 Nutritional Genomics: Nutrigenetics and Nutrigenomics

Abstract: Scientific evidence has proven the strong correlation between gene expression and its response to changes in the nutritional status within individuals. In reality, nutrients have an important role with regards to the entire mechanisms which control gene expression from lower organisms (acellular) to multicellular organisms such as mammals. The gene–diet response phenomenon is well known, such that different genetic make-ups or nutritional status may bring about marked differences in terms of gene expression for different individuals. All these facts point to the possibility that a thorough knowledge of nutritional requirements for different genotypes may lead to innovative nutritive-based protocols in the solution to genetical diseases and disorders. This chapter discusses aspects of nutritional genomics, nutrigenetics and nutrigenomics in depth.

Keywords: gene expression; nutrigenetics; nutrigenomics; nutrition; nutritional genomics

17.1 NUTRITION AND GENE EXPRESSION

Like all other organisms, humans possess genetical variations which mean differences between individuals in terms of response to diets as well as susceptibility to some genetical diseases and disorders such as obesity, cancer and diabetes as well as cardiovascular diseases. These variations in genetical make-up create sharp differences not only phenotypically but also with regard to the performance of food metabolism. The interaction and association of genes, food and the biological outcomes which are manifest healthwise and the ability of individual's body to withstand or overcome diseases also vary from one individual to another.

This reasoning led to nutritional genomics, or nutrigenomics and nutrigenetics, which investigates how food affects or influences the regulation of genetic coding as well as the way an individual's genetic coding influences the metabolism of diets (Scheme 17.1) (Stover 2004, 2006; Stover and Caudill 2008). In simple terms, both diet and genes play important roles in determining the health of an individual and the ability of the body to withstand or fight a certain disease. The important part of nutrigenetic studies is to identify sets of genes that could be regulated by certain diets or nutrition and which are responsible for genetic diseases. This could result in a breakthrough in the process of design and development of diagnostic tools as well as curative mechanisms for problematic genetic diseases.

17.1.1 Nutrigenetics

Nutrigenetics involves the study of the genetic make-ups within and variations between individuals, and how these variations are related to diets and diseases. An aspect of the

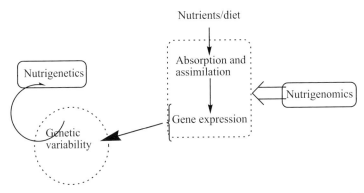

Nutrients/diet

Absorption and assimilation

Nutrigenetics

Nutrigenomics

Genetic variability

Gene expression

Scheme 17.1   The nutrient–genome interactions pattern (Stover 2004, 2006; Stover and Caudill 2008).

discipline is to identify sets of genes which show special responses to components of diets and which may result in increased susceptibility to an association of diet and some particular diseases or disorders. The main advantage of nutrigenetic studies is that, in the final analyses, it can provide plausible and scientific advice and recommendations to individuals with regards to their personal health nutrition.

Ironically, nutrigenetics stems from the tremendous efforts by scientists, medical practitioners, researchers and health specialists to try to explain the causes of a number of hereditary diseases in humans. In essence, reliable and meaningful explanations may not arise from a single piece of information or variable, but rather from a diverse range of causes with different reasons. Scientists and researchers therefore need to adopt integrative approaches towards hereditary and other chronic human diseases (Sing *et al.* 2003).

An example of integrative approaches towards this problem is nutritional genomics; this disciplines examines information which can provide a genetic and molecular understanding for how diet relates to both health and disease, since they tend to influence the gene expression of an individual (Kaput and Rodriguez 2004). It is believed that some dietary ingredients, under certain conditions, are capable of exerting an influence on some individuals either directly or indirectly, thus altering the genomic expression and structure. To those affected, such diets bring serious health risks of dangerous diseases due to the alteration of the portions of the DNA. This fact implies that nutrient components as well as other biologically active ingredients present in foodstuffs have the ability to modulate the genetic expression within the individual.

Gene regulation by dietary components is possible due to the fact that living organisms are capable of altering the expression of amino acid sequences (peptides/proteins) which play key roles in the metabolic pathways. To be more precise, nutrients and other bioactive ingredients in foodstuffs are known to have the ability to alter the gene expression and even impart permanent traits that can be transmitted epigenetically to the next generation (Dennis 2003). Dietary intervention may therefore become meaningful when knowledge of nutritional needs, status and the genetic make-up of an individual is available and can be used to implement strategies to prevent or cure the associated hereditary disease. Moreover, the availability of genomic knowledge and its practices in conjunction with nutrition practices has a bigger potential to actually personalise/individualise nutrition for health benefits. It may prevent some genetic diseases by targeting the genes responsible for these diseases if they are regulated/modulated by dietary components (Stover 2004). In these cases, nutrients and their bioactive components function in the same way as pharmaceuticals by influencing gene

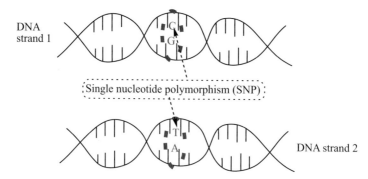

Fig. 17.1 Showing two DNA strands with a difference at a single base-pair location (i.e. a C/T polymorphism; Betoret 2011).

expression, especially where the gene–nutrient interactions have been optimised scientifically to play the role of disease prevention.

The above discussions have set in motion the philosophy behind DNA-driven nutrition (the implication of nutrigenomics and nutrigentics, pertaining to the studies of gene–diet or gene–environmental interactions), especially in chemoprevention disciplines. In this school of thought, the philosophy suggests the possibilities where diet or nutrition may trigger abnormal gene expression (Mathers 2003; Kim 2007; Kornman *et al.* 2007; Lim *et al.* 2007). In addition to this, the diet–gene school of thought advocates the need for personalised diet due to the fact that the effects and/or requirements of diet or nutrition to individuals varies significantly (Stover 2006; Subbiah 2008).

17.1.2 Influence of diet on gene regulation and expression

Although it is true that the bulk of food ingredients are nutrients and are thus metabolised to provide energy, there are certain components in various diets that are known to have the potential to exert an influence on gene expression. These components behave as ligands for transcription factors, thus altering the gene expression directly. The difference between these components and normal nutritious diet ingredients is that the latter have the ability to alter the signal transduction pathways as well as the chromatin structure only by influencing the gene expression, which is what the former (ligands) do but in a more direct fashion (Muller and Kersten 2003; Kaput and Rodriguez 2004). Dietary components and ingredients can therefore interact preferentially with particular sets of susceptibility genes; the resultant effect will either elevate, reduce or counteract a certain disease problem.

The extent of the influence of diet on the gene expression differs between individual due to the genetic differences that exists. This is caused by a gene sequence variation occurring when a single nucleotide, for example, adenine (A), thymine (T), cytosine (C) or guanine (G) in the DNA molecule differs between individuals (i.e. single nucleotide polymorphisms or SNPs; Figure 17.1; Betoret 2011).

17.2 NUTRIGENETIC AREAS OF APPLICATION

A number of papers have reported on the genetic variations that can be linked to the increase in corresponding relations between diet and diseases. The genetic variations available in literature include those associated with body weight, Type 2 diabetes mellitus, obesity, cardiovascular diseases and some autoimmune diseases and cancers.

17.2.1 Body weight

For a long time, obesity has been defined mainly in terms of elementary behavioural aspects such as over eating with little or no exercise either deliberately or from an inability to self-control. Experience has however shown that, together with dieting and exercise, the problem of obesity has been on the rise globally. This may be due to the fact that the whole process of weight loss is difficult and, further, sustainability of weight loss has proved to be a much harder and much difficult task. Obesity and other related clinical conditions are known to be major causes of a number of serious diseases including hypertension, cardiovascular disease, Type 2 diabetes and cancer (National Task Force on the Prevention and Treatment of Obesity 1996; Blum *et al.* 2007).

Body weight and obesity are attributes that are controlled by a number of factors such as the genes, diet and metabolism, behaviour, environment, culture and socioeconomic status. In general, common causes of being overweight and of obesity are known to be polygenic, that is, due to the prevailing gene–gene and gene–environment interactions. In genomics, the term 'environment' refers to all non-genetic aspects that contribute to specific phenotypic traits (Weinsier *et al.* 1998).

17.2.1.1 *Relationship between body weight, obesity and gene expression*

The gene–diet interaction is known to be very dynamic and persists throughout a person's life such that diet is an important factor capable of modulating the gene expression and, given a longer period of time, the phenotype (Leong *et al.* 2003). Among the sets of genes that have been reported to influence the problem of obesity are the 'silent' genes that exert their influence via special interactions with the diet, created by energy-dense and easily available foodstuffs (Weinsier 1998). There are four stages of gene–diet (environment) interactions that are thought to aggravate the problem of obesity; the first two stages are monogenic (involving similar or the same sets of genes) while the other two are polygenic (Loos and Bouchard 2003). The polygenic gene variants play important roles in a number of processes such as energy expenditure, appetite control, fat absorption, obesity-induced insulin resistance, lipid metabolism and the inflammatory response. By identifying the sets of genes which are involved in these processes (lipid and glucose metabolism, inflammatory and immune response, appetite control and detoxification), it is therefore possible to control and manage body weight or obesity problems.

Scientists have recently proposed other explanations for the genetic cause of obesity, linking the problem to gene–diet interaction (Blum *et al.* 2007). In their report, Blum *et al.* (2007) highlighted the concept defined as reward-deficiency syndrome (RDS). They suggested a link between dopaminergic genetic predisposition, glucose metabolism and the gene regulation of dopamine, which could be key to therapeutic remedies for obesity.

RDS disorders are caused by abnormalities in the functioning of the dopamine receptor gene and dopaminergic genes (Blum *et al.* 1996a, b). Dopamine 4-(2-aminoethyl)benzene-1,2-diol (Figure 17.2b), a catecholamine neurotransmitter, plays a very important role in the central nervous system: it is responsible for the neurotransmission of stimuli which regulate emotions in the individual via the interactions between dopamine and other neurotransmitting

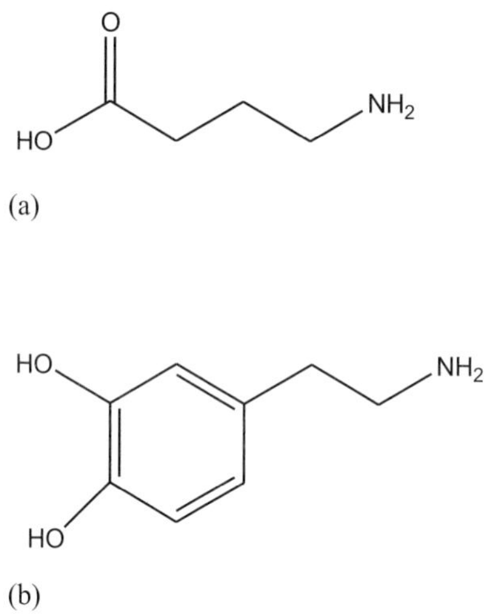

(a)

(b)

Fig. 17.2 Chemical structure of (a) GABA; and (b) dopamine.

agents such as serotonins (linked to depression when levels are low) and opioids (well-being feelings when levels are high) (Blum and Kozlowski 1990). These processes regulate the dopaminergic activity in the central nervous system, and the genes involved can be studied to find a method of escaping the obesity problem. Glucose is also known to regulate dopamine neuronal activity and the gamma aminobutyric acid (GABA) (Figures 17.2a and b) terminal transmitter release (Levin 2001).

17.2.1.2 Nutritional diets for obesity and weight management

The majority of nutritional diets that are recommended for individuals who are obese or over-weight have a lower possibility of converting carbohydrate to fats, facilitate fat oxidation or have ingredients which catalyse the release of serotonin such that weight is controlled. An example of these ingredients in nutritional diets for controlling obesity and overweight include salts of (–)hydroxycitric acid (HCA), potassium and calcium salts of (–)HCA (Figure 17.3; Browns 2005).

Nutritional diets with components capable of reducing stress have been reported to have beneficial effects in the lowering of cortisol, thus cutting down the possibility of fat accumulation in the body (Akhondzadeh et al. 2001). Moreover, food components which stimulate dopamine or elevate the concentration of norepinephrine in the central nervous system, particularly the brain (e.g. phenylalanine; Figure 17.4a), are also recommended (Chen et al. 2004).

The main drawback of phenylalanine is that it has the potential to compete with L-tryptophan (Figure 17.4b) and L-tyrosine (Figure 17.4c), thus hindering the intended purpose of controlling body weight (Milner et al. 1986).

17.2.2 Carcinogenesis

Some diet ingredients have been implicated in development of a number of cancer forms such as breast cancer or gastric cancer, with the evidence of the associated susceptibility genes at the centre of the tumours. Studies have indicated that during heat treatment of some

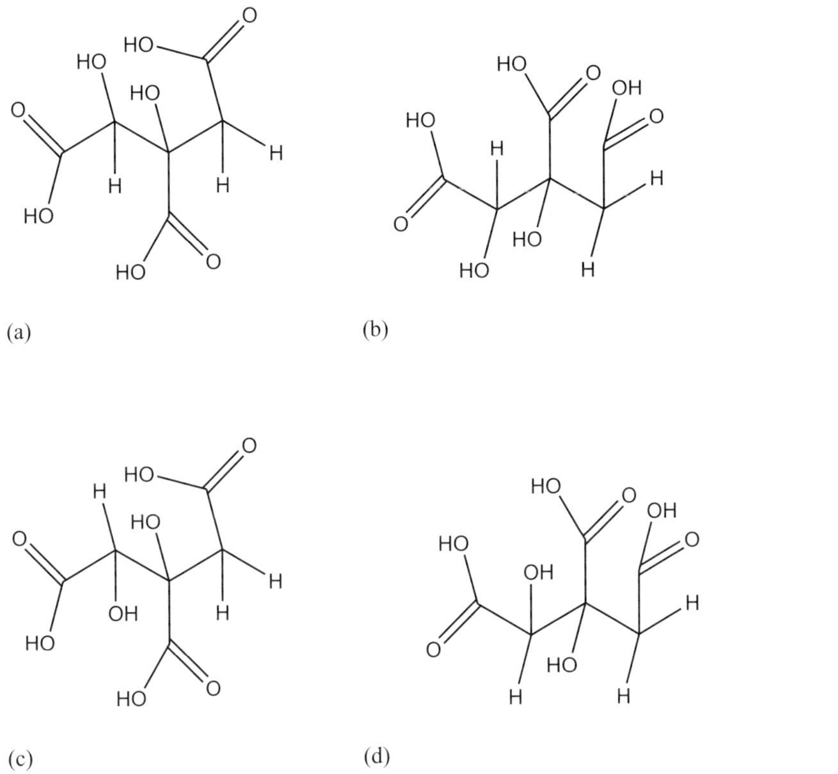

(a) (b)

(c) (d)

Fig. 17.3 Chemical structures of stereoisomers of hydroxycitric acid (Browns *et al.* 2005): (a) (+)–hydroxycitric acid; (b) (–)–hdroxycitric acid; (c) (+)–allo-hydroxycitric; and (d) (–)–allo-hydroxycitric acid.

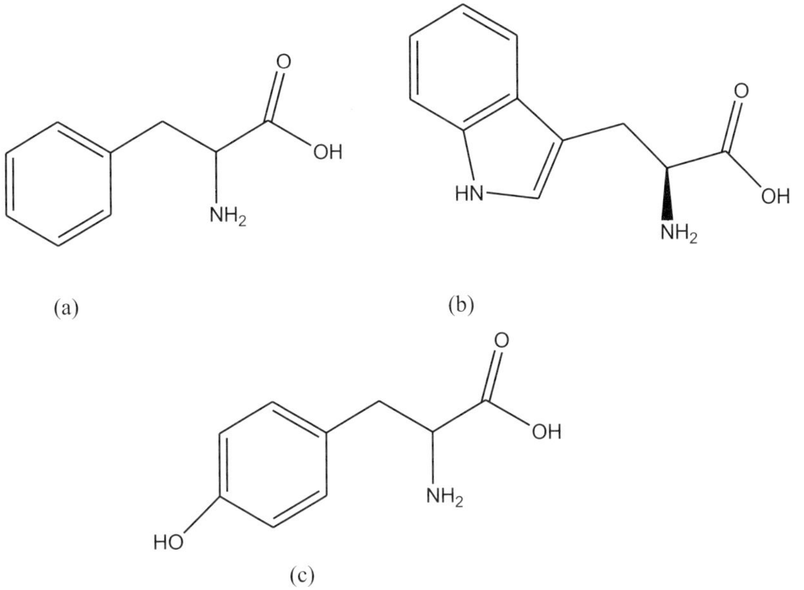

(a) (b)

(c)

Fig. 17.4 Chemical structure of (a) phenylalanine (Chen *et al.* 2004); (b) L-tryptophan; and (c) tyrosine (Milner *et al.* 1986).

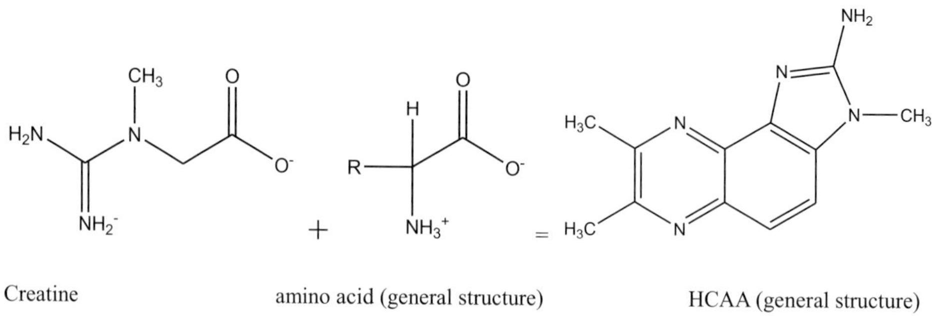

Creatine   amino acid (general structure)   HCAA (general structure)

Fig. 17.5 Reaction of amino acids and creatine to produce heterocyclic aromatic amines (Doll and Peto 1981; Felton and Knize 1990, 1991).

foodstuffs, mainly meat and meat products, there is a high possibility that some mutagenic or carcinogenic compounds may be formed, including polynuclear aromatic hydrocarbons (PAHs) and other heterocyclic nitrogenous compounds such as azaarenes. These nuisance compounds are formed due to protein and fat thermal degradation which occurs at very high temperatures (above 300°C); these temperatures are not attained in normal cooking procedures (Jagerstad *et al.* 1991; Rivera *et al.* 1996; Jagerstad and Skog 2005; Błaszczyk and Janoszka 2008). However, at the ordinary temperatures that are normally used for domestic cooking and preparation of meat, it may still be possible to produce heterocyclic aromatic amines and aminoazaarenes from the ingredients found in meat, such as amino acids, creatine ((2-(Methylguanidino)ethanoic acid) and sugars which occur as by-products of Maillard reactions (Pais *et al.* 1999; Sugimura *et al.* 2004). For example, compounds formed during the heat treatment of red meat through the reaction between amino acids and creatine, which then form heterocyclic aromatic amines (HCAA) through an enzymatic acetylation process catalysed by N-acetyltransferase regulated by the NAT gene, are known to cause cancer when they bind to a particular point on the DNA (Figure 17.5; Doll and Peto 1981; Felton and Knize 1990, 1991).

The high-potent heterocyclic aromatic amines include: 2-amino-3,4-dimethylimidazo(4,5-f)quinoline; 2-amino-3-methylimidazo(4,5-f)quinoline; 2-amino-3,8-dimethyl-imidazo(4,5-f)quinoxaline; 2-amino-3,4,8-trimethylimidazo-(4,5-f)quinoxaline; 2-amino-3,7,8-trimethylimidazo-(4,5-f)quinoxaline; and 2-amino-3-methylimidazo(4,5-f)quinoxaline (Figure 17.6; Nagao 1999; Sugimura *et al.* 2004). Of the above-listed compounds, those most found abundantly in meat and fish when heated are 2-amino-3, 8-dimethyl-imidazo (4,5-f)quinoxaline and 2-amino-1-methyl-6-phenyl-imidazo (4,5-b)pyridine (Skog *et al.* 1995; Pais *et al.* 1999; Zimmerli *et al.* 2001).

Another important phenomenon that may highlight the gene–diet–disease relationship can be seen from dietary lipid biomolecules which form part of the ingredients in most foodstuffs. In addition to them being a source of energy to the body, they are linked to mitogenic processes. These unhealthy mitogenic processes are known to be linked to the development and progression of a number of cancer diseases including colon cancer, colon carcinogenesis, hepatocarcinogenesis, breast cancer and mammary carcinogenesis and metastasis (Anderle *et al.* 2004).

In addition to heterocyclic aromatic amines (HCCA) and polycyclic aromatic hydrocarbons (PAHs), dietary lipids which include long-chain polyunsaturated fatty acids (LC-PUFA)

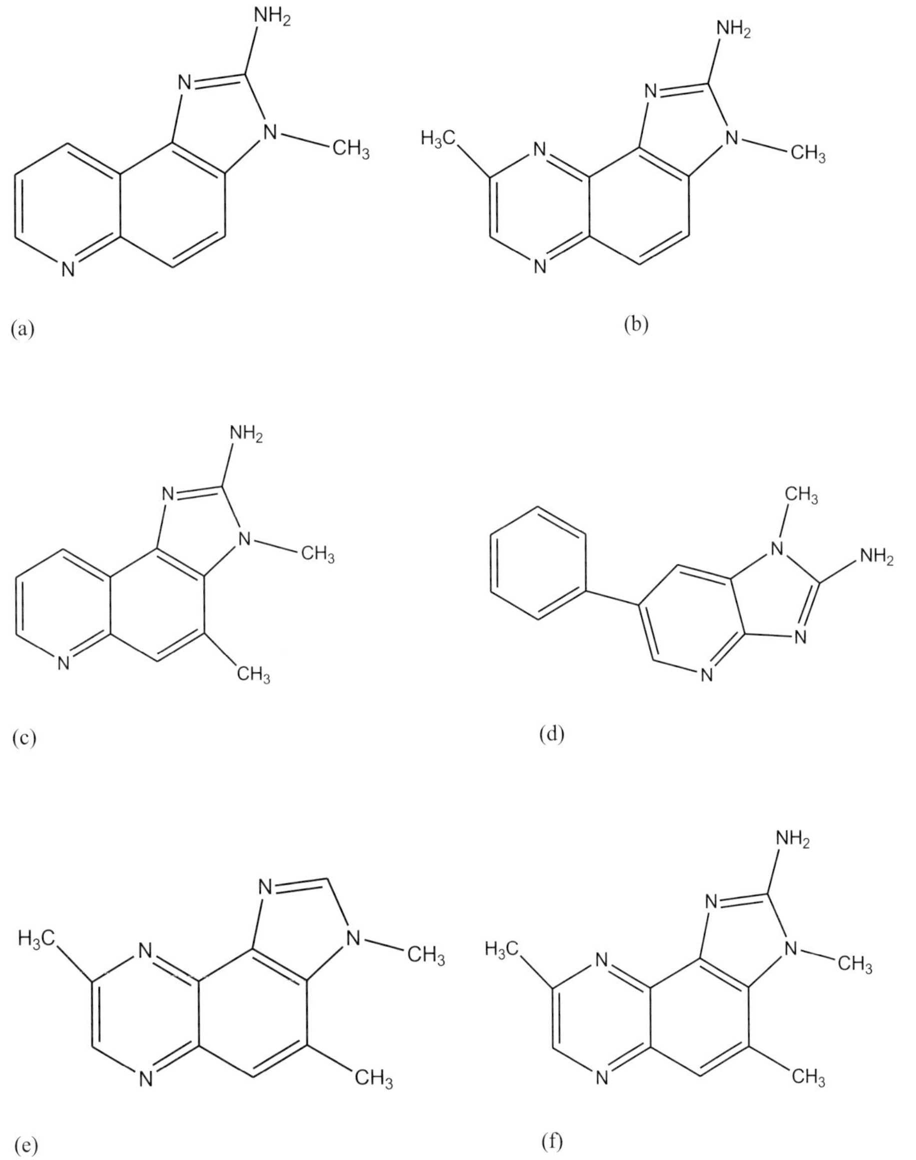

Fig. 17.6 Chemical structures of heterocyclic aromatic amines (Nagao 1999; Sugimura *et. al.* 2004): (a) 2-amino-3-methylimidazo(4,5-f); (b) 2-amino-3,8-dimethyl-imidazo(4,5-f)quinoxaline; (c) 2-amino-3,4-dimethylimidazo(4,5-f)quinoline; (d) 2-amino-1-methyl-6-phenyl-imidazo(4,5-b)pyridine; (e) 2-amino-3,4,8-trimethylimidazo-(4,5-f)quinoxaline; and (f) 2-amino-1,6-dimethylimidazo(4,5-b)pyridine.

are also known to influence carcinogenesis and affect multiple signalling pathways in the body. Long-chain fatty acids which contain unsaturations (e.g. double bonds) in the acyl chain are known as unsaturated fatty acids, and if there are multiple double bonds they are referred to as PUFAs. For example, ω-3 LC-PUFA extracted from fish oil has been reported to induce inhibitory effects, as demonstrated by an azoxymethane-induction of rat colon carcinogenesis model and an HT29 colon tumour growth and metastasis nude mouse model (Calder *et al.* 1998; Chang *et al.* 1998; Willett 2001). Some of the LC-PUFAs, such as arachidonic and eicosapentaenoic acids, are precursors of eicosanoids, biologically active compounds originating from animal products (egg, milk, meat, liver, fish) which regulate physiological processes in the body (Funk 2001). The high level of these ingredients (precursors of eicosanoids) in the diets has been reported to influence health status positively with respect to cardiovascular, immune and inflammatory conditions (Knapp 1999).

In humans and other mammals the biosynthesis of LC-PUFAs is achieved through processes that involve desaturation as well as elongation of the essential fatty acids such as α-linolenic acid (C18:3n-3) or linoleic acid (C18:2n-6) (Figure 17.7) to form much longer-chain and highly unsaturated fatty acids (Sprecher 2000; Nakamura *et al.* 2001), for example arachidonic acid (C20:4n-6), eicosapentaenoic acid and docosahexaenoic acid (C22:6n-3). These processes are catalysed by fatty acyl desaturases (FAD), which is responsible for introducing unsaturations at specific positions in the chain length, and elongases enzyme which assists the condensation of activated fatty acids with malonyl-CoA (Leonard *et al.* 2002; Jakobsson *et al.* 2006).

17.2.3 Inflammatory bowel disease (IBD)

A disease condition known as inflammatory bowel disease (IBD), which occurs due to the inflammation of the intestines caused by abnormalities in the immune responses to the enteric ecosystem, is experienced in one of two main forms: ulcerative colitis (UC) and Crohn's disease (CD) (Lee and Buchman 2009). Normally the medical treatment for IBD prescribed to patients involves the use of 5-aminosalycylate, steroids, immunomodulator and biological therapies. Of these, only the biological therapy which prescribes anti-tumour necrosis factor agents has been reported to selectively block the inflammation, thus treating the disease. The other treatments merely serve to suppress the inflammation (Lee and Buchman 2009). From a genetic point of view, the CD form of IBD has been shown to have a strong link with some genes which play key roles in the innate immunity, while the adaptive immune response genes and specific polymorphism genes are associated with both CD and UC diseases.

Together with these remedies, nutritional supportive options (especially diets rich in ω-free fatty acids such as fish oil, sunflower and corn oils) are preferred therapies for the prevention and recognition roles in the CD cases (Belluzzi *et al.* 1996; Feagan *et al.* 2008). Current research with regards to pathogenesis and functional genetic variations indicates that nutritional supportive diets for IBD should be based on cytoprotective, anti-inflammatory and immunoregulatory nutritional options (Lee and Alan 2009).

17.2.3.1 ω-fatty acids containing dietary nutrition

The fatty acids ω-6 PUFAs and ω-3 PUFAs (e.g. eicosapentaenoic acid and docosahexanoic acid) present mainly as linoleic acid in animal and seed oils are the most important as far as the inflammatory process is concerned. The ω-3 PUFAs play the important role as cofactors for transcriptional factors as well as in the down-expression of pro-inflammatory genes, for example cytokines and cyclo-oxygenase-2 genes (Calder 2008; Roy *et al.* 2007). In some other reports, ω-3 PUFAs have been shown to lessen the severity of UC by down-regulating the cytokine gene expressions. In other studies, ω-3 PUFAs were observed to up-regulate the

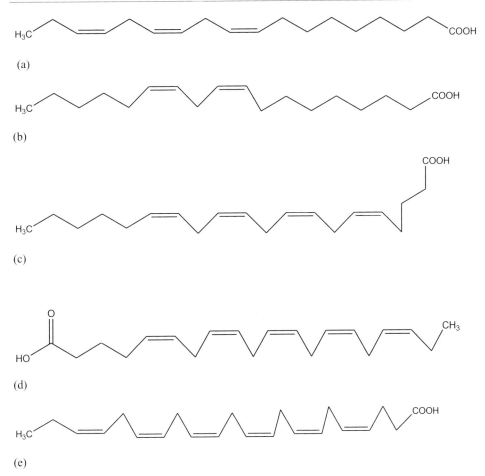

Fig. 17.7 Chemical structures of long-chain polyunsaturated fatty acids (LC-PUFA) (Sprecher 2000; Nakamura *et al.* 2001): (a) α-linolenic acid; (b) linoleic acid; (c) arachidonic acid; (d) eicosapentaenoic acid; and (e) docosahexaenoic acid.

barrier genes, thus enhancing the anti-inflammatory effect (Bassaganya-Riera and Hontecillas 2006; Hudert *et al.* 2006).

17.2.3.2   *Dietary IBD suppression and modification of immune response*

Nutritional diet components and plant extracts play multiple roles simultaneously with regards to contributing to health benefits. Examples of these activities include modulation of immune responses and the suppression of inflammation diseases such as IBD (Philpott *et al.* 2004; Ferguson *et al.* 2007). Those nutrients and phytochemicals which play multi-role tasks in the body system include: arginine and glutamine; the sulphur amino acids (e.g. cysteine or methionine); and vitamins and minerals (e.g. copper, iron, selenium and zinc). As these nutrients and minerals play roles in the immune modulation, they also have an important contribution in reducing the symptoms of IBD disorder in the individual. Some dietary manipulations are known to be of health benefit, for example, it has been reported that decreasing levels of ω-6 PUFAs and increasing levels of ω-3 polyunsaturated acids in the diet of IBD patients can bring some health benefits. This is due to the fact that the manipulation boosts the production of different types of eicosanoids quantitatively. This is advantageous because it will finally replace arachidonic acid inflammatory cell membranes, hence affecting the production of arachidonic acid negatively with a net effect on the up-regulation of the production of pro-inflammatory cytokines (Ferguson *et al.* 2007).

17.3 ANALYTICAL METHODS FOR NUTRIGENETICAL FOOD FUNCTIONS

By their definition, nutrigenomics and nutrigenetics cover many branches of science and hence require a multidisciplinary research approach based on nutritional science, investigating how diet can influence health. Since it is evident that nutritious dietary components can regulate gene expression (thus modulating a number of biochemical processes, transcription factors, protein expression and metabolite production), techniques popular in molecular biology, genetics and biochemistry will have a major role to play in this discipline.

The most commonly used analytical methods and techniques are those for the analyses of biomolecules in complex interactions which, in addition to biochemical techniques, require knowledge on bioinformatics. Under natural conditions, it is known that the regulation of genetic expression process by nutritious diet components may not be a single process; rather, it involves various stages, each of which is regulated by a specific set of genes. Systematic studies which track the flow of molecular and genomic information (e.g. interaction of dietary components with genes, which then influences the expression of specific class of ribonucleic acid, the mRNA which transcribes and translates the genetic information into proteomics and even metabolomics reactions) as they are modulated by dietary components is always performed using a number of techniques (Müller and Kersten 2003).

17.3.1 Northern blot

The techniques used over the years for the study of gene expression by detection of mRNA in biological systems include hybridisation techniques such as the northern blot. This technique uses the principles of electrophoresis to discriminate RNA samples based on their sizes. Visualisation is normally performed with the aid of a hybridisation probe which is composed of a sequence of ribonucleic nucleic acids with a complementary sequence to the RNA of interest being studied (Trayhurn 1996). Northern blot techniques enable scientists to study the way in which cellular activities control the structure and functioning of various processes. The technique also reveals the extent of specific gene expression that occurs during processes such as metabolism, differentiation and even when abnormal processes occur (e.g. mutations or diseases) (Trayhurn 1996; Schlamp *et al.* 2008).

One of the attractive features of northern blotting as a technique is that the membranes used in the blotting can be safely stored; they also provide the possibility of analysis even after a long period of time without losing any information (Streit *et al.* 2009). Another advantage of the northern blotting technique is its reliability. On the other hand, the technique can demonstrate sample degradation due to the action of enzymes, mainly RNases. To control this problem, the use of ascetic techniques such as sterilisation of the working environment or use of RNase inhibiting compounds such as diethylpyrocarbonate, is always encouraged

(Trayhurn 1996). Moreover, the use of toxic chemicals such as ethidium bromide (a known teratogen), diethylpyrocarbonate and formaldehyde and also the use of radioactive material (e.g. ^{32}P) make this technique less popular on safety grounds. Northern blotting has, to a large extent, been replaced by other techniques without these disadvantages, such as real-time polymerase chain reaction (RT-PCR).

17.3.2 Real-time polymerase chain reaction (RT-PCR)

Real-time polymerase chain reaction (RT-PCR) is a technique that amplifies a single or a few copies of a gene into many copies. The principles behind this technique is based on thermal cycling processes in which a gene sequence and short lengths of a gene sequence containing gene sequences complementary to the specific target sequence, known as primers, are enzymatically reacted (using polymerases such as Taq polymerase) in repeated cycles of heating and cooling to induce the amplification of the gene. In the PCR process, the piece of gene sequence that will be generated again is used as a template to continue with the chain reaction such that the process will result in many copies of the desired gene sequence.

Like northern blotting, PCR techniques can provide information about gene expression only for a few sets of genes at a time. This its major limitation as far as nutrigenetics and nutrigenomics is concerned, because it hinders the disclosure of important information relating the relationship between the dietary nutritious components and the genetic effect (Gohil and Chakraborty 2004).

A number of other techniques exist which allow many more sets of genes to be studied and identified, making it possible to effectively relate and link the effect of components in dietary nutritional foodstuffs to the associated metabolism, gene regulation and the respective diseases and disorders (Hu and Kong 2004). Such techniques include that based on DNA microarray principles (Morozova and Marra 2008), as discussed in the following section.

17.3.3 DNA microarray

As the name suggests, microarray refers to a collection of microscopic biochips made up of either oligonucleotides or other bio-probes. The microchips carry numerous gene spots attached to a solid surface such as a glass slide, carefully positioned at specific points within the grid pattern. The principle on which this technique works is based on the hybridisation reaction. This involves exposing two single-stranded DNA molecules to each other and then determining the amount of double-strands that form, yielding the relative quantity of the measure of mRNA in the sample being analysed (Storhoff *et al.* 2005).

The DNA microarray technique is one of the best analytical methods in nutrigenetics and nutrigenomics as it is capable of determining information about multiple genes simultaneously; it is also fast, yielding the required results within a short period of time. Unlike other techniques such as northern blotting, DNA microarrays technique is both environmentally and user-friendly as it is not associated with any toxicity or radioactivity. Moreover, information about the sequence of the DNA is not a prerequisite in the construction of the microarrays, contrary to other techniques used to study gene expression phenomena.

The main drawbacks of this technique are in the acquisition of the equipment as well as the associated costs involved in printing the required number of complementary DNA (cDNA).

17.3.4 Mass spectrometry

Advancements in mass spectrometry have allowed molecular processes, such as those occurring due to the influence of dietary components on the mRNA expression, to be studied. Various mass spectrometry techniques such as Fourier transform ion cyclotron resonance (FTICR) MS; matrix-assisted laser-desorption-ionisation time-of-flight (MALDI-TOF/TOF) MS; electrospray ionisation-quadrupole (ESI-Q/TOF) MS; electrospray-ionisation ion-trap (ESI-IT); orbitrap-MS; and surface-enhanced laser-desorption-ionisation time-of-flight MS (SELDI-TOF-MS); can all reveal useful information about diet–gene–environment interactions (Poon 2007).

17.3.5 Nuclear magnetic resonance (NMR)

Nuclear magnetic resonance (NMR) is another tool that has been used to study the relationship between dietary components and their respective biological responses (Wang *et al.* 2005; Dumas *et al.* 2006). However, the lack of sensitivity of NMR at lower concentrations is its major disadvantage (Fan and Lane 2008).

17.4 CONCLUSION

The contribution of nutrigenetics and nutrigenomics, which link genetic variations to diet–gene interactions and individual nutritional requirements, has created the possibility of personalising nutrition for health. This knowledge can be harnessed to establish diet–gene individual treatments for particular health problems. For this potential to be realised, it is important to continue research into nutritional requirements and their potential to prevent or suppress the development or progression of the diseases associated with diet–gene interactions.

REFERENCES

Akhondzadeh, S., Naghavi, H. R., Vazirian, M., Shayeganpour, A., Rashidi, H. & Khani, M. (2001). Passionflower in the treatment of generalized anxiety: a pilot double-blind randomized controlled trial with oxazepam. *J. Clin. Pharm. Therapeutics* 26 (5), 63–67.

Anderle, P., Farmer, P., Berger, A. & Matthew-Alan, R. (2004) Nutrigenomic approach to understanding the mechanisms by which dietary long-chain fatty acids induce gene signals and control mechanisms involved in carcinogenesis. *Nutrition* 20, 103–108.

Bassaganya-Riera, J. & Hontecillas, R. (2006) Cla and ω-3 PUFA differentially modulate clinical activity and colonic ppar-responsive gene expression in a pig model of experimental ibd. *Clin. Nutr.* 25, 454–465.

Belluzzi, A., Brignola, C., Campieri, M., Pera, A., Boschi, S. & Miglioli, M. (1996) Effect of an enteric-coated fish-oil preparation on relapses in Crohn's disease. *New Engl. J. Med.* 334, 1557–1560.

Betoret, E., Betoret, N., Vidal, D. & Fito, P. (2011) Functional foods development: Trends and technology. *Trends in Food Science and Technology* 22 (9), 498–508.

Błaszczyk, U. & Janoszka, B. (2008) Analysis of azaarenes in pan fried meat and its gravy by liquid chromatography with fluorescence detection. *Food Chem.* 109, 235–242.

Blum, K. & Kozlowski, G. P. (1990) Ethanol and neuromodulator interactions: a cascade model of reward. In: *Alcohol and Behavior*, Ollat, H., Parvez, S. & Parvez, H. (eds), VSP Press, Utrecht, Netherlands, pp. 131–149.

Blum, K., Cull, J. G., Braverman, E. R. & Comings, D. E. (1996a) Reward deficiency syndrome. *The American Science* 84, 132–145.

Blum, K., Sheridan, P. J., Wood, R. C., Braverman, E. R., Chen, T. J. H., Cull, J. G. & Comings, D. E. (1996b) The D2 dopamine receptor gene as a predictor of impulsive-addictive-compulsive behavior: Bayes' theorem. *J. Royal Soc. Med.* 89, 396–400.

Blum, K., Chen, T. J. H., Meshkin, B., Downs, B. W., Gordon, C. A., Blum, S., Mangucci, J. F., Braverman, E. R., Arcuri, V., Deutsch, R. & - Pons, M.-M. (2007) Genotrim™, a DNA-customized nutrigenomic product, targets genetic factors of obesity: Hypothesizing a dopamine–glucose correlation demonstrating reward deficiency syndrome (RDS). *Med. Hypotheses* 68, 844–852.

Browns, B. W., Bagchi, M., Subbaraju, G. V., Shara, M. A., Preuss, H. G. & Bagchi, D. (2005) Bioefficacy of a novel calcium–potassium salt of (-)-hydroxycitric acid. *Mutation Res.* 579 (1–2), 149–162.

Calder, P. C. (2008) Polyunsaturated fatty acids, inflammatory processes and inflammatory bowel diseases. *Mol. Nutr. Food Res.* 52, 885–897.

Calder, P. C., Davis, J., Yaqoob, P., Pala, H., Thies, F. & Newsholme, E. A. (1998) Dietary fish oil suppresses human colon tumour growth in athymic mice. *Clinical Sciences* 94, 303–311.

Chang, W. L., Chapkin, R. S. & Lupton, J. R. (1998) Fish oil blocks azoxymethane-induced rat colon tumorigenesis by increasing cell differentiation and apoptosis rather than decreasing cell proliferation. *J. Nutr.* 28 (3), 491–497.

Chen, T. J., Blum, K., Payte, J. T., Schoolfield, J., Hopper, D., Stanford, M. & Braverman, E. R. (2004) Narcotic antagonists in drug dependence: pilot study showing enhancement of compliance with SYN-10, amino-acid precursors and enkephalinase inhibition therapy. *Medicinal Hypotheses* 63 (3), 538–548.

Dennis, C. (2003) Epigenetics and disease: Altered states. *Nature* 421, 686–688.

Doll, R. & Peto, R. (1981) The causes of cancer: Quantitative estimates of avoidable risks of cancer in the United States today. *J. Natl Cancer Inst.* 66, 1191–1308.

Dumas, M. E., Maibaum, E. C., Teague, C., Ueshima, H., Zhou, B., Lindon, J. C., Nicholson, J. K., Stamler, J., Elliot, P., Chan, Q. & Holmes, E. (2006) Assessment of analytical reproducibility of H-1 NMR spectroscopy based metabonomics for large-scale epidemiological research: the INTERMAP study. *Anal. Chem.* 78 (7), 2199–2208.

Fan, T. W.-M. & Lane, A. N. (2008) Structure-based profiling of metabolites and isotopomers by NMR. *Progress in Nuclear Magnetic Resonance Spectroscopy* 52 (2–3), 69–117.

Feagan, B. G., Sandborn, W. J., Mittmann, U., Bar-Meir, S., D'Haens, G., Bradette, M., Cohen, A., Dallaire, C., Ponich, T. P., McDonald, J. W., Hébuterne, X., Paré, P., Klvana, P., Niv, Y., Ardizzone, S., Alexeeva, O., Rostom, A., Kiudelis, G., Spleiss, J., Gilgen, D., Vandervoort, M. K., Wong, C. J., Zou, G. Y., Donner, A. & Rutgeerts, P. (2008) Omega-3 free fatty acids for the maintenance of remission in Crohn disease: the epic randomized controlled trials. *J. Am. Med. Assoc.* 299, 1690–1697.

Felton, J. S. & Knize, M. G. (1990) Heterocyclic-amine mutagens/carcinogens in foods. In *Chemical Carcinogenesis and Mutagenesis I*, Cooper, C. S. & Grover, P. L. (eds), Springer-Verlag, Berlin, pp. 471–502.

Felton, J. S. & Knize, M. G. (1991) Occurrence, identification, and bacterial mutagenicity of heterocyclic amines in cooked food. *Mutation Res.* 259, 205–218.

Ferguson, L. R., Shelling, A. N., Browning, B. L., Huebner, C. & Petermann, I. (2007) Genes, diet and inflammatory bowel disease. *Mutation Res.* 622, 70–83.

Funk, C. D. (2001) Prostaglandins and leukotrienes: Advances in eicosanoid biology. *Science* 294, 1871–1875.

Gohil, K. & Chakraborty, A. A. (2004) Applications of microarray and bioinformatics tools to dissect molecular responses of the central nervous system to antioxidant micronutrients. *Nutrition* 20, 50–55.

Hu, R. & Kong, A. N. T. (2004) Activation of MAP kinases, apoptosis and nutrigenomics of gene expression elicited by dietary cancer-prevention compounds. *Nutrition* 20, 83–88.

Hudert, C. A., Weylandt, K. H., Lu, Y., Wang, J., Hong, S., Dignass, A., Serhan, C. N. & Kang, J. X. (2006) Transgenic mice rich in endogenous omega-3 fatty acids are protected from colitis. *Proc. Natl Ac. Sci., USA* 103, 11276–11281.

Jagerstad, M. & Skog, K. (2005) Genotoxicity of heat processed foods. *Mutation Res.* 574, 156–172.

Jagerstad, M., Skog, K., Grivas, S. & Olsson, K. (1991) Formation of heterocyclic amines using model systems. *Mutation Res.* 259, 219–233.

Jakobsson, A., Westerberg, R. & Jacobsson, A. (2006) Fatty acid elongases in mammals: their regulation and roles in metabolism. *Prog. Lipid Res.* 45, 237–249.

Kaput, J. & Rodriguez, R. (2004) Nutritional genomics. The next frontier in the postgenomic era. *Physiol. Genomics* 16, 166–177.

Kim, D. H. (2007) The interactive effect of methyl-group diet and polymorphism of methylenetetrahydrofolate reductase on the risk of colorectal cancer. *Mutational Res.* 622, 14–18.

Knapp, H. R. (ed.) (1999) Fatty acids and lipids from cell biology to human disease. *Lipids*, 34 (Suppl), S1–S350.

Kornman, K., Rogus, J., Roh-Schmidt, H., Krempin, D., Davies, A. J., Grann, K. & Randolph, R. K. (2007) Interleukin-1 genotype-selective inhibition of inflammatory mediators by a botanical: a nutrigenetics proof of concept. *Nutrition* 23, 844–852.

Lee, G. & Buchman, A. L. (2009) DNA-driven nutritional therapy of inflammatory bowel disease. *Nutrition* 25, 885–891.

Leonard, A. E., Kelder, B., Bobik, E. G., Chuang, L.-T., Lewis, C. J., Kopchick, J. J., Murkerji, P. & Huang, Y.-S. (2002) Identification and expression of mammalian long-chain PUFA elongation enzymes. *Lipids* 37, 733–740.

Leong, N. M., Mignone, L. I., Newcomb, P. A., Titus-Ernstoff, L., Baron, J. A., Trentham-Dietz, A., Stampfer, M. J., Willett, W. C. & Egan, K. M. (2003) Early life risk factors in cancer: the relation of birth weight to adult obesity. *Intl J. Cancer* 103, 789–791.

Levin, B. E. (2001) Glucosensing neurons do more than just sense glucose. *Int J Obes Related Metabolic Disorders*, Suppl. (5), S68–S72.

Lim, U., Wang, S. S., Hartge, P., Cozen, W., Kelemen, L. E., Chanock, S., Davis, S., Blair, A., Schenk, M., Rothman, N. & Lan, Q. (2007) Gene–nutrient interactions among determinants of folate and onecarbon metabolism on the risk of non-Hodgkin lymphoma: NCI-SEER case-control study. *Blood* 109, 3050–3059.

Loos, R. J. & Bouchard, C. (2003) Obesity: Is it a genetic disorder? *J. Int. Med.* 254, 401–425.

Mathers, J. C. (2003) Nutrition and cancer prevention: diet–gene interactions. *Proc. Nutr. Soc.* 62, 605–610.

Milner, J. D., Irie, K. & Wurtman, R. J. (1986) Effect of phenylalanine on the release of endogenous dopamine from rat striatal slices. *J. Neurochem.* 47, 1444–1448.

Morozova, O. & Marra, M. A. (2008) Applications of next-generation sequencing technologies in functional genomics. *Genomics* 92 (5), 255–264.

Müller, M. & Kersten, S. (2003) Opinion. Nutrigenomics: goals and strategies. *Nature Reviews Genetics* 4, 315–322.

Nagao, M. (1999) A new approach to risk estimation of food-borne carcinogensheterocyclic amines-based on molecular information. *Mutation Res.* 431, 3–12.

Nakamura, M. T., Cho, H. P., Xu, J., Tang, Z. & Clarke, S. D. (2001) Metabolism and functions of highly unsaturated fatty acids: an update. *Lipids* 36, 961–964.

National Task Force on the Prevention and Treatment of Obesity (1996) Long-term pharmacotherapy in the management of obesity. *JAMA* 276, 1907–1915.

Pais, P., Salmon, C. P., Knize, M. G. & Felton, J. S. (1999) Formation of mutagenic/carcinogenic heterocyclic amines in dry-heated model systems, meats, and meat drippings. *J. Agric. Food Chem.* 47, 1098–1108.

Philpott, M. & Ferguson, L. R. (2004) Immunonutrition and cancer. *Mutation Res.* 551, 29–42.

Poon, T. C. W. (2007) Opportunities and limitations of SELDI-TOF-MS in biomedical research: practical advices. *Expert Reviews in Proteomics* 4, 51–65.

Rivera, L., Curto, M. J., Pais, P., Galceran, M. T. & Puignou, L. (1996) Solid-phase extraction for the selective isolation of polycyclic aromatic hydrocarbons, azaarenes and heterocyclic aromatic amines in charcoal-grilled meat. *J. Chromatogr. A*, 731, 85–94.

Roy, N., Barnett, M., Knoch, B., Dommels, Y. & McNabb, W. (2007) Nutrigenomics applied to an animal model of inflammatory bowel diseases: transcriptomic analysis of the effects of eicosapentaenoic acid- and arachidonic acid–enriched diets. *Mutation Res.* 622, 103–116.

Schlamp, K., Weinmann, A., Krupp, M., Maass, T., Galle, P. R. & Teufel, A. (2008) BlotBase: A northern blot database. *Gene* 427 (1–2), 47–50.

Sing, C. F., Stengard, J. H. & Kardia, S. L. (2003) Genes, environment and cardiovascular disease. *Arteriosclerosis, Thrombosis, and Vascular Biology* 23, 1190–1196.

Skog, K., Steineck, G., Augustsson, K. & Jagerstad, M. (1995) Effect of cooking temperature on the formation of heterocyclic amines in fried meat products and pan residues. *Carcinogenesis* 16, 861–867.

Sprecher, H. (2000) Metabolism of highly unsaturated n-3 and n-6 fatty acids. *Biochimica Biophysica Acta* 1486, 219–231.

Storhoff, J. J., Marla, S. S., Garimella, V. & Mirkin, C. A. (2005) Labels and detection methods. In: *Microarray Technology and its Applications*, Müller, U. R. & Nicolau, D. V. (eds), Springer-Verlag, Berlin, Heidelberg, pp. 147–180.

Stover, P. J. (2004) Nutritional genomics. *Physiol. Genomics* 16, 161–165.

Stover, P. J. (2006) Influence of human genetic variation on nutritional requirements. *American Journal of Clinical Nutrition* 83, 436S–442.

Stover, P. J. & Caudill, M. A. (2008) Genetic and epigenetic contributions to human nutrition and health: managing genome-diet interactions. *J. Am. Diet Assoc.* 108 (9), 1480–1487.

Streit, S., Michalski, C. W., Erkan, M., Kleef, J. & Friess, H. (2009) Northern blot analysis for detection of RNA in pancreatic cancer cells and tissues. *Nature Protocols* 4 (1), 37–43.

Subbiah, M. T. (2008) Understanding the nutrigenomic definitions and concepts at the food–genome junction. *OMICS* 12, 229–235.

Sugimura, T., Wakabayayashi, K., Nakagama, H. & Nagao, M. (2004) Heterocyclic amines: Mutagens/carcinogens produced during cooking of meat and fish. *Cancer Science* 95, 290–299.

Trayhurn, P. (1996) Northern blotting. *Proc. Nutr. Soc.* 55, 583–589.

Wang, Y., Tang, H., Nicholson, J. K., Hylands, P. J., Sampson, J. & Holmes, E. (2005) A metabonomic strategy for the detection of the metabolic effects of chamomile (Matricaria recutita L.) ingestion. *J. Agric. Food Chem.* 53 (2), 191–196.

Weinsier, R. L., Hunter, G. R., Heini, A. F., Goran, M. I. & Sell, S. M. (1998) The etiology of obesity: relative contribution of metabolic factors, diet, and physical activity. *Am. J. Med.* 105, 145–150.

Willett, W. C. (2001) Diet and breast cancer. *J. Int. Med.* 249 (5), 395–411.

Zimmerli, B., Rhyn, P., Zoller, O., Schlatter, J. (2001) Occurrence of heterocyclic aromatic amines in the Swiss diet: Analytical method, exposure estimation and risk assessment. *Food Additives and Contaminants* 18, 533–551.

18 Probiotic Foods and Dietary Supplements

Abstract: A concept that is gaining popularity is that of dietary supplements which contain agents with health benefits to humans. One example is probiotic supplements, in which live microbial supplements are introduced in foodstuffs to render health benefits to the host by modulating the gut microbial flora. This supplements nutrient deficiencies as well as corrects the composition of the intestinal balance. The microorganisms supplemented as probiotics are believed to produce antimicrobial compounds which tend to positively influence the gut microbial flora balance. In this chapter, the types and roles of probiotic food additions will be discussed. Note that other dietary supplements such as prebiotics and synbiotics, although similar to probiotics, will be discussed independently in Chapters 19 and 20, respectively.

Keywords: dietary carriers; gut microbial flora; live microbial supplements; probiotics

18.1 MICROBIAL GUT FLORA ACTIVITY

This concept of probiotics began from the observation that the human gut harbours a host of microbial flora which promotes health and the proper functioning of metabolic activities. The existence of indigenous microbes in the gastrointestinal tract has been known of for a long time, although only those microbial flora detrimental to human health were known of in the past. The possibility of health benefits arising from gut microbial flora was therefore neglected for a long time. In the last few decades however, research which involves the manipulation of the indigenous microbial flora so that they become beneficial by improving health has been conducted.

It is currently clear that the intestinal microbial flora contribute significantly to the overall health status of an individual. For example, some microbes which happen to be normally resident in the gut are capable of degrading specific ingredients in foodstuffs to produce beneficial biomolecules such B vitamins and some produce compounds which stimulate the immune system (Holzapfel *et al.* 1998). In addition to this, some indigenous microbes in the intestines are actively involved in the metabolic processes, whereby they transform certain potentially carcinogenic molecules into non-carcinogenic compounds. On the other hand, any disturbance of the gut microbial ecosystem may prove detrimental to the health of an individual.

There are some convincing arguments from some researchers with regard to the efficacy of probiotics for humans, but many other scientists are sceptical of the whole issue due to the lack of controlled studies or even convincing mechanisms which might generate more

Chemistry of Food Additives and Preservatives, First Edition. Titus A. M. Msagati.
© 2013 John Wiley & Sons, Ltd. Published 2013 by John Wiley & Sons, Ltd.

conclusive evidence of the performance of the probiotics in boosting human health (Hose and Sozzi 1991; O'Sullivan *et al.* 1992).

18.2 PROBIOTICS AND NUTRITION

Products containing probiotics of selected microbial strains labelled as 'generally recognised as safe' (GRAS) are normally classified into three groups (Berner and O'Donnell 1998; Sanders and Veld 1999). The three groups are: (1) traditional foodstuffs (mainly non-dairy fermented food items in which active probiotic microorganisms have been incorporated such that the inoculated food items are eaten to provide nutritious health benefits); (2) supplements (mainly fermented dairy products which play the important role of delivering selected active probiotic microbes to the body); and (3) capsule supplement formulations.

The qualities that are always considered for the best strain include reliability, such that there is assurance of the presence of a high enough number of viable microorganism which can last to the specified date of expiry, and efficacy (Pardon *et al.* 1995; Reid 1999; Sanders and Veld 1999). Probiotics are defined as living microorganisms which, upon ingestion, are capable of imparting health benefits to an individual to a greater extent than general nutrition. It therefore follows that the microorganisms used as probiotics have to be alive and not pasteurised. The strains (genera) of probiotic microorganisms that are mainly used include lactobacillus, bifidobacterium and streptococcus.

Characteristics needed for strain or microbial cultures to be considered for use as probiotic materials include the following:

- The strain should be of the indigenous gastro-intestinal tract flora.
- It must have the ability to resist acids and bile attacks, otherwise it will not survive the acidic conditions of the upper gastrointestinal tract.
- It has to be inert in the sense that it should not cause pathogenic, toxic, mutagenic or carcinogenic reactions to the organism, its fermentation products or cell components.
- It should be active against harmful (carcinogenic and pathogenic) microorganisms.
- It must also be genetically stable with no plasmid transfer mechanism.

18.3 PROBIOTICS AND HEALTH

The definition of probiotics is that of live microbial supplements, and whether this definition include dead microorganisms or microbial fragments is under debate (Fuller 1992, 1997). Probiotics have been reported to have a number of health benefits such as the ability to stimulate the immune system, lower blood cholesterol and suppress tumours through their ability to maintain the gut microbial balance properly (Newcomer *et al.* 1983; Fernandes and Shahani 1990; Lee and Salminen 1995; Conway 1996; Fuller 1996). The evidence normally used to prove that probiotics do have health benefits come from experiments on animals where it was shown that when the gut microbial flora is disturbed, a disease condition may result; when an administration of microbial suspension is administered, the healthy condition is restored (Hays 1994).

The general mechanisms by which different types of probiotics work include the ability to compete successfully with the gut pathogens for substrate and for the mucosal binding sites. Probiotics may also be responsible for the stimulation of the immune system under some specific conditions.

18.3.1 Colon cancer

Some reports have highlighted the association between diets high in fat and incidences of colon cancer due to the fact that the fat component in diets stimulates the over secretion of bile acid in colon thus promoting carcinogenesis (Reddy *et al.* 1977). The stimulation of carcinogenesis is driven by fecal microbial enzymes, for example glucuronidase, nitroreductase and azoreductase, which are produced by the autochthonous microflora in the gut (Scheinbach 1998). For example, 3-glucuronidase enzyme has the ability to deconjugate compounds which contain a glucosidic bond, releasing aglycones which are known to be mutagens, and other compounds such as N-hydroxy-N-2-fluorenylacetamide-(3-glucuronide) which results into the formation of compounds such as N-hydroxy-N-2-fluorenylacetamide which are carcinogenic (Weisburger *et al.* 1970).

Nitroreductase and azoreductase enzymes facilitate the conversion of aromatic nitro and azo-compounds into amine derivative compounds such as N-nitroso and N-hydroxy intermediates, which are also carcinogenic (Drasar *et al.* 1972).

Dehydroxylases enzymes catalyse the formation of deoxycholic and lithocholic compounds from bile acids. These cholic derivatives have been implicated in the facilitation of carcinogesis by promoting the binding of carcinogenic molecules to the genetic material (Reddy *et al.* 1977; Autrup *et al.* 1978).

Introducing probiotics as a therapy for colon cancer is considered because probiotics can suppress carcinogenesis by either binding, blocking or even eliminating the problem altogether. Probiotics may also lower the gut pH, thus affecting negatively the activities of the gut microbial population and also enhancing or stimulating the body immune system (McIntosh 1996).

18.3.2 Inflammatory bowel disease (IBD) and colorectal cancer (CRC)

Some probiotics have been recommended as a therapy in dealing with IBD and CRC (Rembacken *et al.* 1999; Guslandi *et al.* 2000; Tuohy *et al.* 2003). In addition to this, a mixed probiotic comprising several lactobacilli genera (*L. acidophilus*, *Lactobacillus bulgaricus*, *Lactobacillus casei* and *Lactobacillus plantarum*), three bifidobacteria genera (*Bifidobacterium breve*, *Bifidobacterium infantis* and *Bifidobacterium longum*) and *Streptococcus thermophilus* were found to effective against another form of IBD, ulcerative colitis as well as Crohn's disease (Campieri *et al.* 2000). These probiotic strains interact with the mucosal regulation of T-cells as well as the regulation of cytokine transcription factors that take place within the mucosa (Shanahan 2000). On the other hand, probiotics that are administered in fermented food items such as dairy products, have been reported to have positive impact as biomarkers of CRC (Burns and Rowland 2000).

18.3.3 Lowering cholesterol

Probiotics (e.g. those comprising lactic acid bacteria) have also been reported to facilitate the reduction of cholesterol in the blood serum. The evidence of this claim comes from a number of possible modes of action that may be explained by the fact that, since cholesterol takes part in the production of bile acids (e.g. cholic and chenodeoxycholic acids), the increased metabolism which goes hand in hand with the increased excretion of bile acids may be considered as a way of lowering cholesterol from the blood serum (Chikai *et al.* 1987). It should be known that the metabolism of bile acids involves their conjugation by amino acids such as glycine and taurine to form the corresponding acidic derivatives (glycocholic). These undergo further hydrolase enzymatic deconjugation and are actually excreted via urine. Apart from these possible ways in which probiotics lowers blood cholesterol, some other reports have discussed the possibility of direct assimilation of cholesterol by probiotics (Gilliland *et al.* 1985).

18.3.4 Lactose maldigestion

Some reports have suggested that people who suffer from lactose maldigestion can be comfortable with the levels of lactose present in fermented food products such as yoghurt as opposed to the higher levels of lactose in raw milk. This can be explained by the fact that these fermented products, as well as probiotic lactic acid bacteria, are rich in lactase enzyme which act on lactose (as a substrate), thus reducing or solving the problem of lactose maldigestion (Marteau *et al.* 2002).

18.3.5 Diarrhoea

Several probiotics have been reported to combat diarrhoea, including *Lactobacillus rhamnosus* GG and *Bifidobacterium bifidum* mixed with *Streptococcus thermophilus* (Isolauri *et al.* 1991). The mode of action of this probiotic strain is believed to involve fortification of the integrity of mucosal linings in addition to the stimulation of the natural immunity via enhanced biomolecules such as the antirotavirus specific immunoglobulin (Ig) A (Saavedra *et al.* 1994).

18.3.6 General mode of action of probiotics

The general mechanisms by which probiotic strains function may involve multifacets or strain specificity. Possible mechanisms include the production of cidal compounds which may contribute to luminal acidification because they produce some volatile fatty acid. They may also play an active role in the competition for nutrients as well as the sticking (attaching) of sites on the gut wall, modulation of the immune response and the processes of regulating colonocyte gene expression (Mack *et al.* 1999; Steer *et al.* 2000; Fooks and Gibson 2002).

18.4 SAFETY AND STABILITY OF PROBIOTICS

From the modes of action of probiotics, assumptions can made that probiotics may cause some disturbances in the body that may cause side effects in certain individuals such as systemic infections, deleterious metabolic activities or even excessive immune stimulation (Marteau 2001; Marteau and Seksik 2004). Among the known cases of infections is one of fungemia by *Saccharomyces boulardii* that involved patients with a catheter (Zunic *et al.* 1991). Otherwise, with regard to the most frequently used probiotic microorganisms such as lactic acid bacteria, safety concerns are very rare in literature (Adams and Marteau 1995). For example, septicaemia and endocarditis infections due to probiotic lactobacilli and bifidobacteria have been reported in some occasions in certain places (Aguirre and Collins 1993; Gasser 1994; Patel *et al.* 1994; Kalima *et al.* 1996; Saxelin 1996; Saxelin *et al.* 1996). The explanation for these rare cases of infections was that the victims had other implants (e.g. a catheter or cardiac valves) and these provided a portal of entry for the infection problem (Horwitch *et al.* 1995). However, the majority of studies have indicated that microbial strains that are normally selected as probiotics such as lactobacilli, bifidobacteria and others have a very good safety record and they are not associated with any of the human infections (Salminen *et al.* 1998; Ishibashi and Yamazaki 2001; Borriello *et al.* 2003).

Bifidobacteria performs other important functions in our bodies such as production of B vitamins for our use, assisting with the functioning of the liver and producing lactic and acetic acids (thus lowering the colon pH which discourages the thriving of harmful microbes, which prefer alkaline environments). Bifidobacteria also discourages the growth of nitrate bacteria in the gut, which have been implicated as carcinogens in the body systems.

For these reasons, these probiotic microorganisms have been listed as GRAS and are therefore safe to be incorporated in foodstuffs; in most cases, they form part of the indigenous microbial flora in human guts (Morgensen *et al.* 2002). This status does not however imply that tests with regards to safety of probiotic strains can be neglcted (Charteris *et al.* 2000; Kheadr *et al.* 2004; Moubareck *et al.* 2005). Since the attribute of microbial resistance to antimicrobial agents is one of the prerequisites of ideal probiotics, evidence of probiotic resistance should be available through proper tests (Herrero *et al.* 1996; Teuber *et al.* 1999).

The issue of probiotic stability is crucial since microbes, just as all other cells, experience a natural decay of their viability with time, affecting their efficacy. The stability of probiotics is generally dependent on a number of factors such as the genus, species, type of strain selected, the formulation, the composition of ingredients and the quality of various components. Other determinants of the probiotic stability are water content, temperature and pH. To ensure the viability of the microbial cell, tests and measurements to confirm the efficacy of probiotics are normally conducted.

Finally, the possibility of triggering allergies for consumers with intolerances to some ingredients or certain microbial strains in probiotics does exist. Such possibilities are normally eliminated by determining the potential allergenic substance and neutralising it.

18.5 SUITABLE DIETARY CARRIERS FOR PROBIOTICS

Many studies have shown the potential of fermented milk products, especially yogurt, as potential products in which probiotics may be incorporated, manipulated and optimised (Hekmat and Koba 2006). Other fermented milk products which have been reported as suitable substrates for the incorporation of probiotics include soy yogurt (Farnworth *et al.* 2007), ice cream (Hekmat and McMahon 1992; Davidson *et al.* 2000; Akin *et al.* 2007) and cheese (Sharp *et al.* 2008).

The benefits of using these cow's milk fermented products as carriers for probiotics are that the pH drops gradually and relatively more slowly than with other dietary media (Farnworth *et al.* 2007). Fermented products such as yogurt are known to have the advantage of excellent maintenance of viable microorganisms, for example, lactobacilli species (Hekmat *et al.* 2009). Moreover, it has been reported that these products provide the best environment for micronutrient fortification where micronutrient additives (e.g. iron, manganese, zinc, molybdenum, chromium and selenium) can be incorporated effectively and without any loss of flavour (Hekmat and Donald 1997; Achanta *et al.* 2006; Zhu *et al.* 2008; Hekmat *et al.* 2009).

18.5.1 Production and evaluation

A number of probiotic microorganisms are used to improve human health. These include both single-strain as well as multi-strain bacteria such as bacillus spores or yeasts species (multi-strain), enterococcus, bacillus, lactobacillus, pediococcus, saccharomyces or bifidobacteria, normally incorporated in fermented foodstuffs such as yoghurts and other milk products. Normally, the production of dietary probiotic additives involves the preparation of the micronutrient blend as well as the probiotic mother culture. Both the micronutrient blend and the type of microbial strain depends on the intended purpose of the probiotic, that is, the type of disease or disorder being targeted.

18.5.2 Fermented dietary products for HIV cases

Micronutrients that are normally considered in probiotics to treat HIV are those which boost the immune response as well as reduce morbidity of the sufferers (by increased CD4 and delayed HIV related mortality), or those which are associated with a delayed progression of HIV to AIDS (Jiampton *et al.* 2003; Fawzi *et al.* 2004; Kaiser *et al.* 2006). Such micronutrients include N-acetyl cysteine (NAC), ω-3 fatty acids, vitamin A (such as β-carotene and palmitate), vitamin B (B1 as thiamin, B6 as pyroxene and B12 as cyanocobalamin), vitamin C (as ascorbic acid), vitamin E (as acetate), niacinamide, folic acid, selenium (as sodium selenite), iron (as ferric pyrophosphate), zinc (as zinc sulphate) and whey protein (Hummelen *et al.* 2010). With regard to the preparation of the probiotic mother culture, strains which will boost both the victim's health and also the host's immune system such as *Lactobacillus rhamnosus* GR-1 or *Lactobacillus rhamnosus* CAN-1 will be considered (Braat *et al.* 2004; Furrie *et al.* 2005; O'Mahony *et al.* 2005).

18.5.3 For diarrhoea cases

Strains suitable for probiotics which are effective in combating diarrhoea diseases are *Lactobacillus reuteri*, *Lactobacillus acidophilus* and *Lactobacillus rhamnosus*. These strains can then be included in the diet in the form of, for example, maize porridge.

18.6 ASSESSMENT OF PROBIOTICS IN FOODSTUFFS AND SUPPLEMENTS

A number of reports have described serious concerns about the quality of probiotic products, especially with regard to the correct count of the viable cell numbers as stipulated in the labels as well as the type of strains (Micanel *et al.* 1997; Shah *et al.* 2000; Weese 2002; Gueimonde *et al.* 2004; Carr and Ibrahim 2005; Lin *et al.* 2006; Maukonen *et al.* 2006; Moreno *et al.* 2006; Al-Otaibi 2009). In addition to this concern, there have been issues about the correct identification of the microbial probiotic species that have been incorporated or used. Some reports have revealed discrepancies between the type of microorganism used as probiotics between that indicated on the label and that actually present in the food products (Canganella *et al.* 1997; Zhong *et al.* 1998). Some concerns have arisen from the possibility of the presence of microbial strains which is not even stated on the label, and which may be potentially harmful (Hamilton-Miller *et al.* 1999). Identifying the presence of another species is important, even for species which belong to the same genus and which normally exert similar probiotic properties (Salminen *et al.* 1998a, b; Spanhaak *et al.* 1998). All these facts point to the need for proper assessment methods to verify the quality as well as measure the potential of the recommended microbial probiotics to render the expected health benefits. Probiotic cultures can be analysed in either dried formulation, liquid formulation or in a frozen state.

18.6.1 Molecular and PCR-based methods

When probiotic microbes such as bifidobacteria or lactobacilli are incorporated in diets, an obvious problem that presents itself is in the detection and counting in order to assess survivals. Another difficult task is distinguishing the probiotic microbes from those endogenous microbes which are indigenous residents in the gut of the host. In most cases, a technique involving 16S ribosomal ribonucleic acid (16S rRNA) probing aided by the polymerase chain reaction (PCR) technique is employed to assess and monitor the strains which contain what is known as the strain-specific signatures within the specified 16S rRNA. In this technique, strain-specific 16S rRNA gene-targeted oligonuclear primers are developed to specifically detect and differentiate the probiotic microbes from those resident in the gut system.

18.6.2 DNA fingerprinting-based methods

Another technique used for probiotic assessment involves the use of DNA fingerprinting, and is also capable of distinguishing probiotic strains from the microbes indigenous in the gut. In this technique, a 16S rDNA restriction fragment length polymorphism (RFLP) analysis is used to assess the fate of probiotic microbes in the gut system after the food containing them has been consumed (Kullen *et al.* 1997). The number of probiotic microbes distinguished from indigenous microbes via 16S rDNA RFLP is however very low. Other approaches related to DNA fingerprinting techniques with higher resolutions have therefore been developed and introduced, including a method known as ribotyping or pulse-field gel electrophoresis (McCartney *et al.* 1996). Moreover, the use of genetically modified probiotic microbes has been introduced to replace gene/DNA probes (Chalfie *et al.* 1994).

18.7 CONCLUSIONS

Despite the fact that the majority of studies and research findings indicate that probiotic microbes in foodstuffs render health benefits to consumers, it seems to be too early to accept such claims and solid proof is still lacking. More work needs to be conducted with regard to establishment of the safety of each class of probiotic microorganisms in the respective food items where they are incorporated, and these studies should (where possible) involve human beings and should not rely on animal experiments for solid and conclusive outcomes. If human studies are unpopular for ethical reasons, then results from animal studies should be translated to humans with great care; there are remarkable physiological differences between animals and humans and also in terms of the indigenous gut flora that inhabit these two groups. Nevertheless, with the current situation and with all the studies that have been conducted so far, experimental results have shown that there are some health benefits associated with the microbial probiotics incorporated in foodstuffs. This should be motivation to conduct more research in the area in order to improve the current findings for the future.

REFERENCES

Achanta, K., Aryana, K. J. & Boeneke, C. A. (2006) Fat free plain set yogurts fortified with various minerals. *LWT* 40, 424–429.

Adams, M. R. & Marteau, P. (1995) On the safety of lactic acid bacteria from food. *International Journal of Food Microbiology* 27, 263–264.

Aguirre, M. & Collins, M. D. (1993) Lactic acid bacteria and human clinical infection. *Journal of Applied Bacteriology* 75, 95–107.

Akin, M. B., Akin, M. S. & Kirmaci, Z. (2007) Effects of inulin and sugar levels on the viability of yogurt and probiotic bacteria and the physical and sensory characteristics in probiotic ice-cream. *Food Chemistry* 104 (1), 93–99.

Al-Otaibi, M. M. (2009) Evaluation of some probiotic fermented milk products from Al-Ahsa markets, Saudi Arabia. *American Journal of Food Technology* 4, 1–8.

Autrup, H., Harris, C. C. & AJeffrey, M. (1978) Metabolism of benzo(a)pyrene and identification of the major benzo(a)pyrene-DNA adducts in cultured human colonocytes. *Cancer Research* 38, 3689–3696.

Berner, L. A. & O'Donnell, J. A. (1998) Functional foods and health claims legislation: applications to dairy foods. *International Dairy Journal* 8, 355–362.

Borriello, S. P., Hammes, W. P., Holzapfel, W., Marteau, P., Schrezenmeir, J., Vaara, M. & Valtonen, V. (2003) Safety of probiotics that contain *Lactobacillus* or *Bifidobacteria*. *Clinical Infectious Diseases* 36, 775–780.

Braat, H., van den Brande, J., van Tol, E., Hommes, D., Peppelenbosch, M. & van Deventer, S. (2004) Lactobacillus rhamnosus induces peripheral hyporesponsiveness in stimulated CD4+T cells via modulation of dendritic cell function. *American Journal of Clinical Nutrition* 80, 1618–1625.

Burns, A. J. & Rowland, I. R. (2000) Anti-carcinogenicity of probiotics and prebiotics. *Current Issues in Intestinal Microbiology* 1, 13–24.

Campieri, M., Rizzello, F., Venturi, A., Poggioli, G., Ugolini, F., Helwig, U., Amadini, C., Romboli, E. & Gionchetti, P. (2000) Combination of antibiotic and probiotic treatment is efficacious in prophylaxis of post-operative recurrence of Crohn's disease: a randomised controlled study vs mesalamine. *Gastroenterology* 118 (4), A781–A781.

Canganella, F., Paganini, S., Ovidi, M., Vettraino, A. M., Bevilacqua, L., Massa, S. & Trovatelli, L. D. (1997) A microbial investigation on probiotic pharmaceutical products used for human health. *Microbiology Research* 152, 171–179.

Carr, J. P. & Ibrahim, S. A. (2005) Viability of bifidobacteria in commercial yogurt products in North Carolina. *Milchwissenschaft*, 60, 414–416.

Chalfie, M., Tu, Y., Euskirchen, G., Ward, W. W. & Prasher, D. C. (1994) Green fluorescent protein as a marker for gene expression. *Science* 263, 802–805.

Charteris, W. P., Kelly, P. M., Morelli, L. & Collins, J. K. (2000) Effect of conjugated bile salts on antibiotic susceptibility of bile salt-tolerant *Lactobacillus* and *Bifidobacterium* isolate. *Journal of Food Protection* 63, 1369–1376.

Chikai, T., Nakao, H. & Uchida, K. (1987) Deconjugation of bile acids by human intestinal bacteria implanted in germ-free rats. *Lipids* 22, 669–671.

Conway, P. L. (1996) Selection criteria for probiotic microorganisms. *Asia Pacific Journal of Clinical Nutrition* 5, 10–14.

Davidson, R. H., Duncan, S. E., Hackney, C. R., Eigel, W. N. & Boling, J. W. (2000) Probiotic culture survival and implications in fermented frozen yogurt characteristics. *Journal of Dairy Science* 83 (4), 666–673.

Drasar, B. S., Renwick, A. G. & Williams, R. T. (1972) The role of the gut flora in the metabolism of cyclamate. *Biochemistry Journal* 129, 881–890.

Farnworth, E. R., Mainville, I., Desjardins, M. P., Gardner, N., Fliss, I. & Champagne, C. (2007) Growth of probiotic bacteria and bifidobacteria in a soy yogurt formulation. *International Journal of Food Microbiology* 116 (1), 174–181.

Fawzi, W., Msamanga, G. I., Spiegelman, D., Wei, R., Kapiga, S., Villamor, E., Mwakagile, D., Mugusi, F., Hertzmark, E., Essex, M. & Hunte, D. J. (2004) A randomized trial of multivitamin supplements and HIV disease progression and mortality. *New England Journal of Medicine* 351, 23–32.

Fernandes, C. F. & Shahani, K. M. (1990). Anticarcinogenic and immunological properties of dietary lactobacilli. *Journal of Food Protection* 53, 704–710.

Fooks, L. J. & Gibson, G. R. (2002) Probiotics as modulators of the gut flora. *British Journal of Nutrition* 88, S39–S49.

Fuller R. (ed.) (1992) *Probiotics 1: The Scientific Basis*. Chapman and Hall, London.

Fuller, R. (1996) Probiotics-panacea or nostrum? *BNF Nutrition Bulletin* 21, 204–208.

Fuller, R. (ed) (1997) *Probiotics 2: Applications and Practical Aspects*. Chapman and Hall, London.

Furrie, E., Macfarlane, S., Kennedy, A. J., Cummings, H., Walsh, S. V., O'Neil, D. A. & Macfarlane, G. T. (2005) Synbiotic therapy (Bifidobacterium longum/Synergy 1) initiates resolution of inflammation in patients with active ulcerative colitis: A randomised controlled pilot trial. *Gut* 54, 242–249.

Gasser, F. (1994) Safety of lactic acid bacteria and their occurrence in human clinical infections. *Bulletin de l'Institut Pasteur* 92, 45–67.

Gilliland, S. E., Nelson, C. R. & Maxwell, C. (1985) Assimilation of cholesterol by *Lactobacillus acidophilus*. *Applied Environmental Microbiology* 49, 377–381.

Gueimonde, M., Delgado, S., Mayo, B., Ruas-Madiedo, P., Margolles, A. & de los Reyes-Gavilan, C. G. (2004) Viability and diversity of probiotic Lactobacillus and Bifidobacterium populations included in commercial fermented milks. *Food Research International* 37, 839–850.

Guslandi, M., Mezzi, G., Sorghi, M. & Testoni, P. A. (2000) Saccharomyces boulardii in maintenance treatment of Crohn's disease. *Digestive Diseases and Sciences* 45, 1462–1464.

Hamilton-Miller, J. M. T., Shah, S. & Winkler, J. T. (1999) Public health issues arising from microbiological and labelling quality of foods and supplements containing probiotic organisms. *Public Health and Nutrition* 2, 223–229.

Hays, S. M. (1994) Natural microbes curb salmonella. *Agricultural Research* 42, 22–26

Hekmat, S. & McMahon, D. J. (1992) Survival of Lactobacillus acidophilus and Bifidobacterium bifidum in ice cream for use as a probiotic food. *Journal of Dairy Science* 75 (6), 1415–1422.

Hekmat, S. & Donald, D. J. (1997) Manufacture and quality of iron-fortified yogurt. *Journal of Dairy Science* 80, 3114–3122.

Hekmat, S. & Koba, L. (2006) Fermented dairy products: Knowledge and consumption. *Canadian Journal of Dietetic Practice and Research* 67 (4) 199–201.

Hekmat, S., Soltani, H. & Reid, G. (2009) Growth and survival of Lactobacillus *reuteri* RC-14 and *Lactobacillus rhamnosus* GR-1 in yogurt for use as a functional food. *Innovative Food Science and Emerging Technologies* 10, 293–296.

Herrero, M., Mayo, B., Gonzalez, B. & Suarez, J. E. (1996) Evaluation of technologically important traits in lactic acid bacteria isolated from spontaneous fermentations. *J. Applied Bacteriology* 81, 565–570.

Holzapfel, W. H., Haberer, P., Snel, J., Schillinger, U. & Huis in't Veld, J. H. (1998) Overview of gut flora and probiotics. *International Journal of Food Microbiology* 41, 85–101.

Horwitch, C. A., Furseth, H. A., Larson, A. M., Jones, T. L., Olliffe, J. F. & Spach, D. H. (1995) Lactobacillemia in three patients with AIDS. *Clinical Infectious Diseases* 21, 1460–1462.

Hose, H. & Sozzi, T. (1991) Probiotics, fact or fiction. *Journal of Chemical Technology and Biotechnology* 51, 540–544.

Hummelen, R., Hemsworth, J. & Reid, G. (2010) Micronutrients, N-Acetyl cysteine, probiotics and prebiotics, a review of effectiveness in reducing HIV progression. *Nutrients* 2 (6), 626–651.

Ishibashi, N. & Yamazaki, S. (2001) Probiotics and safety. *American Journal of Clinical Nutrition* 73, S465–S470.

Isolauri, E., Juntunes, M., Rautanen, T., Sillanaukee, P., Koivula, T. (1991) A human Lactobacillus strain (*Lactobacillus casei sp.* Strain GG) promotes recovery from acute diarrhea in children. *Pediatrics* 88, 90–97.

Jiampton, S., Pepin, J., Suttent, R., Filteau, S., Mahakkanukrauh, B., Hanshaowaorakuf, W., Pongsakdig, C., Puanb, S., Prakashh, S. & Shabbara, J. (2003) A randomized trial of the impact of multiple micronutrient supplementation on mortality among HIV-infected individuals living in Bangkok. *AIDS* 17, 2461–2469.

Kaiser, J. D., Adriana, M. C., Ondercin, J. P., Gifford, S. L., Pless, R. F. & Baum, M. K. (2006) Micronutrient supplementation increases CD4 count in HIV-infected individuals on highly active antiretroviral therapy: A prospective, double-blinded, placebo controlled trial. *Journal of Acquired Immune Deficiency Syndromes* 42, 523–528.

Kalima, P., Masterton, R. G., Roddie, P. H. & Thomas, A. E. (1996) *Lactobacillus rhamnosus* infection in a child following bone marrow transplant. *Journal of Infections* 32, 165–167.

Kheadr, E., Bernoussi, N., Lacroix, C. & Fliss, I. (2004) Comparison of the sensitivity of commercial strains and infant isolates of bifidobacteria to antibiotics and bacteriocins. *International Dairy Journal* 14, 1041–1053.

Kullen, M. J., Amann, M. M., O'Shaughnessy, M. J., O'Sullivan, D. J., Busta, F. F. & Brady, L. J. (1997) Differentiation of ingested and endogenous bifidobacteria by DNA fingerprinting demonstrates survival of an unmodified strain in the gastrointestinal tract of humans. *Journal of Nutrition* 1997, 89–94.

Lee, Y.-K. & Salminen, S. (1995) The coming age of probiotics. *Trends in Science and Technology* 6, 241–245.

Lin, W. H., Hwang, C. F., Chen, L. W. & Tsen, H. Y. (2006) Viable counts, characteristic evaluation for commercial lactic acid bacteria products. *Food Microbiology* 23, 74–81.

Mack, D. R., Michail, S., Wei, S., McDougall, L. & Hollingsworth, M. A. (1999) Probiotics inhibit enteropathogenic *E. coli* adherence *in vitro* by inducing intestinal mucin gene expression. *American Journal of Physiology* 276, G941–G950.

Marteau, P. (2001) Safety aspects of probiotic products. *Scandinavian Journal of Nutrition/Näringsforskning* 45, 22–30.

Marteau, P. & Seksik, P. (2004) Tolerance of Probiotics and Prebiotics. *Journal of Clinical Gastroenterology* 38 (2), S67–S69.

Marteau, P., Seksik, P. & Jian, R. (2002) Probiotics and intestinal health: a clinical perspective. *British Journal of Nutrition* 88, S51–S57.

Maukonen, J., Alakomi, H., Nohynek, L., Hallamaa, K., Leppämäki, S., Mättö, J. Saarela, M. (2006) Suitability of the fluorescent techniques for the enumeration of probiotic bacteria in commercial non-dairy drinks and in pharmaceutical products. *Food Research International* 39, 22–32.

McCartney, A. L., Wenzhi, W. & Tannock, G. W. (1996) Molecular analysis of the composition of the bifidobacterial and lactobacillus micro flora of humans. *Applied Environmental Microbiology* 62, 4608–4613.

McIntosh, G. H. (1996) Probiotics and colon cancer prevention. *Asia Pacific Journal of Clinical Nutrition* 5, 48–52.

Micanel, N., Haynes, I. N. & Playne, M. J. (1997) Viability of probiotic cultures in commercial Australian yogurts. *Australian Journal of Dairy Technology* 52, 24–27.

Moreno, Y., Collado, M. C., Ferrus, M. A., Cobo, J. M., Hernandez, E. & Hernandez, M. (2006) Viability assessment of lactic acid bacteria in commercial dairy products stored at 4°C using LIVE/DEAD BacLigh staining and conventional plate counts. *International Journal of Food Science and Technology* 41, 275–280.

Morgensen, G., Salminen, S., O'Brien, J., Ouwehand, A., Holzapfel, W., Shortt, C., Fonden, R., Miller, G. D., Donohue, D., Playne, M., Crittenden, R., Bianchi Salvadori, B. & Zink, R. (2002) Inventory of microorganisms with a documented history of use in food. *Bulletin International Dairy Federation* 377, 10–18.

Moubareck, C., Gavini, F., Vaugien, L., Butel, M. J. & Doucer-Popularie, F. (2005) Antimicrobial susceptibility of Bifidobacteria. *Journal of Antimicrobial Chemotherapy* 55, 38–44.

Newcomer, A. D., Park, H. S., O'Brien, P. C. & McGill, D. B. (1983) Response of patients with irritable bowel syndrome and lactase deficiency using unfermented acidophilus milk. *American Journal of Clinical Nutrition* 38, 257–263.

O'Mahony, L., McCarthy, J., Kelly, P., Hurley, G., Luo, F., Chen, K., O'Sullivan, G. C., Kiely, B., Collins, J. K., Shanahan, F. & Quigley, E. M. (2005) Lactobacillus and bifidobacterium in irritable bowel syndrome: Symptom responses and relationship to cytokine profiles. *Gastroenterology* 128, 541–551.

O'Sullivan, M. G., Thornton, G., O'Sullivan, G. C. & Collins, J. K. (1992) Probiotic bacteria: myth or reality? *Trends in Food Science and Technology* 3, 309–314.

Pardon, G., Alvarez, S., Rachid, M., Aquero, G. & Gobbato, N. (1995) Probiotic bacteria for humans: clinical system for evaluation of effectiveness-immune system stimulation by probiotics. *Journal of Dairy Science* 78, 1597–1606.

Patel, R., Cockerill, F. R., Porayko, M. K., Osmon, D. R., Ilstrup, D. M. & Kenting, M. R. (1994) Lactobacillemia in liver transplant patients. *Clinical Infectious Diseases* 18, 207–212.

Reddy, B. S., Watanabe, K., Weisburger, J. H. & Wynder, E. L. (1977) Promoting effect of bile acids in colon carcinogenensis in germ-flee and conventional F344 rats. *Cancer Research* 37, 3238–3242.

Reid, G. (1999) The efficacy of probiotics. In: Tannock, G. W. (ed.), *Probiotics: A Critical Review*. Horizon Scientific Press, Wymondham, UK, pp. 129–133.

Rembacken, B. J., Snelling, A. M., Hawkey, P. M., Chalmers, D. M. & Axon, A. T. (1999) Non-pathogenic *Escherichia coli* versus mesalizine for the treatment of ulcerative colitis: a randomised trial. *Lancet* 354, 635–639.

Saavedra, J. M., Bauman, N. A., Oung, I., Perman, J. A. & Yolken, R. H. (1994) Feeding of *Bifidobacterium bifidum* and *Streptococcus thermophilus* to infants in hospital for prevention of diarrhea and shedding of rotavirus. *Lancet* 344, 1046–1049.

Salminen, S., Ouwehand, A. C., Isolauri, E. (1998a) Clinical applications of probiotic bacteria. *International Dairy Journal* 8, 563–572.

Salminen, S., von Wright, A., Morelli, L., Marteau, P., Brassart, D., de Vos, W. M. & Valtonen, V. (1998b) Demonstration of safety of probiotics-a review. *International Journal of Food Microbiology* 44, 93–106.

Sanders, M. E. & Veld, J. H. (1999) Bringing a probiotic-containing functional food to the market: microbiological, product regulatory and labelling issues. *Antonie van Leeuwenhoek* 76, 293–315.

Saxelin, M. (1996) Colonization of the human gastrointestinal tract by probiotic bacteria. *Nutrition Today* 31, 5S–8S.

Saxelin, M., Chuang, N. H., Chassy, B., Rautelin, H., Mäkelä, P. H., Salminen, S. & Gorbach, S. L. (1996) Lactobacilli and bacteremia in Southern Finland 1989–1992. *Clinical Infectious Diseases* 22, 564–566.

Scheinbach, S. (1998) Probiotics: Functionality and commercial status. *Biotechnology Advances*, 16 (3), 581–608.

Shah, N. P., Ali, J. F. & Ravula, R. R. (2000) Populations of Lactobacillus acidophilus, Bifidobacterium spp. and Lactobacillus casei in commercial fermented milk products. *Bioscience and Microflora* 19, 35–39.

Shanahan, F. (2000) Probiotics and inflammatory bowel disease: is there a scientific rationale? *Inflammation Bowel Diseases* 6, 107–115.

Sharp, M. D., McMahon, D. J. & Broadbent, J. R. (2008) Comparative evaluation of yogurt and low-fat cheddar cheese as delivery media for probiotic *Lactobacillus casei*. *Journal of Food Science* 73 (7), M375–M377.

Spanhaak, S., Havenaar, R. & Schaafsma, G. (1998) The effect of consumption of milk fermented by Lactobacillus casei strain Shirota on the intestinal micro flora and the immune parameters in humans. *European Journal of Clinical Nutrition* 52, 899–907.

Steer, T., Carpenter, H., Tuohy, K. & Gibson, G. R. (2000) Perspectives on the role of the human gut microbiota and its modulation by pro- and prebiotics. *Nutrition Research Reviews* 13, 229–254.

Teuber, M., Meile, L. & Schwarz, F. (1999) Acquired antibiotic resistance in lactic acid bacteria from food. *Antonie Van Leeuwenhoek* 76, 115–137.

Tuohy, K. M., Probert, H. M., Smejkal, C. W. & Gibson, G. R. (2003) Using probiotics and prebiotics to improve gut health. *Therapeutic Focus* 8 (15), 693–700.

Weese, J. S. (2002) Microbiologic evaluation of commercial probiotics. *Journal of the American Veterinary Medical Association* 220, 794–797.

Weisburger, J. H., Grontham, R. E. & Weisburger, E. K. (1970) Metabolism of the carcinogenic N-hydroxy-N-2-fluorenylacetamide in germ-free rats. *Biochemical Pharmacology* 19, 151–162.

Zhong, W., Millsap, K., Bialkowska-Hobrzanska, H. & Reid, G. (1998) Differentiation of Lactobacillus species by molecular typing. *Applied Environmental Microbiology* 64, 2418–2423.

Zhu, L., Miller, D., Nelson, D. & Glahn, R. (2008) Soluble ferric pyrophosphate: A novel iron source for individuals with high iron needs. *The FASEB Journal* 22, 678.

Zunic, P., Lacotte, J., Pegoix, M., Buteux, G., Leroy, G., Mosquett, B., Molin, M. & Fongérnie, Á. (1991) *Saccharomyces boulardii*. *Therapie* 46, 497–501.

19 Prebiotics

Abstract: In the past few decades, scientists have developed a great interest in nutrition and its relation to health. Moreover, nutritionists and dieticians have intensified the research activities related to dietary modulation of the human gut. The motivation for this interest is drawn from knowledge of the presence of hosts of microbial population in the gastrointestinal tract. The previous chapter dealt with probiotics; this chapter will focus on prebiotics, the non-viable food ingredients selectively metabolised by the beneficial gut microbial flora and with diet modulation of gut microbes, thus bringing health benefits to the host individual. The modulation catalyses or stimulates an increase either in microbial populations or in microbial activities, thus enhancing resistance towards harmful microbes and the immune response. This chapter discusses the role of prebiotics in foods and their contribution to the improvement of human health.

Keywords: microbial population; prebiotics; saccharides

19.1 PREBIOTICS AND HEALTH

The purpose of foods such as prebiotics, probiotics and synbiotics lies somewhere between foods and medicines (drugs). While probiotics are live microorganisms incorporated in foodstuffs, prebiotics are non-digestible food ingredients intended for selective stimulation of growth and/or activity of gut microbial flora in the digestive system and colon to induce health benefits in individuals (Gibson and Roberfroid 1995). The definition of prebiotics has recently been extended such that it now refers to the selective fermented food components which support specific changes, both in the composition and/or activity, in the gastrointestinal microbiota that will result in potential health benefits to individuals (Gibson *et al.* 2004).

In reality, prebiotics exert their influence in an indirect manner due to the fact that they are being fed to a specific gut strain or a few gut microbial strains such as indigenous bifidobacteria and lactobacilli (Teitelbaum and Walker 2002). As a result, they will induce selective modification of the microbial flora within the individual digestive system. It is these modifications that actually provide health benefits for the host, rather than the actual prebiotics themselves.

Further Thinking

Prebiotics are selectively fermented, dietary ingredients that induce specific changes both in the composition and/or activity of the gastrointestinal microbial flora, thus providing health benefit(s) to the host. Prebiotics targets the microbial flora indigenous in the colon, that is, microbes which are already present within the gut system. They provide a source of substrate to these microbial gut microbial populations, positively affecting their activities and numbers.

Generally, functional food components referred to as prebiotics are soluble fibres. Structurally, they are actually carbohydrates such as oligosaccharides and inulin, although there are some which are non-carbohydrates. As the name suggests, oligosaccharides, are composed of sugar molecules with chain lengths made up of short-chain polysaccharide of up to 20 sugar units. Some oligosaccharides are naturally occurring and can be extracted from their sources, such as fruits or vegetables, using traditional extraction techniques involving organic solvents. Other oligosaccharides are synthetic and can be produced commercially through hydrolysis of polysaccharides.

19.2 FACTORS THAT INFLUENCE THE ACTIVITY AND EFFECTIVENESS OF PREBIOTICS

Prebiotic oligomers which are mainly carbohydrate in nature possess features and attributes that influence their performance as functional foods, enhancing the health status of their hosts. These oligomers have features such as glycosidic linkages which join the monosaccharide monomers, providing the selectivity needed for the processes of fermentation as well as digestion in the intestines. For example, the fermentation process of fructo-oligosaccharides by bifidobacteria is selective due to the fact that this oligomer is hydrolysed by a corresponding enzyme known as β-fructofuranosidase which targets the glycosidic bonds.

Moreover, the majority of known oligosaccharide prebiotics contain saccharides such as glucose, fructose and xylose. Apart from these oligosaccharides, there have been no reports of successful prebiotics with other monosaccharides. The structural chemistry of these oligomers must therefore be crucial to the prebiotic activity and effectiveness.

Another important factor contributing to effectiveness is molecular weight. This can be seen from the fact that oligomers are prebiotic active while polysaccharides are not known to have prebiotic properties (De Leenheer 1994). The contribution of molecular weight factor in prebiotics is crucial and a determinant factor, as demonstrated by the fact that xylan is not selective while xylo-oligosaccharides are (Okazaki *et al.* 1990; Jaskari *et al.* 1998).

The degree of polymerisation is another contributing factor. The majority of carbohydrate compounds in inulin with degree of polymerisation of ≤ 25 are known to be prebiotic active; those which have a degree of polymerisation > 25 are not (De Leenheer 1994).

19.3 TYPES OF OLIGOSACCHARIDES

A number of oligosaccharides have potential as prebiotics, including fructo-oligosaccharides, gluco-oligosaccharides, pectic-oligosaccharides, lactosucrose, the sugar alcohols, levans, inulin, resistant starch, galacto-oligosaccharides, xylo-oligosaccharides, lactulose, lactosucrose, palatinose, isomalto-oligosaccharides and soybean oligosaccharides (Gibson *et al.* 2000). The oligosaccharides are synthesised by a number of species of bacteria, fungi and plants and they are grouped into different classes based on their glycosidic linkages. Inulins contain $\beta(2\text{-}1)$ linkages and levans $\beta(2\text{-}6)$ linkage. Others, such as xylo-oligosaccharides are formed by chains of xylose molecules linked by $\beta(1\text{-}4)$ linkages and they consist mainly of xylobiose, xylotriose and xylo-tertaose.

Further Thinking

Prebiotics do not normally undergo hydrolysis or absorption in the upper part of the gastrointestinal tract but only in the lower part, where they selectively stimulate the growth and/or activity of desirable bacteria in the colon. Members of microbial flora in the colon, especially lactobacillus and bifidobacteria, are the main target for increased growth and/or activity due to prebiotics. These microbes are essential as they protect the host by competing for nutrients and space with other microbes such as pathogenic bacterial or fungal strains and thus boosting the immune system. Moreover, the microbial fermentation activities of prebiotics in the gut results in the production of short-chain fatty acids such as acetic, propionic and butyric acids, which are important energy sources for the mucosal cells.

19.3.1   Fructo-oligosaccharides

Fructo-oligosaccharides are short chains of fructose sugar molecules. These molecules have been used as dietary supplements for a long time and remain popular today as prebiotics, where they are used as substrates for the gut colon microbial flora in order to improve the health of the hosts. They have also been reported to perform a number of important functions in the digestive system such as promotion of calcium absorption and lowering of gut pH. Moreover, fructo-oligosaccharides are regarded as dietary fibres; they therefore have low calorific value and their fermentation yields acids and gases thus lowering the gut pH. The low pH is beneficial because it enhances the solubilisation of calcium, a phenomenon with positive health implications for an individual (van den Heuvel *et al.* 1999; Zafar *et al.* 2004). The lowering of gut pH also encourages the growth of lactic acid bacteria, which in turn promotes butyrate and lactate production (Cherbut *et al.* 2003).

Further Thinking

The ideal prebiotics should be resistant to degradation by acids and hydrolytic enzymes present in the digestive system; they should also be fermentable by microbes in the colon. By definition, they must be capable of selective stimulation to positively promote both the growth and/or activity of the gut microbial flora, to render health benefits to the host.

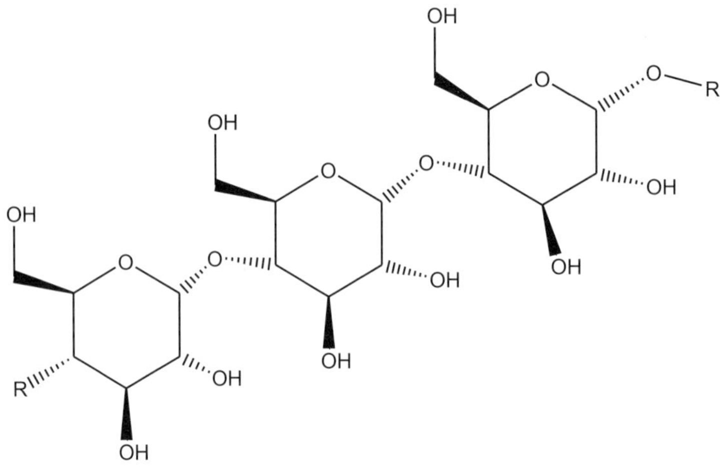

Fig. 19.1 General structure of gluco-oligosaccharide (Wichienchot *et al.* 2006)

19.3.2 Gluco-oligosaccharides

Gluco-oligosaccharides (Figure 19.1), also known as gluco-oligosides, are products resulting from an incomplete enzymatic degradation of starch. They have the general chemical formula $(O-\alpha-D-gucopyranosyl)_n$, where n is an integer in the range 2–10. These molecules are synthesised enzymatically using an enzyme known as glycosyl-transferase from leuconostoc mesenteroides, which facilitate the transfer of glucose monomers from sucrose to maltose. Some reports have indicated that gluco-oligosaccharides with a molecular weight in the range 7.8–65.6 kDa are prebiotic active (Wichienchot *et al.* 2006).

19.3.3 Isomalto-oligosaccharides

Isomalto-oligosaccharides (Figure 19.2) are a mixture of sugars such as isomaltose, isomaltotriose, panose and isomaltotetraose (Kohmoto *et al.* 1991; Hsiao-Ling *et al.* 2001). They are non-digestible low-calorie health sweeteners which also support the growth and/or activity of microbial flora indigenous in the colon, especially the bifidobacteria and lactic acid bacteria. Isomalto-oligosaccharides are composed of glucose monomers linked by $\alpha(1-6)$ glucosidic bonds. They are enzymatically synthesised from starch and other oligomers with $\alpha(1-6)$ glucosidic bonds.

Isomalto-oligosaccharides are known to have the ability to maintain lactic acid bacteria while, at the same time, supporting the generation of butyrate, an attribute which is highly desirable for an effective prebiotic agent (Kohmoto *et al.* 1991; Hsiao-Ling *et al.* 2001).

19.3.4 Inulin fibre

Inulin or inulin fibre is a class of oligosaccharides comprising short chains of sugar molecules which are indigestible in the human gut. In nature, inulin can be found in some edible plants including many vegetables such as chicory tubers and wild yams. It is an effective prebiotic because of the fact that it is not digested by the digestive enzymes in the stomach and in the small intestines; it is therefore present as a good energy source and food for the microbial

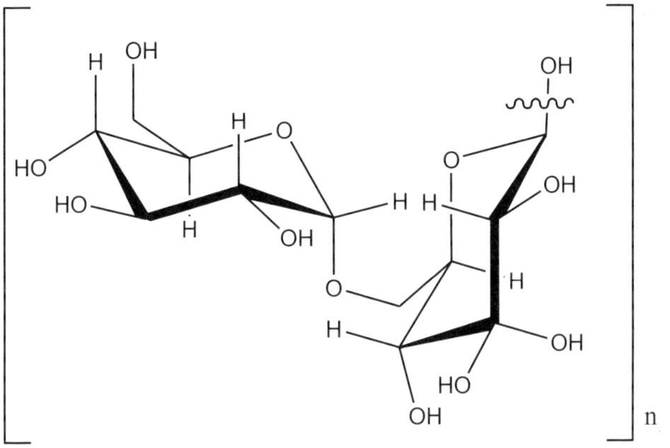

Fig. 19.2 The general structural formula for isomaltose (Kohmoto *et al.* 1991; Hsiao-Ling *et al.* 2001)

flora in the gut. Inulin therefore has the positive effect of catalysing the multiplication of beneficial microbes such as bifidobacteria in the gut (Gibson *et al.* 1995; Kleeson *et al.* 1997; Roberfroid *et al.* 1998; Jenkins *et al.* 1999).

Inulin is found in smaller amounts in garlic, onion, bananas, wheat, oats, soybean and rye. Large quantities of these foods would have to be consumed before any noticeable benefits. Fortification procedures utilising known prebiotics in foods, especially fermentable foods such as dairy products, is therefore necessary.

19.4 QUALITY ASSESSMENT OF PREBIOTICS

For any material to be a suitable prebiotic agent, it must demonstrate certain quality attributes. These features include the property of non-digestibility, in the sense that it has the necessary resistance to the action of gastric acidity, resistance to hydrolysis by digestive system enzymes and resistance to gastro-intestinal absorption phenomena. An ideal prebiotic agent must also be capable of undergoing fermentation by the gut microbial flora and also induce selective stimulation of growth and/or activity of intestinal microbes.

Methods normally used to assess the quality of prebiotics includes the use of the quantitative score test known as prebiotic index (PI), which assesses whether a particular food component is a prebiotic material. PI is generally defined as the increase in bifidobacteria (given as a measure of the absolute number of new colony-forming units in cfu/g of faeces) divided by the daily dose (in grams) of prebiotic ingested, that is:

$$PI = \frac{Bif - Bac + Lac - Clos}{Total}$$

where Bif represents the number of bifidobacteria; Bac refers to the number of bacteria; Lac is the number of lactobacilli; Clos refers to the number of clostridia; and Tot is the total bacterial number (all at sample time divided by the number at inoculation). The higher the number of microbes at the end, the better the prebiotic agent (and vice versa).

19.5 CONCLUSIONS

Knowledge of indigenous microbial flora in the gut and intestines has proved to be essential for issues related to health and nutrition. This can lead to improvements in methods for the modulation of microflora through diets incorporated with prebiotic agents, providing health benefits. This technology however needs to be supported by establishment of the actual health benefits associated with prebiotic intake. This calls for more research in the area of prebiotics, especially into the mechanisms by which prebiotics exert their health benefits.

REFERENCES

Cherbut, C., Michel, C. & Lecannu, G. (2003) The prebiotic characteristics of fructooligosaccharides are necessary for reduction of TNBS-induced colitis in rats. *Journal of Nutrition* 133, 21–27.

De Leenheer, L. (1994) Production and use of inulin: industrial reality with a promising future. In: Carbohydrates as Organic Raw Materials, *III*, van Bekkum, H., Roper, H. & Voragen, A. G. J. (eds), VCH, Weinheim.

Gibson, G. R. & Roberfroid, M. B. (1995) Dietary modulation of the human colonic microbiota: Introducing the concept of prebiotics. *Journal of Nutrition* 125, 1401–1412.

Gibson, G. R., Probert, H. M., Van Loo, J., Rastall, R. A. & Roberfroid, M. B. (2004) Dietary modulation of the human colonic microbiota: Updating the concept of prebiotics. *Nutrition Research Reviews* 17, 259–275.

Gibson, G. R., Beatty, E. R. , Wang, X. & Cummings, J. H. (1995) Selective stimulation of bifidobacteria in the human colon by oligofructose and inulin. *Gastroenterology* 108, 975–982.

Gibson, G. R., Ottaway, P. B. & Rastall, R. A. (2000) *Prebiotics: New Developments in Functional Foods.* Chandos Publishing Limited, Oxford.

Hsiao-Ling, C., Yu-Ho, L., Jiun-Jr, L & Lie-Yon, K. (2001) Effectes of isomaltooligosaccharides on bowel functions and indicators of nutritional status in constipated elderly men. *Journal of American College of Nutrition* 20 (1), 44–49.

Jaskari, J., Kontula, P., Siitonen, A., Jousimies-Somer, H., Mattila-Sandholm, T. & Poutanen, K. (1998) Oat β-glucan and xylan hydrolysates as selective substrates for Bifidobacterium and Lactobacillus strains. *Applied Microbiology and Biotechnology* 49, 175–181.

Jenkins, D. J., Kendall, C. W. & Vuksan, V. (1999) Inulin, oligofructose and intestinal function. *Journal of Nutrition* 129, 1431S–1433S.

Kleessen, B., Sykura, B., Zunft, H. J. & Blaut, M. (1997) Effects of inulin and lactose on fecal microflora, microbial activity and bowel habit in elderly constipated persons. *American Journal of Clinical Nutrition* 65, 1397–1402.

Kohmoto, T., Fujui, F., Takaku, H. & Mitsuoka, T. (1991) Dose-response test of isomaltooligosaccharides for increasing fecal Bifidobacteria. *Agricultural and Biological Chemistry* 55, 2157–2159.

Okazaki, M., Fujikawa, S. & Matsumoto, N. (1990) Effects of xylooligosaccharide on growth of bifidobacteria. *J.* Japanese Society of Nutrition and Food Sciences 43, 395–401.

Roberfroid, M. B., Jan A. E., Loo, V. & Gibson, G. R. (1998) The bifidogenic nature of chicory inulin and its hydrolysis products. *Journal of Nutrition* 128, 11–19.

Teitelbaum, J. E. & Walker, W. A. (2002) Nutritional impact of pre- and probiotics as protective gastrointestinal organisms. *Annual Review of Nutrition* 22, 107–138.

van den Heuvel, E. G. H. M., Muys, T., van Dokkum, W. & Schaafsma, G. (1999) Oligofructose stimulates calcium absorption in adolescents. *American Journal of Clinical Nutrition* 69 (3), 544–548.

Wichienchot, S., Prasertsan, P., Hongpattarakere, T., Gibson, G. R. & Rastall, (2006) In vitro three-stage continuous fermentation of gluco-oligosaccharides produced by *Gluconobacter oxydans* NCIMB 4943 by the human colonic microflora. *Current Issues in Intestinal Microbiology* 7, 13–18.

Zafar, T. A., Weaver, C. M., Zhao, Y., Martin, B. R. & Wastney, M. E. (2004) Non-digestible oligosaccharides increase calcium absorption and suppress bone resorption in ovariectomized rats. *Journal of Nutrition* 134 (2), 399–402.

20 Synbiotics

Abstract: Probiotics and prebiotics have been discussed in Chapters 18 and 19, respectively. In this chapter, a combination of these used in foods to enhance health benefits, referred to as synbiotics, is discussed. Currently, a number of probiotic strains have been designed or developed with particular characteristics for specific health benefits. It is a plausible idea to target prebiotics at these probiotic strains to enhance the health benefit outcome. This is possible from the fact that prebiotics capitalise from the use of non-dietary ingredients or indigestible substrates to improve gut health. The range of foods in which prebiotics can be incorporated outnumbers those in which probiotics can be added, thus providing further advantages in nutrition.

Keywords: antibiotic agents; antimicrial agents; dietary symbiotic formulation; indigestible substrates; probiotics–prebiotics synergy

20.1 SYNBIOTIC FOODS AND HEALTH

Further Thinking

Prebiotics provide the necessary food and energy source for probiotics (live microbes), enabling them to acquire important properties such as greater intolerance of oxygen, low pH, and temperature. Synbiotics are essential in nutrition because without prebiotics, probiotics may not survive and multiply in the colon.

Synbiotics are functional foods that combine both probiotics and prebiotics (Gibson and Roberfroid 1995). Although probiotics are technically not stimulated by prebiotics, because probiotics are mainly active in the small intestines and prebiotics are active in the colon a combination of the two has enhanced health benefits to the host (Gibson and Roberfroid 1995). However, the synergetic effects of synbiotics are only realised in cases where prebiotic components selectively favour probiotic strains; otherwise, there is no synergy that can be ascribed to the presence of both pre- and probiotics at the same time.

Synbiotics play the important roles of boosting the immune response of the host and also improving the survival of the probiotic microbes crossing the upper part of the gastrointestinal tract (hence enhancing their effect in the colon, because their specific substrate is now readily

---

Chemistry of Food Additives and Preservatives, First Edition. Titus A. M. Msagati.
© 2013 John Wiley & Sons, Ltd. Published 2013 by John Wiley & Sons, Ltd.

available). The fermentation process provides health benefits to the host, to which both the probiotics and prebiotics contribute. Another major benefit of synbiotics is that they result in an overall increased positive effect with regards to the persistence of the probiotic microbes in the gastrointestinal tract.

Further Thinking

To promote the action of probiotic microbes over other harmful bacteria competing for a limited food source, prebiotics must be incorporated and function together with probiotics to offer a reliable health benefit to the host. Apart from supporting microbial growth in the colon, prebiotics also provide the necessary attachment platform to support an increased growth rate of beneficial microbes, automatically discouraging the growth of harmful microbes.

Examples of synbiotics (probiotic + prebiotic): bifidobacteria + fructo-oligosaccharide; lactobacilli + lactitol; bifidobacteria + gluco-oligosaccharide; llactobacillus + inulins; and bifidobacteria + inulin.

20.2   HEALTH BENEFITS OF SYNBIOTICS

In order to realise the health benefits of synbiotics, it is imperative that a particular probiotic is appropriately matched with a prebiotic in order to target specific effects. Since synbiotics are composed of two parts, the health-related benefits will be realised from the contribution of each part independently rather than the synergetic contribution. For example, the antimicrobial property of synbiotics comprises independent contributions from both probiotics and prebiotics.

20.2.1   Disease prevention

Probiotics play the important role of boosting the immune responses while the prebiotics provide a platform for the attachment of probiotic microbes along the walls of the gut, at the sites where harmful pathogenic microbes would have attached. For example, bacterial strains which cause conditions such as diarrhoea tend to attach themselves on the walls of the gut linings; prebiotics may dislodge them and prevent the occurrence of diarrhoea. In a similar manner, synbiotics are capable of inhibiting many other harmful processes in the gut ecosystem.

20.2.2   Anti-tumour activity

Synbiotics also demonstrate anti-tumour activity. This is possible since prebiotic agents, which are mostly oligosaccharides (sugars), support the elevation of the levels of calcium and magnesium in the colon. Since these are essential elements needed by microbes in the colon, they play an important role in the probiotic microbial cell multiplication. However, some bacterial strains in the gut may convert magnesium and calcium into harmful by-products such as insoluble bile or fatty acid salts. The probiotic microbes in the gut have the ability to bind and neutralise some carcinogens, thus inhibiting the growth of some tumours, as well as other bacterial strains that may convert matter into carcinogenic compounds.

Prebiotic oligosaccharides (mainly those with 3-6 sugars) can bind calcium and magnesium in the ilium but later release them in the colon where the absorption process is more efficient. This phenomenon may provide health benefits in terms of prevention of diseases such as osteoporosis.

20.2.3 Cancer therapy

It has been reported that the probiotic part of the synbiotics only initiates the anti-cancer process and not the promotion of carcinogenesis events; the prebiotic component however plays an active role in the prevention of the development of the growth of the tumour (Burns and Rowland 2000). Some research findings have shown that synbiotics generate better results at the onset of cancer rather than at later stages (Burns and Rowland 2000). Since synbiotics contain both probiotics and prebiotics, they are both present during the onset of the tumourigenesis, even to the subsequent stages where the tumour is halted or prevented.

Synbiotics therefore functions as both preventive (prophylaxis) as well as treatment agents (therapeutic) for cancer (Fotiadis *et al.* 2008). These findings indicate that a mixture of probiotics and prebiotics can be the best weapon to fight cancer diseases (Burns and Rowland 2000). The presence of probiotics and prebiotics is also important for the growth of microbes indigenous in the gut as well the probiotic strain themselves, because of the symbiotic relationship exists (Liong 2008).

20.2.4 Boosting of immune system

Synbiotics modulate the functioning of the immune system since probiotic microbes are an important factor in boosting the levels of immunoglobulin A (IgA) as well as that of the white blood phagocytic cells. They may also assist in cases of allergy due to the fact that probiotic microbes may occupy spaces in the gut walls, eliminating the possibility of antigens responsible for allergies to attach themselves.

20.3 MECHANISM OF ACTION OF SYNBIOTICS

Synbiotic foods are becoming increasingly popular. The performance of synbiotics is always highly dependent upon the quality of probiotics in the sense that the chosen probiotic microbes must be capable of surviving the intestinal environments (e.g. pH, enzymatic action) and they must also be able to attach themselves to the gut walls. In addition to these factors, the selection of prebiotics which are compatible with the chosen probiotic microorganisms is also crucial; a mismatch will result in the failure of the synbiotic agent.

The mode of action of synbitoics involves synergy of the mechanisms by which probiotics and prebiotics act (Liong 2008). When a matching pair of probiotic and prebiotic species microbes is utilised, the following synergetic modes of action are possible:

- A prebiotic component in a given symbiotic formulation stimulates the corresponding probiotic, which will in turn modify the metabolic activities of gut microbial flora and hence provide health benefits to the host.
- A prebiotic will modify the gene expression, eliminating harmful pathogenic microbes in the process. Although there exists the possibility of carcinogenic by-products, the probiotic component of the synbiotic will bind and degrade the carcinogenic compounds formed.
- Prebiotics increase the levels of micronutrients such as calcium and magnesium in the large intestines while the probiotic alters the physicochemical environment in the large intestines; the combined result is health benefits to the host.
- When nutrients are available, probiotic microbes produce short-chain fatty acids (e.g. acetic acid, propionic acid and butyric acid) which will create pH conditions in which pathogens cannot grow, thus eliminating them.

20.4   THE FUTURE OF SYNBOTIC FOODS

The research into synbiotics is gaining momentum; current problem regarding antibacterial resistance is their major limitation (Burns and Rowland 2000). On the other hand, synbiotics are showing very promising results in terms of treating and/or preventing diseases such as growth of tumours (Fotiadis *et al.* 2008). Synbiotics have also been reported to stimulate immunity and prevent infection by other harmful microorganisms; they may therefore be very useful and low-cost therapies against many disease conditions. There are many other areas where synbiotics are considered to play a positive role, including infectious inflammatory diseases and some types of allergic conditions (Geier *et al.* 2007).

Despite the progress in research activities with regards to synbiotics, which has so far been performed on animals, very little if any research has been conducted on human being; published findings and observations could be considered as being of little use. Further synbiotics studies need to involve human being as biomarkers for diseases such as cancer.

REFERENCES

Burns, A. J. & Rowland, I. R. (2000) Anti-carcinogenicity of probiotics and prebiotics. *Current Issues in Intestinal Microbiology* 1 (1), 13–24.

Fotiadis, C. I., Stoidis, C. N., Spyropoulos, B. G. & Zografos, E. D. (2008) Role of probiotics, prebiotics and synbiotics in chemoprevention for colorectal cancer. *World Journal of Gastroenterology* 14 (42), 6453–6457.

Geier, M. S., Butler, R. N. & Howarth, G. S. (2007) Inflammatory bowel disease: Current insights into pathogenesis and new therapeutic options; probiotics, prebiotics and synbiotics. *International Journal of Food Microbiology* 115, 1–11.

Gibson, G. R. & Roberfroid, M. (1995) Dietary modulation of the human colonic microbiota: introducing the concept of prebiotics. *Journal of Nutrition* 125, 1401–1412.

Liong, M. T. (2008) Roles of probiotics and prebiotics in colon cancer prevention: postulated mechanisms and in-vivo evidence. *International Journal of Molecular Sciences* 9, 854–863.

21 Microencapsulation and Bioencapsulation

Abstract: Some products of the food industry undergo changes due to physico-chemical processes if the food items are not protected or are stored for some time before being consumed. In addition, there is a growing demand from consumers for processed foods to maintain their bioactivity properties until introduced to the digestive system. Microencapsualtion and bioencapsulation are techniques currently in use to address these concerns. Unlike the other chapters in this book which deal with food components or ingredients, microcapsules and biocapsules refers to products of processes in which food components are coated (encapsulated) by inert shells to protect them from the external environment; such products are also used in the controlled release of particular food components. In this chapter, we discuss various microencapsulation and bioencapsulation techniques suitable for various molecules and biological species.

Keywords: bioencapsulation; food coating; food packaging; microencapsulation

21.1 INTRODUCTION TO MICROENCAPSULATION AND BIOENCAPSULATION

Microencapsulation is a technique used for the packaging of different types and forms of products such as solids, liquids and even gaseous substances in very small sealed coatings (capsules), which allow the release the encapsulated materials at controlled rates and under specific sets of conditions (Anal and Stevens 2005; Kailasapathy and Masondole 2005; Anal *et al*. 2006). From this definition, it follows that a microcapsule is composed of a semi-permeable or strong membrane surrounding a core. It can vary in terms of shape, configuration and even diameter; generally the diameter can range from several micrometres to at least 1 millimetre (Figures 21.1a, b and 21.2).

Microencapsulation finds application in the food as well as in biopharmaceutical industry (Smidsrød and Skjak-Braek 1990). The technique is attractive because of its ability to entrap valuable ingredients, referred to as the core material, internal phase or fill. The protective polymer material which encapsulates the active ingredients is known as the encapsulating material, wall material, shell, coating or membrane (Hogan *et al*. 2001; Figure 21.2). The entrapped ingredients are released when and where needed without any compromise regarding their integrity. When at the target, a microcapsule can be opened to release the core in various ways, for example, the use of heat to fracture the shell, diffusion, solvation and also pressure. In general, encapsulation finds wide application in the food industry such as in stabilising the core, preventing oxidation within the core, masking of colour, flavour or

Chemistry of Food Additives and Preservatives, First Edition. Titus A. M. Msagati.
© 2013 John Wiley & Sons, Ltd. Published 2013 by John Wiley & Sons, Ltd.

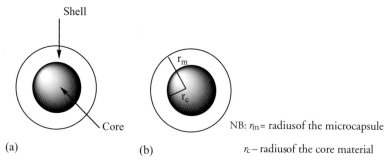

Fig. 21.1 Schematic representation of (a) core surrounded by the shell and (b) cross-section of a micro-capsule (Anal and Stevens 2005; Kailasapathy and Masondole 2005; Anal et al. 2006).

odour, and preventing the loss of any nutritive ingredients through various physico-chemical changes.

The principle behind microencapsulation is based on the creation of the microcavity in the capsule by a polymer material used for encapsulation. The formed microcavity governs and controls the mechanisms involving the interactions between the core and the environment surrounding the core. Microencapsulation is needed in food industries to incorporate flavours and colours as well as additives such as antioxidants, nutraceuticals, probiotics, prebiotics and synbiotics. The process of microencapsulation takes into consideration a number of factors such as the stability of the bioactive ingredients and aims to discourage any adverse interactions within the shell during the processing or storage, until the active ingredient is released within the consumer's gut. Core materials are therefore protected from any detrimental conditions that would result in the deterioration of its integrity.

Due to differences in terms of the shape, structures and configuration of core materials, microcapsules are designed with different shapes. Variations include encapsulating multicores, multiwalls, smooth and regular shapes and irregular shapes (Figure 21.2).

Generally, food-grade microcapsules are designed to either be permeable (especially when intended to be used for controlled release); semi-permeable (where they are selectively

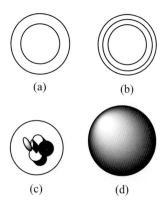

Fig. 21.2 Configurations and shapes of microcapsules: (a) simple microcapsule; (b) multiwall capsule; (c) multicore capsule; and (d) matrix micocapsule (Anal and Stevens 2005; Kailasapathy and Masondole 2005; Anal et al. 2006).

permeable to some specific molecules and effectively impermeable to the external environment); or impermeable (where rupturing of the shell is required to release the core).

On the other hand, bioencapsulation deals mainly with the encapsulation of cells, tissues and active biomolecules using semi-permeable membranes. Biomaterials are safely enclosed and protected from any external factors that may result in potential damage (de Vos *et al.* 2009). As for microencapsulation, bioencapsulation finds application in the food industry (probiotics and nutraceuticals), pharmaceutical industry and in biotechnology.

21.2 COMMONLY USED FOOD-GRADE MICROCAPSULES

The attributes normally needed for materials suitable for food microcapsules are mainly low viscosity in the presence of high contents of solids and good solubility. Moreover, materials used for capsules should form films which are cohesive with the polymer materials, be inert in the sense that they should not react chemically with the core material and be compatible with the core material. The core materials must have properties appropriate for the desired use such as physical and mechanical strength, impermeability, stability to chemical attack (e.g. stability to the action of enzymes), flexibility and optical qualities. For these reasons, materials such as alginates, chitosan, gums (acacia gums, xanthan-gelan blends, gelatin, carrageenans), starches, cellulose acetate phthalate, maltodextrins and also corn syrup (Sections 21.2.1–21.2.7) have been widely reported as effective microcapsules (Anandaraman and Reineccius 1986; Bangs and Reineccius 1988; Thevenet 1988; Kenyon 1995).

Materials used in the synthesis of coating material can be classified as one of four different groups: (1) water insolubles (e.g. cellulosic materials such as ethylcellulose, cellulose nitrate, polymethacrylate and silicones); (2) water solubles (starch, caroboxymethylcellulose, hydroxyethylcellulose and gumsgelatine); (3) enteric resins (e.g. cellulose acetate phthalate); and (4) waxes and lipids (e.g. beeswax, paraffin, stearic acid, glyceryl stearates and stearyl alcohol).

As well as being used as encapsulating agents in their original form, the above-listed materials can undergo surface modification in order to impart some specific surface-active function (Moreau and Rosenberg 1993, 1996; Young *et al.* a, b; Kim and Morr 1996). Another reason for modification is that a single type of encapsulating material rarely possesses all the required attributes needed, necessitating the blending of more than one material (e.g. a carbohydrate material and a proteinous material; Bangs and Reineccius 1988; Sankarikutty *et al.* 1988; Trubiano and Lacourse 1988; Bhandari *et al.* 1992; Moreau and Rosenberg 1993; Young *et al.* a, b; Faldt and Bergenstahl 1994; Sheu and Rosenberg 1995; Dian *et al.* 1996). The chemistry of the most commonly used microcapsules in the food industry is discussed in the following sections.

21.2.1 Alginates and alginate derivatives

Alginate is a naturally occurring linear polysaccharide. It is a polymer made of monomers of β-D-mannuronic and α-L-guluronic acids linked by a 1,4-glycosidic bond (Figures 21.3a–c; Ault *et al.* 1935).

For food applications, alginate derivatives made by cross-linking with divalent cations such as calcium in the structures are mostly used. These derivatives (such as calcium alginates)

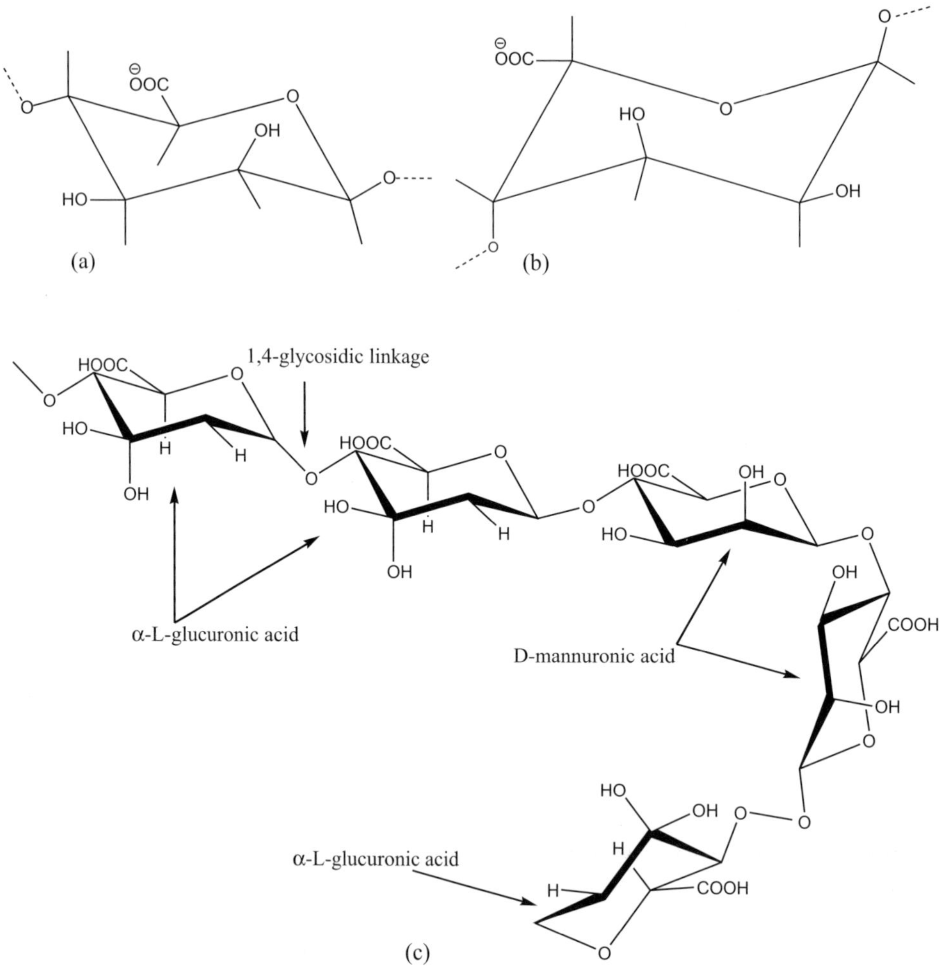

Fig. 21.3 Chemical structure of (a) D-mannuronic acid and (b) L-guluronic acid; (c) arrangements of glucuronic and mannuronic acid monomers in alginates (Ault *et al.* 1935).

have been reported in a number of applications, such as for the encapsulation of probiotic microbes and other organic acids (Steenson *et al.* 1987; Shah and Ralura 2000; Sultana *et al.* 2000; Truelstrup-Hansen *et al.* 2002).

Alginates are attractive as microcapsules due to their safety record: that they are not toxic, are easily available and can easily be fabricated and handled (Dimantov *et al.* 2003; Chandramouli *et al.* 2004; Gouin 2004). However, alginate microcapsules are also known to have some drawbacks due to their instability in acidic media and their mechanical instability especially when used in lactic acid media (Mortazavian *et al.* 2007). Moreover, where there is a demand for industrial application, alginates always prove to be expensive; they also tend to form cracked surfaces as well as porous beads, undesirable at a large-scale production (Gouin 2004).

To counter these shortcomings, alginate capsules are blended with other polymeric materials (Krasaekoopt *et al.* 2003). Polymeric materials blended with alginates include starch,

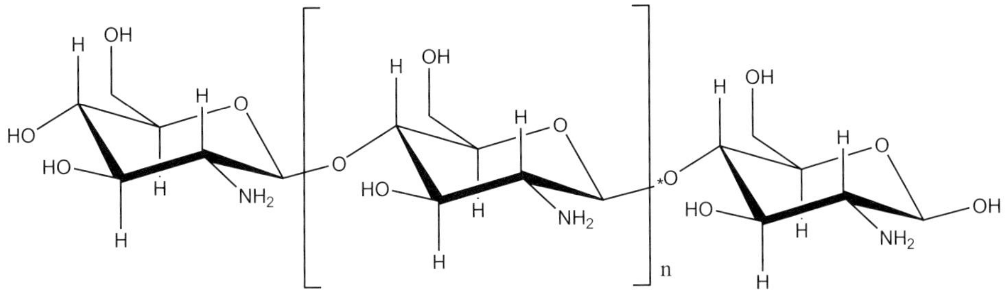

Fig. 21.4   Structure of chitosan, demonstrating how the units are linked together (Klien *et al.* 1983).

glycerol, poly-amino acids (e.g. poly-*L*-lysine) and chitosan. It should be noted that in the process of coating chitosan on alginate capsules, calcium chloride is first added to chitosan to supply Ca^{2+} ions, important in enhancing the chitosan–alginate coating mechanisms (Krasaekoopt *et al.* 2003).

Further Thinking

The rule of the selection protocol that applies for wall materials used for microcapsules stipulates that the oil-soluble core polymers are compatible with wall materials that are water-soluble. Likewise, core polymers which are water-soluble must be matched with wall materials that are oil-soluble.

21.2.2   Chitosan

Chitosan is a linear polymer of D-glucosamine and N-acetyl-D-glucosamine (acetylated unit) saccharide units connected by β-(1-4) links (Figure 21.4). The D-glucosamine in chitosan exists in the deacetylated form while the N-acetyl-D-glucosamine exists in the acetylated form. Chitosan is known to undergo ionotropic gelation, which gives it a gel structure in the same way as for gelatin. It is also a negatively charged linear polysaccharide which is soluble at acidic pH ranges (especially for pH $<$ 6) and it has the ability to increase the polymer chain (polymerise) by cross-linking processes catalysed by some specific monoanions and polyanions (Klien *et al.* 1983). As well as being used in the coating of alginates (Zhou *et al.* 1998), chitosan has been blended with other polymers such as hexamethylene diisocyanate and glutaraldehyde, and the blends have been reported to be of superior quality as compared to chitosan capsule alone (Groboillot *et al.* 1993).

21.2.3   Xanthan–gelan blends

There exists an optimal ratio of xantan–gelan blends (Figures 21.5a and b) needed for it to be suitable for use as food-grade capsule, dependent upon its use. For example, Sun and Griffiths (2000) reported that the ratio of xanthan: gelan in the blend for optimal use in probiotics or prebiotics should be 1:0.75. The main advantages of the xanthan–gelan gum

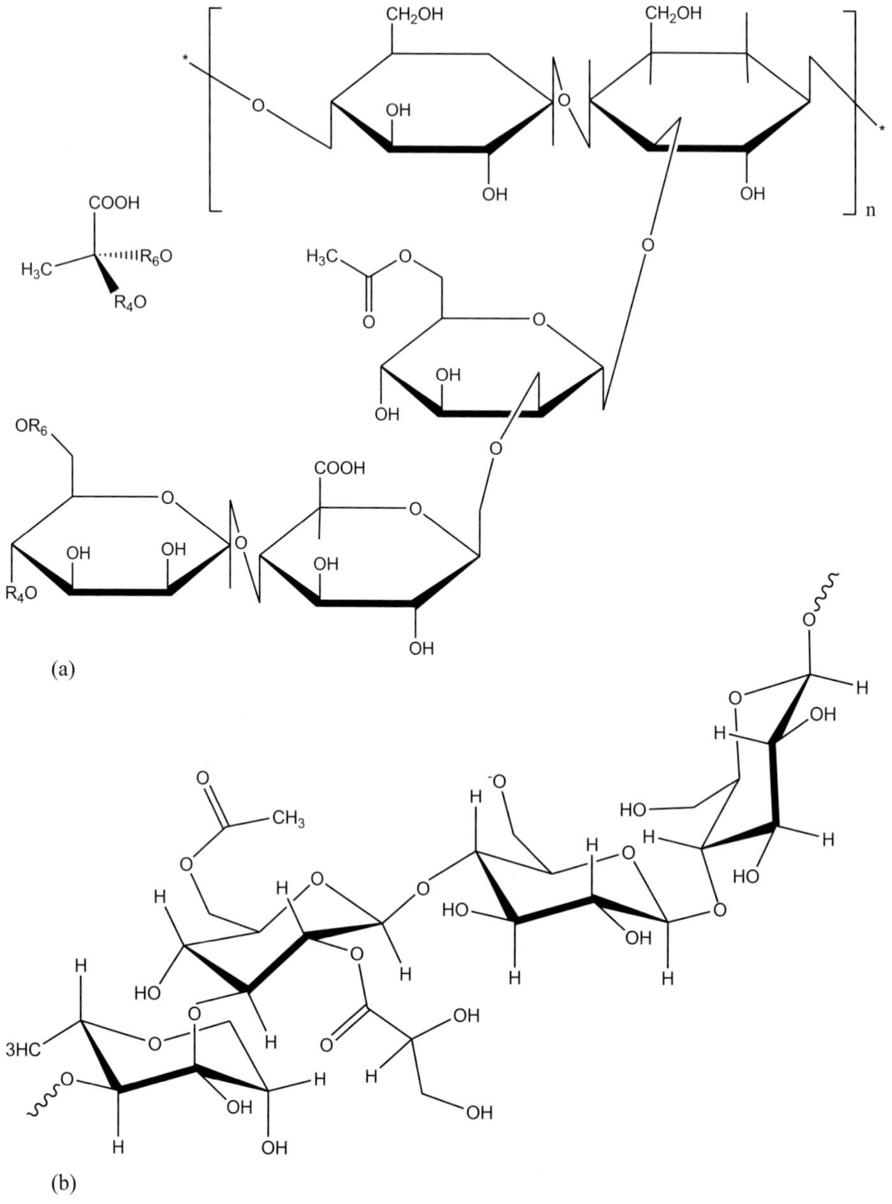

(a)

(b)

Fig. 21.5 Chemical structure of (a) xanthan and (b) gelan (Sun and Griffiths 2000).

blend is that it does not decompose even when the pH of the matrix is low. The gum blend is normally stabilised with calcium ions, retaining its safety status (Sanderson 1990).

21.2.4 Starch materials

Starch (amylose and amylopectin; Figure 21.6a and b; David and Cox 2008) is used as capsules in a diverse way. For example, it is used for the coating of alginate capsules; to

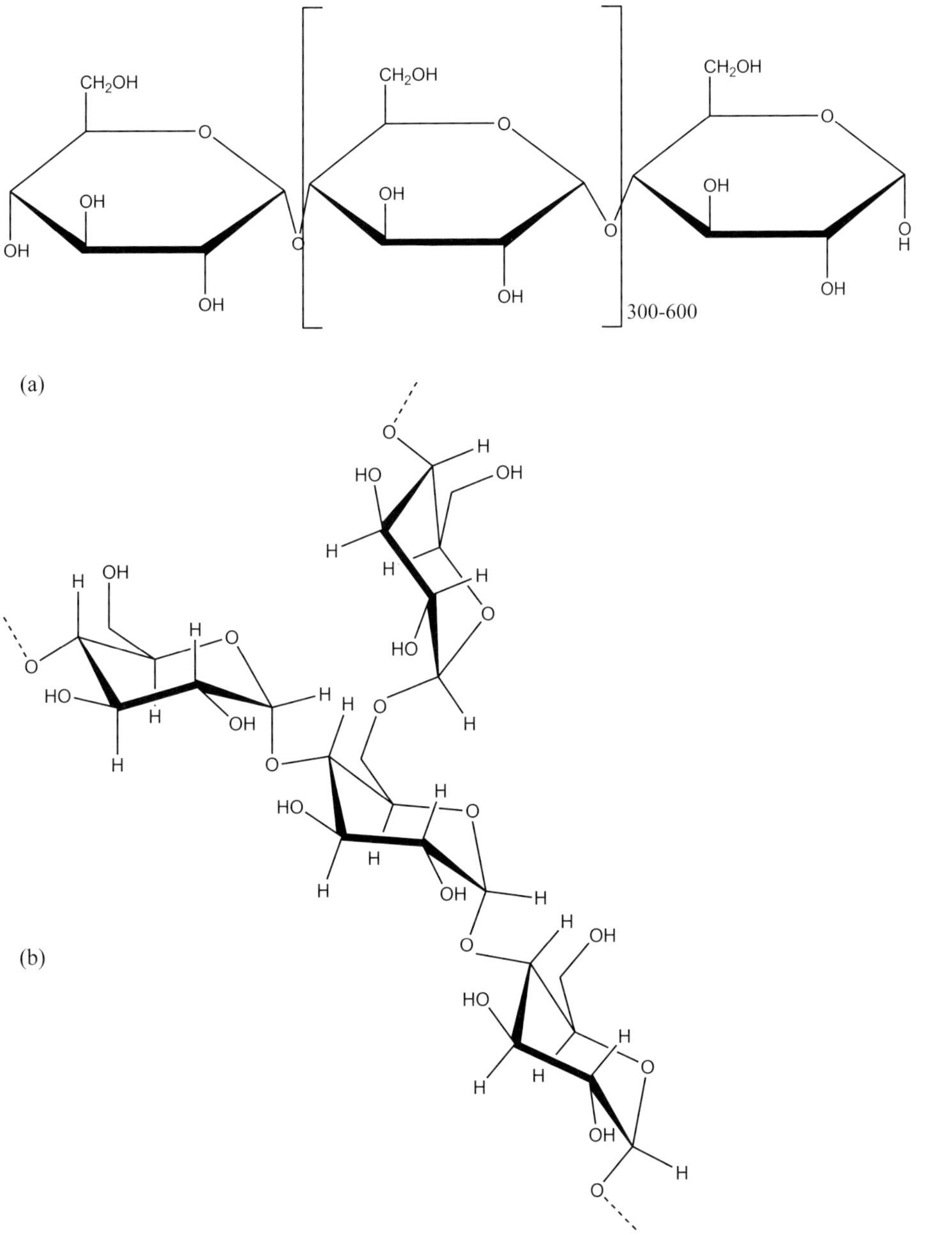

Fig. 21.6 Chemical structure of (a) amylose and (b) amylopectin (David and Cox 2008).

boost the functions of coat formation (Dimantov 2003) when used in the form of high-amylose corm starch; or, when corn starch has been lyophilised, it can be applied as a coat-forming raw material. Use of lyophilised corn starch is limited however, due to the fact that it is unstable in the presence of pancreatic amylase enzymes (present in the digestive system; Fanta *et al.* 2001). To counter this shortcoming, a resistant form of starch is used instead.

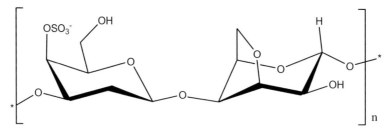

Fig. 21.7 Chemical structure of kappa carrageenan.

21.2.5 Kappa-carrageenan and other carrageenan derivatives

Kappa-carrageenan, a neural polysaccharide polymer (Figure 21.7), is known to dissolve at high temperature; monovalent cations and anions such as K^+Cl^- (potassium chloride) are normally used to stabilise it (Krasaekoopt *et al.* 2003). In cases where kappa-carrageenan is used for probiotics, the use of K^+Cl^- has been reported to be detrimental to microbes such as lactic acid bacteria (normally used in dairy products as it inhibits the activity of these microbes; Audet *et al.* 1988). For this reason, other monovalent ions such as Cs^+Cl^-, Rb^+Cl^- or $NH_4^+Cl^-$ have been used instead, which do not result in any of the inhibitory effects observed with K^+Cl^-. The performance of kappa-carrageenan as an efficient food capsule is improved when blended with a galactomannan vegetable gum known as either locust bean gum (LBG) or carobin at ratios of 1:2 (kappa-carrageenan:locust); this produces a capsule with notable stability in the presence of food acids (Miles *et al.* 1984; Audet *et al.* 1988).

21.2.6 Gelatin and its blends

Gelatin is a complex substance formed by peptides and proteins which are produced during the partial hydrolysis of collagen. This material, which is actually a proteinous gum, has a number of attractive features which makes it suitable for use as food microcapsule. These features include its ability to form thermo-reversible gels and it is amphoteric, thus a suitable candidate for blending with anionic polysaccharide gum polymers. These blends can provide additive effects (Hyndman *et al.* 1993). Other known gelatin blends that are used as food capsules include the gelain-toluene diisocyanate blend, which forms good capsules in terms of their resistance to cracking and breaking due to an efficient cross-link mechanism between the two polymer gums. A blend of gelatin gum and gum arabic has also demonstrated good performance, especially in the production of capsules suitable for application with soybean oils (Truelstrup-Hansen *et al.* 2002).

21.2.7 Cellulose acetate phthalate

The presence of phthalate moiety makes the cellulose acetate phthalate negatively charged. One of the striking features of this polymer (Figure 21.8) capsule is that it has great solubility at pH 6 but not at pH 5 (Malm *et al.* 1951). Capsules of this polymer have been known to be very safe, thus have a wide application not only in the food industry but also for pharmaceutical products (Krasaekoopt *et al.* 2003). Coating wax on cellulose acetate phthalate has also been reported to add quality (Rao *et al.* 1989).

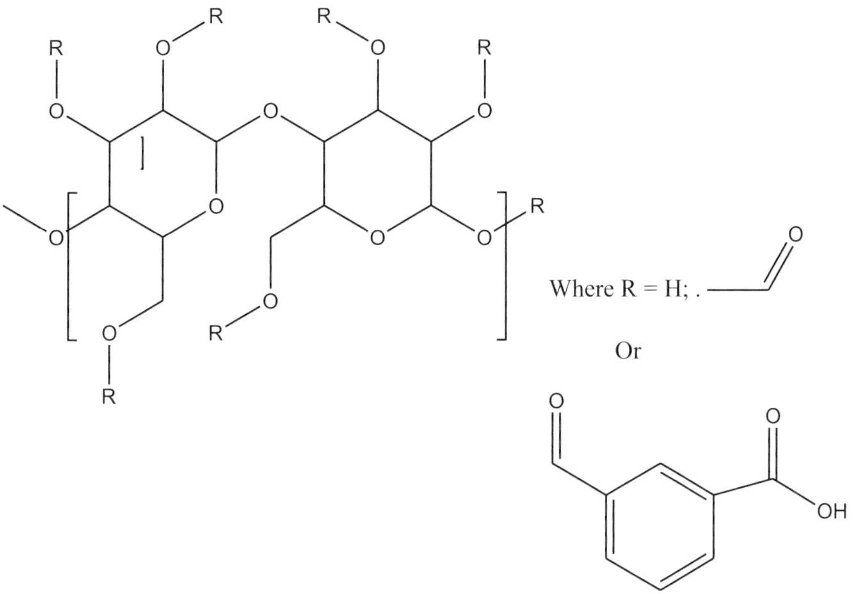

Fig. 21.8 Chemical structure of cellulose acetate phthalate.

21.3 METHODS OF FOOD MICROENCAPSULATION

There are several techniques that are normally used for encapsulation in food as well as other areas (Ranney 1969; Benita 1996; Arshady 1999). Each of these techniques has particular benefits and attributes, making them suitable for application to systems with specific qualities in terms of (1) physico-chemical, (2) chemical and (3) physico-mechanical properties (Ranney 1969; Benita 1996; Arshady 1999). Some techniques from each of these three categories, especially those which are widely used, are discussed in the following sections.

Further Thinking

The choice of the best encapsulation technique is governed by factors such as the permeability of the coating wall, the type of polymer or core material or the size of the required particle. These priorities will guide the choice and tailoring of the preferable process which will result in the best capsule.

21.3.1 Physico-chemical

From Scheme 21.1 which shows the various the classifications of the encapsulation techniques, it is clear that the various classes are discriminated from each by the underlying principles that governs each of the processes in various groups. Even within the same classes, there may be different principles for each subdivision (Ranney 1969; Benita 1996; Arshady

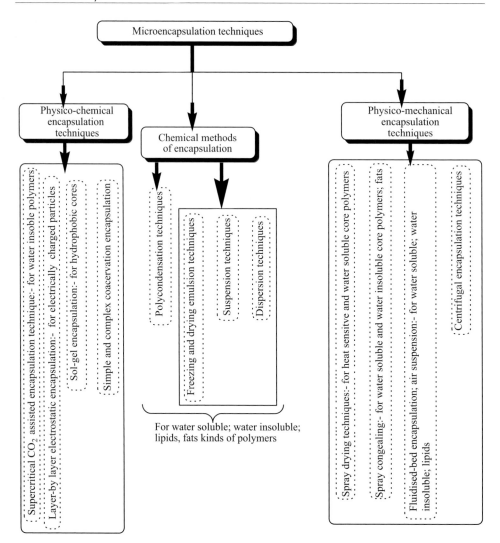

Scheme 21.1 Microencapsulation techniques (Ranney 1969; Benita 1996; Arshady 1999).

1999). There are a number of subdivion which fall under physico-chemical encapsulation techniques each of these subgroups will be discussed separately.

21.3.1.1 *Supercritical fluid-assisted*

Depending on the supercritical fluid used, these techniques can be suitable for both water-soluble and water-insoluble core polymers. By definition, supercritical fluids possess properties intermediate of liquids and gases. Supercritical fluids such as supercritical CO_2, modified supercritical CO_2, supercritical nitrous oxide, supercritical water and others are generally attractive for use in encapsulation techniques because of their properties. These include solubility in a variety of solutes and compatibility with other gases such as nitrogen or hydrogen. The densities of supercritical fluids change greatly whenever there is a small variation in terms of pressure or temperature, and this provides the possibility of experimenting with properties of supercritical fluids. Supercritical fluids are safe (not toxic), economical, non-flammable and are in many cases available in high-purity form.

Supercritical fluids have been used for encapsulation processes for food products such as vitamins, colouring agents and flavouring agents (Liu *et al.* 2002; Chambon *et al.* 2004). In such applications, the capsules materials must be of the type which do not dissolve biomolecules such as proteins or carbohydrates but which can dissolve hydrocarbons. Supercritical-fluid-assisted encapsulation techniques include rapid expansion of supercritical solution, gas anti-solvent and particles from gas-saturated solution.

21.3.1.2 *Coacervation*

In coacervation, acids (mineral acids such as sulphuric acid, hydrochloric acid or organic acid) are employed to alter the pH of the coating material in order to lower its solubility. This will cause the coating material to precipitate from the solution (e.g. gelatin or gum arabic solution), thus depositing around the core (polymer) particles.

21.3.2 **Chemical**

21.3.2.1 *Polycondensation*

Polycondensation phenomena applied in encapsulation techniques are based on the Schotten–Baumann reaction, whereby either amides are synthesised from the reactions of amines and acid chlorides or esters are obtained from the reactions of acid chlorides and alcohols (Schotten 1884; Baumann 1886; Frantz *et al.* 2002). Schotten–Baumann reactions generally take place in the presence of water and an organic solvent such as dichloromethane or diethyl-ether (Scheme 21.2; Schotten 1884; Baumann 1886; Frantz *et al.* 2002).

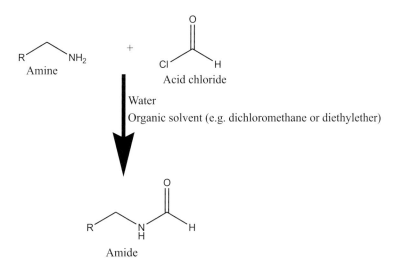

Scheme 21.2   Polycondensation encapsulation reactions based on Schotten–Baumann reaction (Schotten 1884; Baumann 1886; Frantz *et al.* 2002).

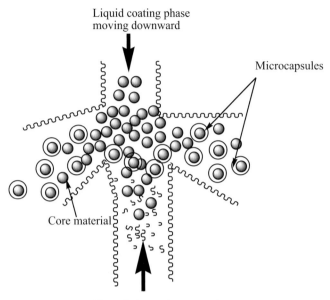

Liquid coating phase
moving downward

Microcapsules

Core material

Heated air moving upward

Scheme 21.3   Formation of coatings by air suspension technique (Ganesan *et al.* 2003; Werner 2005).

21.3.2.2 *Air suspension*

In this technique, emulsions, dispersions, solutions (solvent solutions, aqueous solution) or hot melts are used for encapsulation, thus making it easier to control the optimal coating parameters. The process involves coating particles which are suspended in a perpendicular position with a stream of heated air moving upwards while the solution is moving downwards (Scheme 21.3; Ganesan *et al.* 2003; Werner 2005). The movement of heated air upwards becomes diverted and slows down as it reaches the top, allowing particles to settle on the surface of the core material. This process will repeat for several cycles, thus forming a coat or capsule (Ganesan *et al.* 2003).

21.3.2.3 *Rotating disc suspension*

Another suspension encapsulation technique is known as rotating disc suspension, and operates in a slightly different manner. In this suspension method, hot suspensions of core materials are put into a spinning disc (Scheme 21.4). As a result of rotation and centrifugal forces, the polymer substance (the core) becomes shelled (coated) and the shell is hardened by cooling (Werner 2005). This technique is attractive due to its simplicity, low cost, speed and high efficiency.

21.3.2.4 *Dispersion*

In this technique, the polymer (core) is dispersed in the solution which is being stirred and the particles are deposited on the surface of the core. The characteristics of the coating are determined by the experimental parameters that control the dispersion; these include the viscosity of the solution, the speed of stirring and the surface tension of the solution.

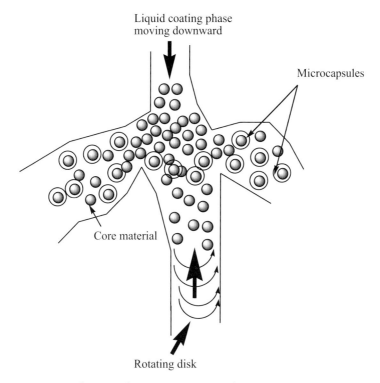

Liquid coating phase
moving downward

Microcapsules

Core material

Rotating disk

Scheme 21.4   Process of rotating disc suspension encapsulation (Werner 2005).

21.3.3 Physico-mechanical

21.3.3.1 *Spray drying*

This technique is mostly suitable for water-soluble core polymers since other types of solvent systems tend to generate unpleasant odours and they may not be very environmentally friendly. The process of spray drying starts by dissolving the active substance in a solution containing the polymer to be used until it is stuck in the polymer solid (Figure 21.9; Jafari *et al.* 2008).

The most attractive feature of spray drying is the brief contact time, which makes it possible for labile materials to be handled properly and efficiently and the process is cheap in terms of operating costs. The spray-drying encapsulation technique finds applications in flavourings, fragrances and oils (Jafari *et al.* 2008).

21.4 MICROENCAPSULATION FOR FOOD COLOURANTS

Due to an increased demand for natural food colourants versus synthetic colourants, which are governed by strict guidelines and legislations, the search for ways to preserve natural dyes for use in foods has intensified (Giusti and Wrolstad 1996). The need for better preserving techniques for food colourants is due to the number of factors that lead to the disappearance or fading of colours, including pH, light, enzymes, air (O_2), concentration, temperature, the presence of other colouring agents or the presence of metallic species (Mazza and Miniati

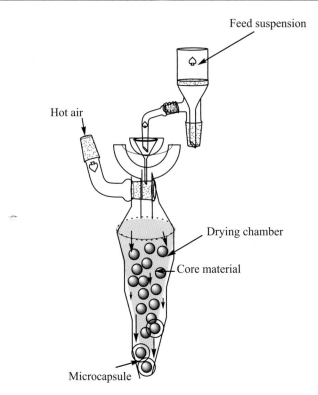

Fig. 21.9 The spray-drying encapsulation technique (Jafari *et al.* 2008).

1993; Rodriguez-Saona *et al.* 1999). To safeguard the integrity and stability of these colouring active ingredients, reliable techniques need to be developed so that colours are preserved and maintained.

Microencapsulation is one of the very promising techniques for the preservation and release of food colours when and where needed. A number of microencapsulation techniques for various food colours have been reported. For example, Ersus and Yurdagel (2007) have reported the use of spray drier as the microencapsulation technique for the anthocyanin pigment extracts from black carrot (*Daucuscarota* L.). Stability of anthocyanin extracts, as determined from the spray-dried microencapsulated powders, were measured after optimising the storage temperature and light. After a period of more than two months, the colour had decreased by a third of the original colour when the storage temperature was 25°C. At 4°C however, the loss was about a tenth of the original colour. With regard to the wall material, three types of maltodextrins (Stardri 10, Glucodry 210 and MDX 29) were used both as carriers and also as coating agents. Results suggested that Glucodry 210 performed better as a wall material than the other maltodextrins used.

A number of other researchers have reported on microencapsulation of food colourants after the extraction process. For example, Ge *et al.* (2009) reported the microencapsulation of red rose pigments after they had extracted them from a hybrid rose. Since wall materials for the extracted red rose pigments are hydrophobic (and thus oil soluble) while the core polymer is hydrophilic (hence water soluble), the microencapsulation for the red rose pigment was performed with the hydrophobic oil/lipid-soluble wall materials, mainly beeswax

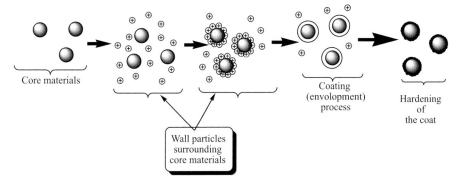

Scheme 21.5   Microencapsulation stages for food colourants (Ersus and Yurdagel 2007; Ge *et al.* 2009).

and/or stearic acid to enable the embedding process of the active ingredient of the pigment. Scheme 21.5 depicts the various stages of microencapsulation for food colourants (Ersus and Yurdagel 2007; Ge *et al.* 2009).

21.5 BIOENCAPSULATION FOR PROBIOTICS

As defined previously in Chapter 18, probiotics are live microbes that are capable of bringing health benefits to their host when administered in sufficient amounts. However, probiotics are only useful for their intended purpose if their viability is well protected from the production stages all the way to storage and administration. For this reason, the search for suitable carriers which can withstand a chemical and enzymatic environment and which are capable of delivering such biomolecules/biomaterials/biospecies is on-going.

A number of naturally occurring polymer materials with excellent compatibility and biodegredability have been used as carriers and as systems to deliver intended food ingredients in a controlled fashion. These polymeric materials include alginates, which offer many advantages such as excellent resilience, controlled delivery and release. The form of alginates used most often in bioencapsulation is calcium alginate gel beads. The major limitation to alginate capsules is that they offer a very limited stability (Krasaekoopt *et al.* 2004; Mandal *et al.* 2006).

On the other hand, protein polymers have been reported to be more attractive than the alginates due to their nutritive value. The presence of a polypeptide functionality also provides the wide possibility for encapsulation as well as reverse binding of other active species before releasing the core to the specified target (Chen *et al.* 2006; Chen and Subirade 2008). Moreover, the encapsulation process used for proteinous materials is performed using enzymatic hydrolysis, thus with the potential to produce bioactive petide compounds that may bring *in vivo* health benefits to the host (Kilara and Panyam 2003; Korhonen and Pihlanto 2003). Protein polymer materials reported as carriers for food ingredients, probiotics and other different molecules in bioencapsulation procedures include: whey protein micro-beads (Doherty *et al.* 2011), casein, collagen, albumin (Rossler *et al.* 1995; Kuijpers *et al.* 2000; Latha *et al.* 2000; Beaulieu *et al.* 2002; Picot and Lacroix 2004).

21.6 CONCLUSIONS

Microencapsulation and bioencapsulation have shown great potential as viable carriers for important food ingredients and their controlled released to the desired *in vivo* targets. With further research in this area, this technology may play an important role in the production of foods in which ingredients with health benefits have been incorporated. In instances where some food ingredients show reactivity towards other food molecules, and produce undesirable by-products which may lead to deterioration of food quality, encapsulation is the technology to protect important food ingredients from such phenomena.

REFERENCES

Anal, A. K. & Stevens, W. F. (2005) Chitosan-alginate multilayer beads for controlled release of ampicillin. *International Journal of Pharmaceutics* 290, 45–54.

Anal, A. K., Stevens, W. F. & Remuñan-Lopez, C. (2006) Ionotropic cross-linked chitosan microspheres for controlled release of ampicillin. *International Journal of Pharmaceutics* 312, 166–173.

Anandaraman, S. & Reineccius, G. A. (1986) Stability of encapsulated orange peel oil. *Food Technology* 40, 88–93.

Arshady, R. (1999) *Microspheres, Microcapsules & Liposomes: Medical & Biotechnology Applications*. Citus Books, London.

Audet, P., Paquin, C. & Lacroix, C. (1988) Immobilized growing lactic acid bacteria with ê-carrageenan-locust bean gum gel. *Applied Microbiology and Biotechnology* 29, 11–18.

Ault, P. G., Haworth, W. N. & Hirst, E. L. (1935) Preparation of d-mannuronic acid and its derivatives. *Journal of Chemical Society* (0), 517–518.

Bangs, W. E. & Reineccius, G. A. (1988) Corn starch derivatives. In: *Flavor Encapsulation*, Risch, S. J. & Reineccius, G. A. (eds), ACS symposium series, American Chemical Society, Washington, DC, volume 370, pp. 12–28.

Baumann, E. (1886) Ueber eine einfache method der darstellung von benzoësäureäthern. *Berichte der deutschen chemischen Gesellschaft* 19 (2), 3218–3222.

Beaulieu, L., Savoie, L., Paquin, P. & Subirade, M. (2002) Elaboration and characterization of whey protein beads by an emulsification/cold gelation process: application for the protection of retinol. *Biomacromolecules* 3, 239–248.

Benita, S. (1996) *Microencapsulation Methods and Industrial Application*. Marcel Dekker, Inc., New York.

Bhandari, B. R., Dumoulin, E. D., Richard, H. M. J., Noleau, I. & Lebert, A. M. (1992) Flavor encapsulation by spray drying: Application to citral and linalyl acetate. *Journal of Food Science* 57 (1), 217–221.

Chambon, P., Cloutet, E. & Cramail, H. (2004) Synthesis of core-shell polyurethane-poly(dimethylsiloxane) particles in supercritical carbon. *Macromolecules* 37, 5856–5859.

Chandramouli, V., Kalasapathy, K., Peiris, P. & Jones, M. (2004) An improved method of microencapsulation and its evaluation to protect *Lactobacillus* spp. in simulated gastric conditions. *Journal of Microbiological Methods* 56, 27–35.

Chen, L. & Subirade, M. (2008) Food-protein-derived materials and their use as carriers and delivery systems for active food components. In: *Delivery and Controlled Release of Bioactives in Foods and Nutraceuticals*, Garti, N. (ed.), Woodhead Publishing Ltd., UK, pp. 251–278.

Chen, L. Y., Remondetto, G. E. & Subirade, M. (2006) Food protein-based materials as nutraceutical delivery systems. *Trends in Food Science and Technology* 17 (5), 272–283.

David, N. & Cox, M. M. (2008) *Principles of Biochemistry*, 5th edition. Freeman and Company, New York, USA.

de Vos, P., Bucko, M., Gemeiner, P., Navratil, M., Svitel, J. & Faas, M. (2009) Multiscale requirements for bioencapsulation in medicine and biotechnology. *Biomaterials* 30 (13), 2559–2570.

Dian, N. L. H. M., Sudin, N. & Yusoff, M. S. A. (1996) Characteristics of microencapsulated palm-based oil as affected by type of wall material. *Journal of the Science of Food and Agriculture* 70, 422–426.

Dimantov, A., Greenberg, M., Kesselman, E. & Shimoni E. (2003) Study of high amylase corn starch as food grade enteric coating in a microcapsule model systems. *Innovations in Food Science, Engineering and Technology* 5, 93–100.

Doherty, S. B., Gee, V. L., Ross, R. P., Stanton, C., Fitzgerald, G. F. & Brodkorb, A. (2011) Development and characterisation of whey protein micro-beads as potential matrices for probiotic protection. *Food Hydrocolloids* 25, 1604–1617.

Ersus, S. & Yurdagel, U. (2007) Microencapsulation of anthocyanin pigments of black carrot (Daucuscarota L.) by spray drier. *Journal of Food Engineering* 80, 805–812.

Faldt, P. & Bergenstahl, B. (1994) The surface composition of spray dried protein-lactose powders. *Colloids and Surfaces A: Physicochemical and Engineering Aspects* 90, 183–190.

Fanta, G. F., Knutson, C. A., Eskins, K. S. & Felker, F. C. (2001) Starch microcapsules for delivery of active agents. US patent 6,238,677.

Frantz, D. E., Weaver, D. G., Carey, J. P., Kress, M. H. & Dolling, U. H. (2002) Practical synthesis of Aryl Triflates under aqueous conditions. *Organic Letters* 4, 4717–4718.

Ganesan, M., Pal, T. K. & Jayakumar, M. (2003) Pellet coating by air suspension technique using a mini-model coating unit. *Bollettino Chimico Farmaceutico* 142 (7), 290–294.

Ge, X., Wan, Z., Song, N., Fan, A. & Wua, R. (2009) Efficient methods for the extraction and microencapsulation of red pigments from a hybrid rose. *Journal of Food Engineering* 94, 122–128.

Giusti, M. M. & Wrolstad, R. E. (1996) Radish anthocyanin extract as a natural red colorant for maraschino cherries. *Journal of Food Science* 61 (4), 688–694.

Gouin, S. (2004) Microencapsulation-industrial appraisal of existing technologies and trend. *Trends in Food Science and Technology* 15, 330–347.

Groboillot, A. F., Champagne, C. P., Darling, G. D. & Poncelet, D. (1993) Membrane formation by interfacial cross-linking of chitosan for encapsulation of *Lactobacillus lactis*. *Biotechnology and Bioengineering* 42, 1157–1163.

Hogan, S. A., McNamee, B. F., Dolores O'Riordan, E. & O'Sullivan, M. (2001) Emulsification and microencapsulation properties of sodium caseinate/carbohydrate blends. *International Dairy Journal* 11, 137–144.

Hyndman, C. L., Groboillot, A., Poncelet, D., Champagne, C. & Neufeld, R. J. (1993) Microencapsulation of Lactococcus lactis with cross-link gelatin membranes. *Journal of Chemical Technology and Biotechnology* 56, 259–263.

Jafari, S. M., Assadpoor, E., He, Y. & Bhandari, B. (2008) Encapsulation efficiency of food flavours and oils during spray drying. *Drying Technology* 26, 816–835.

Kailasapathy, K. & Masondole, L. (2005) Survival of free and microencapsulated Lactobacillus acidophilus and Bifidobacterium lactis and their effect on texture of feta cheese. *Australian Journal of Dairy Technology* 60, 252–258.

Kenyon, M. M. (1995) Modified starch, maltodextrin, and corn syrup solids as wall materials for food encapsulation. In: *Encapsulation and Controlled Release of Food Ingredients*, Risch, S. J. & Reineccius G. A. (eds), ACS Symposium Series, American Chemical Society, Washington, DC, volume 590, pp. 42–50.

Kilara, A. & Panyam, D. (2003) Peptides from milk proteins and their properties. *Critical Reviews in Food Science and Nutrition* 43, 607–633.

Kim, Y. D. & Morr, C. V. (1996) Microencapsulation properties of gum arabic and several food proteins: Spray dried orange oil emulsion particles. *Journal of Agricultural Food Chemistry* 44, 1314–1320.

Klien, J., Stock, J. & Vorlop, K. D. (1983) Pore size and properties of spherical calcium alginate biocatalysts. *European Journal of Applied Microbiology and Biotechnology* 18, 86–91.

Korhonen, H. & Pihlanto, A. (2003) Food-derived bioactive peptides-opportunities for designing future foods. *Current Pharmaceutical Design* 9, 1297–1308.

Krasaekoopt, W., Bhandari, B. & Deeth, H. (2003) Evaluation of encapsulation techniques of probiotics for yoghurt. *International Dairy Journal* 13, 3–13.

Krasaekoopt, W., Bhandari, B. & Deeth, H. (2004) The influence of coating materials on some properties of alginate beads and survivability of microencapsulated probiotic bacteria. *International Dairy Journal* 14 (8), 737–743.

Kuijpers, A. J., van Wachem, P. B., van Luyn, M. J., Brouwer, L. A., Engbers, G. H. & Krijgsveld, J. (2000) In vitro and in vivo evaluation of gelatin-chondroitin sulphate hydrogels for controlled release of antibacterial proteins. *Biomaterials* 21, 1763–1772.

Latha, M. S., Lal, A. V., Kumary, T. V., Sreekumar, R. & Jayakrishnan, A. (2000) Progesterone release from glutaraldehyde cross-linked casein microspheres: in vitro studies and in vivo response in rabbits. *Contraception* 61, 329–334.

Liu, X. D., Yu, W. Y., Zhang, Y., Xue, W. M., Yu, W. T., Xiong, Y., Ma, X. J., Chen, Y., Yuan, Q. (2002) Characterization of structure and diffusion behavior of Ca-alginate beads prepared with external or internal calcium sources. *Journal of Microencapsulation* 19, 775–782.

Malm, C. J., Emerson, J. & Hiatt, G. D. (1951) Cellulose acetate phthalate as enteric coating material. *Journal of the American Pharmacists Association* 10, 520–522.

Mandal, S., Puniya, A. K. & Singh, K. (2006) Effect of alginate concentrations on survival of microencapsulated Lactobacillus casei NCDC-298. *International Dairy Journal* 16 (10), 1190–1195.

Mazza, G. & Miniati, E. (1993) *Anthocyanins in Fruits, Vegetables and Grains.* CRC Press, London.

Miles, M. J., Morris, V. J. & Carroll, V. (1984) Carob gum kappa-carrageenan mixed gels-mechanical-poperties and X-ray fiber diffraction studies. *Macromolecules* 17, 2443–2445.

Moreau, D. L. & Rosenberg, M. (1993) Microstructure and fat extractability in microcapsules based on whey proteins or mixtures of whey proteins and lactose. *Food Structure* 12, 457–468.

Moreau, D. L. & Rosenberg, M. (1996) Oxidative stability of anhydrous milk-fat microencapsulated in whey proteins. *Journal of Food Science* 61, 39–43.

Mortazavian, A., Razavi, S. H., Ehsani, M. R. A. & Sohrabvandi, S. (2007) Principles and methods of microencapsulation of probiotic microorganisms. *Iranian Journal of Biotechnology* 5 (1), 1–18.

Picot, A. & Lacroix, C. (2004) Encapsulation of bifidobacteria in whey protein-based microcapsules and survival in simulated gastrointestinal conditions and in yoghurt. *International Dairy Journal* 14 (6), 505–515.

Ranney, M. W. (1969) *Microencapsulation Technology*, Noyes Development Corp Publishers, Beaverton, Orlando.

Rao, A. V., Shiwnarin, N. & Maharij, I. (1989) Survival of microencapsulated Bifidobacterium pseudolongum in simulated gastric and intestinal juices. *Canadian Institute of Food Science and Technology Journal* 22, 345–349.

Rodriguez-Saona, L. E., Giusti, M. M. & Wrolstad, R. E. (1999) Color and pigment stability of red radish and red fleshed potato anthocyanins in juice model systems. *Journal of Food Science* 64, 451–456.

Rossler, B., Kreuter, J., Scherer, D. (1995) Collagen microparticles: preparation and properties. *Journal of Microencapsulation* 12, 49–57.

Sanderson, G. R. (1990) Gellan gum. In: *Food Gels*, Harris, P. (ed.) Springer, pp. 201–233.

Sankarikutty, B., Sreekumar, M. M., Narayanan, C. S. & Mathew, A. G. (1988) Studies on microencapsulation of cardamon oil by spray drying technique. *Journal of Food Science and Technology* 25 (6), 352–356.

Schotten, C. (1884) Ueber die oxidation des piperidins. *Berichte der deutschen chemischen Gesellschaft* 17 (2), 2544–2547.

Shah, N. P. & Rarula, R. R. (2000) Microencapsulation of probiotic bacteria and their survival in frozen fermented dairy desserts. *Australian Journal of Dairy Technology* 55, 139–144.

Sheu, T.-Y. & Rosenberg, M. (1995) Microencapsulation by spray drying ethyl caprylate in whey protein and carbohydrate wall systems. *Journal of Food Science* 60 (1), 98–103.

Smidsrød, O. & Skjak-Braek, G. 1990. Alginate as immobilization matrix for cells. *Trends in Biotechnology* 8, 71–78.

Steenson, L. R., Klaenhammer, T. R. & Swaisgood, H. E. (1987) Calcium alginate-immobilized cultures of lactic streptococci are protected from attack by lytic bacteriophage. *Journal of Dairy Science* 70, 1121–1127.

Sultana, K., Godward, G., Reynolds, N., Arumugaswamy, R., Peiris, P. & Kailasapathy, K. (2000) Encapsulation of probiotic bacteria with alginate-starch and evaluation of survival in simulated gastrointestinal conditions and in yoghurt. *International Journal of Food Microbiology* 62, 47–55.

Sun, W. & Griffiths, M. W. (2000) Survival of bifidobacteria in yogurt and simulate gastric juice following immobilization in gellanxanthan beads. *International Journal of Food Microbiology* 61, 17–25.

Thevenet, F. (1988) Acacia gums: Stabilisers for flavor encapsulation. In *Flavor Encapsulation*, Risch, S. J. & Reineccius, G. A. (eds), ACS Symposium Series, American Chemical Society, Washington, DC, volume 370, pp. 37–44.

Trubiano, P. E. & Lacourse, N. L. (1988) Emulsion stabilizing starches. In *Flavor Encapsulation*, Risch, S. J. & Reineccius, G. A. (eds), ACS Symposium Series, American Chemical Society, Washington, DC, volume 370, pp. 45–54.

Truelstrup-Hansen, L., Allan-wojtas, P. M., Jin, Y. L. & Paulson, A. T. (2002) Survival of free and calcium-alginate microencapsulated *Bifidobacterium* spp. in simulated gastro-intestinal conditions. *Food Microbiology* 19, 35–45.

Werner, S. R. L. (2005) Air suspension coating of dairy powders: A micro-level process approach. PhD thesis, Massey University, Palmerston North, New Zealand.

Young, S. L., Sarda, X. & Rosenberg, M (1993a) Microencapsulating properties of whey proteins. 2. Combination of whey proteins with carbohydrates. *Journal of Dairy Science* 76, 2878–2885.

Young, S. L., Sarda, X. & Rosenberg, M. (1993b) Microencapsulating properties of whey proteins. 1. Microencapsulation of anhydrous milk fat. *Journal of Dairy Science* 76, 2868–2877.

Zhou, Y., Martins, E., Groboillot, A., Champagne, C. P., Neufeld, R. J. (1998) Spectrophotometric quantification of lactic bacteria in alginate and control of cell release with chitosan coating. *Journal of Applied Microbiology* 84, 342–348.

General Conclusions

In this book, different aspects of the chemistry of a number of food additives have been discussed. It is anticipated that the contents of this book will generate more interest in research areas on food additives and preservatives. This book has highlighted the need for more focused research into food additives to all stakeholders in the areas of food chemistry and food technology. To avoid negative consequences to consumers from toxic additives or or their metabolites, the process of monitoring and quality control must be a continuous one. Analysts and policy makers need to be on alert, especially in cases where little research has been conducted on some of the additives.

The discussion of food additives and preservatives in this book has shown that, in many instances, more than one additive is added to a particular food product and also that one food additive may play more than one role. For example, some organic acids serve as acidulants, flavourings and preservers. Some food emulsifying agents also serve as stabilisers as well as sweeteners. Proper regulatory mechanisms must be enforced and should be adhered to so that the correct ratio of these multi-tasking additives is used for each intended purpose without violating the limits set for their use.

For some additives which have been reported to affect human health by causing some disease conditions and sicknesses, or for those consumers who have been diagnosed with particular sensitivities/allergies to some food additives, it is important that appropriate warnings be included in the packaging of additives.

A discussion of recently introduced technologies which combine aspects of food and health, for example nutraceuticals, nutrigenetics, probiotics, prebiotics and synbiotics, is included so that researchers are aware of the latest developments in the field.

Several processes that are important in food processing, such as microencapsulation and bioencapsulation, are also discussed in this book due to their importance in the food industry. The development of analytical methods to monitor residues of food additives in food products is very encouraging; where regulations and restrictions have been imposed on a certain class of additives, they can be monitored easily and with high certainty.

With the continuous introduction of new food additives, information is constantly needed to understand the chemistry of these newly introduced food additives, how they work, possible metabolites and, most importantly, their safety to consumers.

Index

Chemistry of Food Additives and Preservatives, First Edition. Titus A. M. Msagati.
© 2013 John Wiley & Sons, Ltd. Published 2013 by John Wiley & Sons, Ltd.